Improving standards and certification in agri-food supply chains

Ensuring safety, sustainability and social responsibility

It is widely recognised that agriculture is a significant contributor to global warming and climate change. Agriculture needs to reduce its environmental impact and adapt to current climate change whilst still feeding a growing population, i.e. become more 'climate-smart'. Burleigh Dodds Science Publishing is playing its part in achieving this by bringing together key research on making the production of the world's most important crops and livestock products more sustainable.

Based on extensive research, our publications specifically target the challenge of climate-smart agriculture. In this way we are using 'smart publishing' to help achieve climate-smart agriculture.

Burleigh Dodds Science Publishing is an independent and innovative publisher delivering high quality customer-focused agricultural science content in both print and online formats for the academic and research communities. Our aim is to build a foundation of knowledge on which researchers can build to meet the challenge of climate-smart agriculture.

For more information about Burleigh Dodds Science Publishing simply call us on +44 (0) 1223 839365, email info@bdspublishing.com or alternatively please visit our website at www.bdspublishing.com.

Related titles:

Advances in ensuring the microbiological safety of fresh produce
Print (ISBN 978-1-80146-268-6); Online (ISBN 978-1-80146-269-3, 978-1-80146-270-9)

Developing smart agri-food supply chains
Print (ISBN 978-1-78676-749-3); Online (ISBN 978-1-78676-751-6, 978-1-78676-752-3)

Frontiers in agri-food supply chains
Print (ISBN 978-1-80146-271-6); Online (ISBN 978-1-80146-272-3, 978-1-80146-273-0)

Key issues in agricultural ethics
Print (ISBN 978-1-80146-313-3); Online (ISBN 978-1-80146-314-0, 978-1-80146-315-7)

Understanding and fostering soil carbon sequestration
Print (ISBN 978-1-78676-969-5); Online (ISBN 978-1-78676-971-8, 978-1-78676-972-5)

Chapters are available individually from our online bookshop:
https://shop.bdspublishing.com

climate•SMART•publishing
IN AGRICULTURAL SCIENCE

BURLEIGH DODDS SERIES IN AGRICULTURAL SCIENCE
NUMBER 148

Improving standards and certification in agri-food supply chains

Ensuring safety, sustainability and social responsibility

Edited by Professor Louise Manning, University of Lincoln, UK

burleigh dodds
SCIENCE PUBLISHING

Published by Burleigh Dodds Science Publishing Limited
82 High Street, Sawston, Cambridge CB22 3HJ, UK
www.bdspublishing.com

Burleigh Dodds Science Publishing, 1518 Walnut Street, Suite 900, Philadelphia, PA 19102-3406, USA

First published 2024 by Burleigh Dodds Science Publishing Limited

Library of Congress Control Number: 2024940546

British Library Cataloguing in Publication Data
A catalogue record for this book is available from the British Library

ISBN 978-1-80146-451-2 (print)
ISBN 978-1-80146-453-6 (PDF)
ISBN 978-1-80146-452-9 (ePub)
ISSN 2059-6936 (print)
ISSN 2059-6944 (online)

DOI: 10.19103/AS.2024.0133

Typeset by Deanta Global Publishing Services, Dublin, Ireland

Contents

Part 2 The expanding scope of agri-food chain standards

Series list

Title	Series number
Achieving sustainable cultivation of maize - Vol 1 From improved varieties to local applications Edited by: Dr Dave Watson, CGIAR Maize Research Program Manager, CIMMYT, Mexico	001
Achieving sustainable cultivation of maize - Vol 2 Cultivation techniques, pest and disease control Edited by: Dr Dave Watson, CGIAR Maize Research Program Manager, CIMMYT, Mexico	002
Achieving sustainable cultivation of rice - Vol 1 Breeding for higher yield and quality Edited by: Prof. Takuji Sasaki, Tokyo University of Agriculture, Japan	003
Achieving sustainable cultivation of rice - Vol 2 Cultivation, pest and disease management Edited by: Prof. Takuji Sasaki, Tokyo University of Agriculture, Japan	004
Achieving sustainable cultivation of wheat - Vol 1 Breeding, quality traits, pests and diseases Edited by: Prof. Peter Langridge, The University of Adelaide, Australia	005
Achieving sustainable cultivation of wheat - Vol 2 Cultivation techniques Edited by: Prof. Peter Langridge, The University of Adelaide, Australia	006
Achieving sustainable cultivation of tomatoes Edited by: Dr Autar Mattoo, USDA-ARS, USA and Prof. Avtar Handa, Purdue University, USA	007
Achieving sustainable production of milk - Vol 1 Milk composition, genetics and breeding Edited by: Dr Nico van Belzen, International Dairy Federation (IDF), Belgium	008
Achieving sustainable production of milk - Vol 2 Safety, quality and sustainability Edited by: Dr Nico van Belzen, International Dairy Federation (IDF), Belgium	009
Achieving sustainable production of milk - Vol 3 Dairy herd management and welfare Edited by: Prof. John Webster, University of Bristol, UK	010
Ensuring safety and quality in the production of beef - Vol 1 Safety Edited by: Prof. Gary Acuff, Texas A&M University, USA and Prof. James Dickson, Iowa State University, USA	011
Ensuring safety and quality in the production of beef - Vol 2 Quality Edited by: Prof. Michael Dikeman, Kansas State University, USA	012
Achieving sustainable production of poultry meat - Vol 1 Safety, quality and sustainability Edited by: Prof. Steven C. Ricke, University of Arkansas, USA	013
Achieving sustainable production of poultry meat - Vol 2 Breeding and nutrition Edited by: Prof. Todd Applegate, University of Georgia, USA	014

Introduction

Setting and enforcing standards to ensure minimum levels of safety and quality in agri-food supply chains has a long history. However, the complexity and geographic dislocation of modern agri-food supply chains, with a host of intermediaries separating farmers from the final consumer, often located in differing jurisdictions, has significantly complicated the process of standard setting as well as applying governance structures such as monitoring and ensuring compliance. The thematic scope of standards has also now expanded from its original narrow focus to cover issues such as social responsibility and ethical trading, animal welfare, environmental sustainability, biodiversity loss and nature-related risks. Finally, the range of stakeholders involved in setting and monitoring standards influencing agri-food supply chains has proliferated, from official national, regional (e.g. the EU) and international bodies such as the Codex Alimentarius Commission (CAC) to non-government organisations (NGOs), the food industry itself and, more recently the banking and investment sector concerned to ensure their clients meet environmental, social and governance (ESG) standards and manage climate-related investment risks. Focussing on generic and product specific approaches, this collection reviews the range of standards and certification schemes that have been developed and the ways to optimise their contribution to achieving safe, sustainable and socially responsible agricultural and post-harvest food production.

Part 1 Setting standards

Part 1 reviews the array of standard-setting bodies and standards, ranging from the international frameworks developed by the Food and Agriculture Organisation (FAO) and the CAC to the more specific initiatives developed by non-governmental organisations (NGOs).

Chapter 1 begins by reviewing the origins and development of the CAC, established by the United Nations (UN) Food and Agriculture Organisation (FAO) and the World Health Organisation (WHO). It describes the Codex organisational structure and standard setting process. The chapter goes on to review the main achievements of Codex in promoting global food safety and quality, as well as the current and future challenges facing the organisation.

Since 1963 Codex has set commodity safety and quality standards for all major food commodities as well as developing 79 Guidelines dealing with general issues such as food import and export inspection and certification systems and 54 Codes of Practice addressing, for example, guidance on preventing contamination (chemical, physical or microbiological) of food products. In addition, the Commission has created a comprehensive system of thousands of maximum residue levels (MRLs) for food additives, pesticide and veterinary drugs and maximum limits for important contaminants. These standards underpin global traded agreements for agri-food products so common product and production standards can be accepted by all parties. A signature Codex achievement is the widespread adoption hazard analysis critical control point (HACCP) principles and guidance on the development of management systems for safe food production. The General Principles of Food Hygiene (which include the crucial HACCP system), the General Standard for Food Additives and Contaminants, as well as related codes to reduce contamination, are important and globally-recognised texts on food safety. The Codex Food Labelling System provides nations with all the necessary advice to ensure that customers are given accurate information about products and are not misled. The Codex Risk Analysis System offers a comprehensive approach for members to protect their populations based on sound science. Codex nutrition and labelling standards have helped inform both governments and consumers in the area of nutrition and foods for special uses.

More recently, Codex has developed guidelines for rapid risk analysis following detection of contaminants in food where there is no regulatory level. Areas for future development include traceability where innovations in tracking technologies, traceability systems, electronic documentation and labelling technologies have made it easier to track ingredients and products from farm to fork, despite the complexity of globalised agri-food supply chains (a theme explored in more detail in Chapter 13). Other opportunities include new partnerships e.g. with the United Nations Environment Programme (UNEP) to strengthen the way Codex addresses sustainability issues. As the chapter indicates, achieving consensus between member nations in some areas, speed of standard setting, and efficient and effective participation of all members are still challenging issues, as is the financing of the scientific advice underlying Codex and capacity building for developing countries, together with their effective participation in Codex meetings

As Chapter 2 indicates, agri-food supply chains have been linked to a number of negative environmental and social externalities, including environmental pollution, biodiversity loss, greenhouse gas (GHG) emissions and labour abuses, with growing societal demands that such negative impacts should be addressed. This chapter considers the role of supranational policy, and regional and national policy in driving sustainability standards and improvement across agri-food supply chains. The chapter also reviews developments in forms of hybrid governance and uses the example of the role of green finance to show how agri-food supply chain strategies are being reshaped.

Global governance architectures relevant to food production include: The World Trade Organisation (WTO), the FAO and the CAC (as discussed in Chapter 1). More recently, they have been extended to include The UN

Framework Convention on Climate Change (UNFCCC) and their annual conferences of parties (COPs), as well as The Intergovernmental Panel on Climate Change (IPCC) which provides the scientific findings informing COP discussions. These governance architectures directly influence global, regional and national public policy. An example is the supranational recognition of sustainability issues which has led to the subsequent enshrining in law of net zero carbon policies in individual countries with major implications for the food industries in those countries. Green finance strategies have also been driven through the same governance mechanism.

However, because these policy initiatives typically rely on mediated consensus, supranational institutions can be slow to act. They, and the policies they create, can also be inflexible, technocratic and lacking in accountability to individual sovereign powers and the citizens that elect them, can fail to reflect regional needs and concerns, and can lack the ability to enforce agreed changes on recalcitrant countries. These limitations and the quasi-public nature of 'public' goods has encouraged a growing interest in public-private partnerships (PPPs) as a way of funding and managing the achievement of sustainability outcomes. PPPs allow hybrid governance with the potential to deliver innovative, flexible and dynamic governance when compared to purely regulatory or market controls

Within PPPs, the role of governments is to establish frameworks and standards, together with incentives and sanctions, to steer corporate players to collectively and individually improve performance within their local operations in areas such as the sustainability of their operations. Private aspects of agri-food governance include individual or collective retailer-led and transnational corporation-led food governance systems, often enacted through the development of private third-party standards and certification schemes. These private standards are based on prerequisite programmes such as good agricultural practices (GAP), good manufacturing practices (GMP) and good hygienic practices (GHP) (as set out by Codex) or environmental and social governance (ESG) strategies which are increasingly being extended to encompass sustainable food production.

Governments have also moved to create new frameworks, requirements and standards for the finance sector to take account of environmental issues, as well as new environmental and social reporting standards for major companies such as the EU Corporate Sustainability Reporting Directive. The former has stimulated the growth of green finance which has been seen as a key catalyst for public, private and public/private approaches to improving sustainability. Green finance can provide both public and private financial flows (banking, micro-credit, insurance, and investment) to promote the delivery of the UN Sustainable Development Goals (SDGs), and enable green growth and the adoption of climate mitigation strategies.

Building on the previous chapter, Chapter 3 examines the development of voluntary sustainability initiatives, standards and certification systems. It addresses the challenges of coordination across national, regional, multilateral and private industry initiatives, as well as difficulties in comparing them and monitoring their performance. The chapter starts by reviewing successive phases in the development of sustainability standards prompted by catalysts such as the 1983 Brundtland Commission, the 1992 Earth Summit, as well as the 2000 launch of the UN Millennium Development Goals and the UN Global Compact with its 10 principles to guide business operations towards greater sustainability, together with the UN Sustainable Development Goals (SDGs) launched in 2016. The Global Compact, for example, has now become the world's largest corporate sustainability initiative, with more than 13,000 companies in over 170 developing and developed countries.

As the focus on sustainability grew over the last thirty years, companies began to develop their own corporate social responsibility (CSR) initiatives. By their very nature, these are corporate strategic responses to the particular regulatory and market pressures they face, whether from customers, suppliers and business partners, government authorities or NGOs. These CSR initiatives sought to protect company reputations (as well as sales and profits) and stave off a rise in prescriptive regulation. Today these standards are at the heart of many corporate sustainability actions. Often created by industry consortia, NGOs or multistakeholder initiatives, voluntary sustainability standards go beyond minimum legal standards imposed by legislators (often termed the regulatory baseline). These private, voluntary standards are usually limited to a well-defined sustainability theme, value chain, specific sector or product.

The UN has estimated that there are in the region of 450 voluntary sustainability standards, mostly covering agricultural commodities. However, valuable as these individual initiatives have been, the piecemeal nature of their development means there is a lack of a connective structure (governance architecture) to ensure interoperability or compatibility between national, multilateral, regional and private industry perspectives on sustainability. To assist users of sustainability standards, the International Trade Centre (ITC) Trade for Sustainable Development (T4SD) programme has developed the Standards Map. It provides the world's only global repository of such initiatives and standards, including references to codes of conduct, audit protocols, certification programmes, sustainability reporting frameworks, sustainable finance indicators and international standards. As a response to the complicated landscape of sustainability initiatives, the Standards Map focuses on making such information more accessible and transparent for everyone, as a global public good. The Map provides a publicly-available inventory covering over 300 sustainability initiatives, applicable to more than 80 sectors and 190 countries, making it easier to compare standards against more than 1,600 criteria and assess levels of compliance.

At the same time, the ITC has become the convener of a community of sustainability experts, including technical development agencies, international agencies, industry associations, business support institutions, civil society organisations and governments that promote sustainability practices and work to align sustainability frameworks. This complements other initiatives such as the International Social and Environmental Accreditation and Labelling (ISEAL) Alliance established in 2002 with the aim of creating a solid foundation for collaborative working between third party certification and standards organisations, creating credibility tools and platforms for shared learning. Founder members included Fairtrade International, the Rainforest Alliance and IFOAM, the body representing the global organic movement. ISEAL has e.g. developed a Sustainability Good Practice Guide to inform sustainable standard organisations and improve the rigour and credibility of sustainability standard systems.

Multinational corporations have also joined forces e.g. through the Consumer Goods Forum setting up the Global Social Compliance Programme (GSCP) in 2009 with the 'aim of harmonising existing efforts and deliver a common, consistent and global approach across sectors for the continuous improvement of working and environmental conditions in global supply chains'. A long-standing grouping of food and agriculture companies with a focus on pre-competitive topics is the Sustainable Agriculture Initiative Platform (known as the SAI Platform). Established in 2002, with roots in discussions between Danone, Nestlé and Unilever, its original focus was sustainable agricultural practices in arable in the EU and United States (US), given that they were already collaborating through the emerging Roundtables on tropical products such as oil palm and soy. Membership has grown to over 150 of the largest food and agriculture companies across 30 countries, covering a vast array of production systems. Like many other collective initiatives, its focus is the development and sharing metrics covering both good practices and outcome measures.

Looking ahead, and as discussed in Chapter 2, policymakers may increasingly recognise and use voluntary standards as a form of 'earned recognition', signalling a shift away from the 'voluntary' part of such sustainability standards when they are being used to fully or partially demonstrate regulatory compliance. The advantage of this approach is a reduction in the surveillance of food business operators and farmers where private inspections and audits replace, in part or wholly, regulatory inspections and audits. Many farmers and FBOs highlight the cost of demonstrating compliance to multiple standards and requirements both in terms of time and economic costs. Policymakers are starting to use these private standards in free trade agreements and public procurement policies. As noted in earlier chapters (and discussed in Chapter 13), standards and supply chain actors are now also joining forces with technology providers to track and/or trace products to their origin, and provide data to support sustainability claims down to the farmer level. Supply chains are being digitalised through the adoption of a range of mobile and internet technologies. Procurement professionals and standard setters increasingly use artificial intelligence, the Internet of Things (IoT) and blockchain technologies to track and trace products at each stage of the supply chain.

Given the non-profit, independent nature of NGOs, they are in a key position to improve and enhance the uptake of sustainability practices and standards. Chapter 4 explores the role of NGOs in improving sustainability standards in the agri-food supply chain. Key roles include lobbying, collaboration, providing subject matter expertise, facilitating knowledge exchange, setting standards, providing incentives, communication to a range of stakeholders as well as strengthening accountability and credibility.

As the chapter shows, one of a NGO's most commonly-acknowledged roles is that of lobbying and enforcing accountability, using leverage over government and business and trying to influence policy and practice to identify a particular challenge, whether it be social injustices in the agri-food supply chain or damaging impacts on the environment. However, NGOs have also played a proactive role in not only identifying but helping to resolve these challenges. One of the most important roles that NGOs play is collaborating with other NGOs, food supply chain stakeholders, farmers and governments to develop collective solutions. By developing and providing subject matter expertise NGOs can also support supply chains to improve environmental and social standards through advice and assistance with designing, implementing, monitoring, recording and reporting. The role NGOs play as subject matter experts is undoubtedly related to their role in facilitating knowledge exchange. Within this role, NGOs provide specialist training, resources, support and technical expertise, both to companies and to individual farmers and others delivering change on the ground. A key role played by NGOs is communication in generating and supporting change towards more sustainable and resilient farming systems. This role is particularly important where there may be insufficient demand for change from consumers.

Finally, NGOs have played a central role in developing robust, independent standards and certification schemes. Particularly where such standards are independently verified, they provide robust verification of practices or outcomes in order to meet ESG commitments, environmental pledges and other targets. Many of these standards have now been taken up by major companies in the agri-food supply chain and made mandatory for suppliers. They provide both a route to practice change and incentives to make this change e.g. through price premiums for certified produce or improved market access. As indicated in earlier chapters, governments are increasingly considering incorporating sustainability standards into their own policies or using them as a templates for future public regulation.

The chapter illustrates these varied contributions made by NGOs with a range of case study examples, most notably from the UK-based NGO Linking

Environment and Farming (LEAF). LEAF works with farmers to promote Integrated Farm Management (IFM), a whole farm business approach which emphasises a holistic view of the farming system. IFM has 9 interconnected sections covering themes such as business planning, soil health, plant health, habitat and biodiversity enhancement, energy and waste management, water management, livestock management and engaging society. LEAF manages LEAF Marque, which is the standard that farming businesses are assessed against by third party organisations (certification bodies) to achieve LEAF Marque certification. In 2020 there were 27 countries with LEAF Marque certified businesses

Finally, the chapter highlights key ongoing challenges faced by NGOs, including the risk of greenwashing by agri-food supply chain actors making claims that are not reflected in actual change on the ground. This risk can be aggravated by the plethora of different sustainability certification schemes which risk unnecessary duplication (and waste of resources), competition and confusion in the market. NGOs continue to differ in the amount of leverage they have with governments and key corporate players as well as the financial and other resources needed to fulfil their role. Finally, it can be difficult to design policy and standards that are suitable across a wide variety and complexity of contexts given the differing (and sometimes conflicting) agendas of different agri-food supply chain actors. There can be challenging potential trade-offs in meeting the strategic needs of the agri-food supply chain and wider society.

Part 2 The expanding scope of agri-food chain standards

Chapters in Part 2 explore the growing range of coverage of standards in contemporary agri-food supply chains.

Building on the chapters in Part 1, Chapter 8 provides a critical assessment of the current landscape of sustainability standards, focussing on who has developed them, their content and rationale. It also explores how companies select the standards they use, and how this is changing. Recent choices made by brands and retailers accentuate some of the challenges associated with standards for some agri-food supply chain actors such as smallholders.

The most visible standards in agriculture and land management since the mid-1990s have been those developed by multi-stakeholder organisations, often dubbed voluntary sustainability standards (VSS) usually associated with third-party certification processes, for example, Fairtrade International or Rainforest Alliance. These are often associated with the processes of governing *through* standards in which civil society, often in partnership with business, creates a standard that they then lobby for business to adopt, and offer independent verification, and usually a consumer-facing label or certificate (business to consumer or B2C). Whilst VSS are developed with companies or industry bodies as part of a broad coalition of stakeholders, their origins lie in a critique of both the lack of effective transnational regulation and the limitations of private sector regulation or standards developed by companies themselves.

Also technically voluntary, but compulsory for producers wishing to sell to the particular company, are private company standards, so-called 'do it yourself' standards where a company has its own standard that it applies to its suppliers, thereby exercising governance *of* its supply chain. This kind of standard often remains a business-to-business (B2B) communication. These standards may be developed in collaboration with civil society, but they are proprietary instruments, sometimes with an associated trademark. An example is Nespresso's AAA Sustainable Quality programme launched in 2003, initially focussing on improving product quality but, with growing concern about climate change and other environmental changes affecting yields, now broadened to encompass environmental and social sustainability issues, drawing on the expertise of NGOs such as the Rainforest Alliance.

The choice by companies about whether to commit to a sustainability agenda, their strategic priorities and the tools to use to deliver those priorities, as well as how well they implement these elements, will vary according to the mix and strength of the external pressures on the business and their internal drivers. The sustainability standards landscape has been characterised as a market in which companies choose standards not necessarily on what is technically the 'best' but what meets their needs at a particular time and in relation to specific internal and external business pressures. There is conflicting evidence as to whether, in practice, the nature of this standards market leads to a race to the bottom or to the top.

Some of the problems this generates can be seen in the way sustainability standards affect smallholders. Sustainability standards are by definition exclusionary: they set out what is necessary to merit a particular certificate or label with compliance being made a requirement to sell produce to a particular buyer. This can be problematic for smaller producers, who often lack the resources to comply, and for development, especially where expectations of compliance are not accompanied by economic or social support to achieve compliance. Moreover, certification to meet sustainability standards can even erode producer incomes and resilience (e.g. by increasing costs and entry requirements to access some markets). Realising the associated benefits requires a supportive institutional context, both in the country of production and of consumption and along the supply chain. Often the issue is not whether the implementation of standards-driven criteria themselves might lead to positive outcomes, but who is implementing them, how they work with smallholders and whether they listen to their concerns.

More broadly, whether a company uses a VSS with third party certification, participates in a collective industry initiative or uses its own private standard, the link to the company's values and its over-arching sustainability strategy is an important indicator of its potential to initiate and deliver longer-term change. Scrutiny is therefore needed not only of the specific standards but also a company's approach to sustainability more broadly, if it is integrated within their overarching business mission, goals and strategy and the extent to which it is linked to operational activities, organisational transformation and a vision of future sustainable development.

Chapter 6 looks at a key challenge in achieving more sustainable food production: how to achieve 'net zero'. Net zero refers to the goal of achieving 'a balance between anthropogenic emissions by sources and removals by sinks' as outlined in the 2015 Paris Climate Change Agreement. The chapter reviews differing potential solutions, starting with outlining the principles of 'carbon farming', a whole farm approach that seeks to optimise carbon capture (sequestration in carbon sinks) by implementing practices that improve the rate of carbon dioxide (CO2) removal from the atmosphere and storing CO2 in plant material and/or soil organic matter. Other definitions expand the definition of carbon farming to include both practices for sequestering atmospheric carbon and/or for reducing the production of GHG emissions at farm level. This approach links all GHG emissions to a carbon dioxide equivalence in terms of global warming potential or CO2e. The chapter reviews a range of potential practices that allow carbon farming to both achieve 'net zero' in the balance of GHG emissions/capture whilst optimising food production, therefore contributing to more resilient food production systems.

However, the chapter also highlights the significant barriers to the adoption of carbon farming practices. Whilst there is a wealth of research on the mechanisms of soil carbon sequestration and on how particular agronomic practices can enhance sequestration, improve soil health and deliver other ecosystem services, this has not always been translated into specific, practical guidance tailored to individual farms. In addition to a lack of knowledge in some cases, farmers are concerned about the costs of transitioning to new practices, where these cannot be passed on through agri-food supply chains to the end consumer, and there remain concerns about how far these practices affect may affect farm-level yields. A particular concern at farm level is how to effectively measure the effectiveness of carbon farming practices.

With this in mind, the chapter looks at the current range of model-based carbon calculation tools available to farmers (including AgreCalc, Cool Farm Tool, Farm Carbon Calculator and Sandy in the UK). As highlighted in this chapter (and discussed in more detail in Chapter 14), these tools have evolved separately, have different assumptions and model structures and can, therefore, report conflicting results which undermines their credibility. The chapter discusses relevant standards and protocols to test and improve model robustness and comparability. It also reviews the role of more established techniques such as life cycle assessment can play, as well as the emergence of a carbon trading market.

More recently, the concept of 'regenerative agriculture' has been promoted strongly by civil society, NGOs and multi-national corporations within the food supply chain. As with organic agricultural practices, regenerative agriculture is associated with low external input-based agriculture i.e. limiting artificial nutrient use and the use of crop protection products. The theoretical foundation for regenerative agriculture is agroecology which views agriculture as an adapted form of ecosystem in which an understanding of fundamental biogeochemical flows/cycles and functional interactions between organisms within this type of ecosystem can used to optimise its ability to deliver a range of ecosystem services, including food production. Examples of practices promoting agroecosystem function include optimising the physical, chemical and biological health of soils to promote crop growth and health (reducing the need for external inputs such as fertilisers and pesticides), promoting biodiversity (e.g. to encourage beneficial species such as pollinators or predators of crop pests), and ensuring diversity in agricultural landscapes (to optimise beneficial functional interactions between species).

However, despite the growing focus on the concept, there remains no generally-accepted definition of 'regenerative agriculture' with a range of overlapping (and sometimes conflicting) process and outcome-based definitions. This has led to a differentiation in the emergent standards and certification schemes. As the chapter points out, in the short term, the flexibility of the framing of regenerative agricultural practices may be a strength in that it allows a wider range of actors to choose the particular set of regenerative practices that suit the needs of a particular farming system and to gain widespread engagement from a range of farming communities. In the long term, if not effectively implemented and monitored, this very diversity could dilute the intended impact. Research and better knowledge exchange (by both scientists and farmers (since the latter have often pioneered regenerative techniques independently on their own farms) is key to establishing a clear foundation of authoritative knowledge on how particular regenerative agricultural practices work, how to measure their effects and optimise their implementation.

There are signs that greater coherence is emerging. A recent, and potentially influential, initiative to define regenerative agriculture has been made by the Sustainable Agriculture Initiative Platform (SAI Platform). SAI Platform represents leading global food processing and other agri-food businesses, including Cargill, Coca-Cola, Danone, Diageo, Kellogg's, Kraft Heinz, Nestlé and Unilever. The SAI Platform has defined regenerative agriculture as: 'an outcome-based farming approach that protects and improves soil health, biodiversity, climate and water resources while supporting farming business development.' It has outlined a framework with four impact areas: soil health, water, biodiversity and climate. The framework sets out material criteria which determine how these four impact areas are affected, and key outcomes which optimise impact area function (e.g. soil health, water resources, biodiversity and mitigation of climate change) and a set of key principles and practices to achieve these outcomes. This framework is designed to allow farmers in different contexts to assess the specific environmental and production risks they face, identify outcomes to mitigate those risks, prioritise farm management actions and establish metrics for measuring performance.

A second major initiative is emerging from GLOBALG.A.P., a standards owning organisation with a group of farm assurance certification schemes. From its initial foundations GLOBALG.A.P has grown to provide some of the most respected and internationally-recognised standards supporting the global trade of farmed products, particularly fruits and vegetables, flowers and aquaculture products. GLOBALG.A.P. standards have been adopted by over 200,000 certified producers in more than 135 countries farming over 4 million hectares. In addition to its flagship Integrated Farm Assurance (IFA) standard focussing on quality, GLOBALG.A.P. has developed a number of add-on or 'bolt on' standards that are focused particularly on environmental sustainability and social responsibility, including GLOBALG.A.P. Farm Sustainability Assessment (GGFSA), BioDiversity, Sustainable Programme for Irrigation and Groundwater Use (SPRING) and Risk Assessment on Social Practice (GRASP). Whilst GLOBALG.A.P.'s flagship IFA embeds elements of economic, environmental and social sustainability, there is recognition of the need for a dedicated environmental sustainability standard. An Environmental Sustainability Working Group (ESWG) was established in 2022 and tasked to develop a new standard, the Environmental Sustainability Solution (ESS), for launch in 2025. Like SAI, this promises to provide a broad-ranging standard to better define and codify regenerative agricultural practices.

In 2022 the Kunming-Montreal Global Biodiversity Framework (the "Paris Agreement for Nature") called for "innovative" financing initiatives to secure greater levels of investment in nature recovery. New finance mechanisms and markets for nature are emerging, some of which offer alternative income-generating opportunities for farmers and land managers. Chapter 7 provides an overview of some of the options that may exist for farmers wishing to engage in more sustainable land management and generate enhancements in nature to earn additional income beyond food production.

Agri-environment Schemes (AES) are an important and long-established mechanism for tackling biodiversity decline in the farmed landscape and are essentially based on "Payment for Ecosystem Services" (PES). In Europe and the UK, for example, farmers receive public money in return for undertaking food production using more sustainable methods or implementing management with specific environmental benefits. Despite some successes, however, it is evident that to date, these schemes have failed to achieve sufficient reductions in rates of biodiversity decline in farmed landscapes. Whilst AES can be expected to continue to play an important role in nature recovery efforts, other "innovative" financing initiatives called for under the Paris Agreement are needed to achieve the scale of transformative changes in land use that are needed to tackle the biodiversity crisis.

An increasingly wide range of environmental credits are being traded and in a variety of ways. As discussed in Chapter 6, carbon markets dominate, but there is growing interest in biodiversity and other nature-credit markets. The chapter provides an overview of the spectrum of markets and finance mechanisms that has developed. Biodiversity offsets are an example of such an approach, now used routinely to compensate for biodiversity losses that would otherwise occur post-infrastructural development, despite actions to avoid and limit damage to ecosystems in-situ. Biodiversity offsets typically come into play in quite subscribed circumstances in which obvious, direct, and irreversible losses of biodiversity occur, for example as a result of a mining, road building, or other infrastructure development project. Biodiversity offsets require the achievement of tightly defined outcomes, with a strict, formal requirement to demonstrate ecological equivalence between what is lost and what is gained. There may be opportunities to generate and sell biodiversity units through statutory/ national biodiversity net gain or offset schemes (such as Biodiversity Net Gain (BNG) in England) or to engage directly with businesses seeking to compensate for their own impact from infrastructural development. Many farmers, if they engage, are likely to opt to work through habitat banks which take responsibility for designing and implementing the specialist management that might be needed and administer any regulatory provisions for tracking outcomes.

A biodiversity credit is a tradable security representing a unit of biodiversity improvement, which is brought about by some form of land-management change. Credits can be traded in biodiversity or nature credit markets, designed "to deliver equitable, nature positive outcomes". Oversight and verification are important to ensure that biodiversity or nature do benefit in tangible ways, commensurate with the traded value. Australia, for example, has put a national regulatory framework in place for biodiversity credit markets to support this.

Markets for nature offer farmers attractive opportunities to diversify their businesses into farming for nature and provision of natural capital in a variety of ways. The options and schemes are proliferating with farmers needing guidance about the best options to choose, though a new generation of advisors and advice is emerging. For example, the Green Finance Initiative has developed a "Farmers' toolkit for assessing nature market opportunities" which provides free advice to UK farmers to brief them on options to consider and guide them in the process of engaging in nature markets, should they decide it is right for them and their business.

To illustrate the practicalities involved, the chapter includes a case study on the The Wendling Beck Environment Project in Norfolk (England), looking to transform land use from arable farming to the generation of biodiversity and a variety of natural capital benefits whilst still providing a sustainable way of living for the four landowners involved and creating employment for others. This approach has identified the sale of biodiversity units as a key revenue stream, alongside government payments for delivering other ecosystem services and revenues from regenerative crop and livestock farming.

As the chapter discusses, various efforts have been made to establish principles and standards to promote or ensure good practice in both offset schemes and nature credit markets. The Business and Biodiversity Offset Programme (BBOP) e.g. has developed principles on biodiversity offsets with wide collaboration and "buy in" from the full spectrum of players in the offset process, which are considered to embody good international practice. Although widely accepted, these principles can be challenging to apply in practice, indicating that more work on robust standards is needed as these markets develop and mature.

As Chapter 8 points out, it is surprising that, despite the urgency of addressing the environmental impact and social issues related to livestock production, research and practice relating to standards in this sector is still in its infancy. Sustainability standards and programmes in livestock production are, however, now developing rapidly and continuously to address the multitude of challenges in the sector, partly in response to the high level of societal debate about livestock production and environmental pressure from NGOs. This trend, however, has resulted in a multiplicity of standards, with no consensus on a generic framework, making it difficult to comprehensively compare standards.

In comparing on-farm implementation of standards, the chapter highlights it can also be difficult to attribute improvements to specific initiatives. Research comparing compliance scores of different welfare programmes suggests variations are likely due to differences in the design of the standard, rather than differences in animal welfare levels per se. Heterogeneity in standards makes it difficult to compare standards both on paper and in terms of realisable, measurable outcomes on farms.

The chapter uses a particular transition theory model which is applicable to food supply chains: the 'sustainable market transformation model' (SMTM). Echoing themes in Chapters 2, 3 and 6, the SMTM identifies four phases of transition in standards evolution, from inception through competitive advantage to synergy and, finally, institutionalisation.

Caring Dairy, initiated in 2004 by ice-cream maker Ben & Jerry's (now owned by Unilever) for their European production operation in the Netherlands, can be seen as one of the first dairy programmes on sustainability and presents an example of an initiative developing from the inception to the competitive advantage phase. The project has turned into the Caring Dairy programme in which 400 dairy farmers in the Netherlands participate to improve their on-farm sustainability. The growing number of these individual initiatives has led, more recently, to moves towards synergy e.g. the development of the Dairy Sustainability Framework (DSF), linked to the SAI platform. The DSF was created and is led by leading global dairy producers accounting for over 30% of global milk production. The aim is to harmonise sustainability approaches throughout the dairy sector by establishing 11 key sustainability criteria and associated indicators in the framework which allow comparability between standards, reducing the number of individual standards and audits farmers need to comply with. Echoing themes in Chapter 6, there are now move towards the fourth phase of institutionalisation e.g. by the Dutch Council for the Environment and Infrastructure in proposing linking public eco-scheme payments with private sustainability initiatives using performance indicators linked to a points-based system.

The chapter uses two specific dairy sustainability initiatives, Origin Green (in Ireland) and Foqus planet (in the Netherlands, Belgium and Germany) to explore key challenges in standard development and implementation. Meeting the demands of different stakeholders and customers can be challenging and can result in duplication of effort. There are inherent trade-offs in developing and implementing standards such as Origin Green and Foqus planet, where the aim is to increase sustainability across multiple areas of sustainability (e.g. social, environmental, and economic sustainability). Trade-offs include the need to make continuous improvements to standards while minimising the burden on farmers which frequent changes to standard criteria may create (e.g. in reconciling public policy and private sustainability standards). Interviewees for both Origin Green and Foqus planet identified the balancing act involved in keeping standards scientifically valid at the same time as being practical for farmers to implement. An increase in the comprehensiveness of standards potentially comes at the expense of practicality of implementation by farmers.

Other challenges include ensuring effective measurement of performance. While historically target- and practice-based indicators have been commonly used in standards, there has been a more recent shift towards including outcome or performance-based indicators based on direct observations, measurements or calculations. However, a shift to performance-based indicators comes with its own risks. It may require organisations to be more selective in the sustainability issues they want to address in their standards, given the associated time and cost investment involved in measurement. Reports from the DSF show that members prioritise GHG emissions and animal welfare whereas water quality, on-farm waste management and working conditions receive least priority. In addition to performance-based approaches such as the use of indicators and benchmarks, there are value-based approaches that can be used. The development of Participatory Guarantee Systems may be regarded as an example of this approach, in which producers are certified through active participation of stakeholders.

Building on Chapter 1, and developing themes in Chapters 5 and 8, Chapter 9 looks at developments in applying GAP standards across global agri-food supply chains. As the chapter points out, regulators can compel farmers to comply with GAP through either penalties (risk of prosecution), 'the stick" or through 'the carrot' and incentives for compliance. Voluntary market-orientated GAP standards define specific criteria with a requirement for organisations to either comply with or exceed legal minimum standards, often linked to a certification process. Private GAP standards can be B2B, as previously described or B2C where the product packaging includes a logo or symbol identifying directly to consumers that products have been produced to a given standard e.g. the Red Tractor Standard in the UK.

Reflecting on the synergy phase in standards development identified in Chapter 8, Chapter 9 also highlights the forming by a group of international retailers and food manufacturers in 2000 of the Global Food Safety Initiative (GFSI) with a focus on food safety, improving consumer confidence and consumer protection and improving the cost effectiveness of assurance processes from field to fork. GFSI has evolved from a benchmarking organisation for third-party private certification standards to an organisation with the aim of delivering strategic regulatory outcomes, facilitating regulatory compliance for organisations and facilitating improved understanding and application of Codex standards.

However, development of private B2B standards by retailers either individually or collectively through GFSI (e.g. in the drive for standard harmonisation) can result in a potential lack of transparency and a barrier to market access. As noted in Chapter 5, it can also lead to the marginalisation of smallholder producers and workers in developing countries who lack the resources to meet the standards' requirements. As the chapter points out, upgrading individual organisations, or indeed a whole value chain, to embed higher environmental or social standards requires significant changes and investment in infrastructure, knowledge and capacities.

These challenges are illustrated by the experience of VietGAP, launched by the Vietnamese Ministry of Agriculture and Rural Development (MARD) in 2008 as a set of minimal GAP standards and guidelines for Vietnamese agriculture. Covering themes such as food safety and quality, worker welfare and environmental sustainability, VietGAP standards have been developed for major crops like rice, coffee, tea, fruits and vegetables, livestock, and fisheries

Key strengths of VietGAP include the fact that it provides a holistic set of good agricultural practices that extend beyond food safety, that it provides a programme which is customised to local conditions with a stepwise approach to adoption, seeks to secure participatory involvement of farmers and others, and provides strong alignment with comparable international GAP standards supporting access to export markets. However, major weaknesses of the programme include the voluntary nature of participation which limits widespread adoption, insufficient training and auditing support for smallholder farmers within the programme, high costs of compliance for smallholders, and a lack of consumer awareness of the benefits of VietGAP-certified produce, including the lack of a standardised VietGAP logo which has reducing consumer support and demand for certified produce. Other major weaknesses include minimum annual testing regimes for crops and weak external auditing. Currently, sampling and laboratory testing requirements are limited and not strictly enforced. Poor control over the quality and consistency of GAP checklists, auditor competency, verification and validation processes for certified farms are all issues which continue to impact the programme. The limited number of independent external third-party audits also appears to be insufficient to prevent non-compliance (and fraudulent claims) in group certification schemes. All of these aspects can be addressed with appropriate interventions. While VietGAP has made a good start in modernising Vietnamese agricultural methods, most exporters and importers still rely on GFSI benchmarked certifications to access the international supply chain. VietGAP must expand in scale, scope and credibility through greater investment to support adoption, ensure rigorous auditing and build stronger domestic and export markets for certified produce.

Chapter 10 develops a wide range of themes explored e.g. in Chapter 4 on the varied and changing role of NGOs as well as issues relating to consolidation and the shift to more performance-based measurement discussed in Chapter 8, and some of the practical issues involved in standards implementation discussed in Chapter 9. The Sustainable Agriculture Network (SAN) developed as a global collaborative network of 22 NGOs to be better able to pool expertise and resources, with its own SAN Certification Standard has enabled more than one million small, medium, and large-scale producers to develop more sustainable farming techniques.

However, it has moved beyond this to develop the Sustainable Agriculture Framework (SAF), a comprehensive sustainability indicator framework designed to guide other organizations towards more effective standard-setting processes and sustainability programmes. SAF is designed to be flexible (modular), holistic (with a broader thematic and landscape focus), fully participatory and outcome-based. The chapter shows how SAF has been used in practice to develop integrated landscape programmes, as well as projects focussed in on-farm biodiversity and integrated pest management and multifunctional field margins.

Chapter 11 also picks up on challenges in developing effective sustainable certification schemes. As it discusses, the initial premise of the Rainforest Alliance was simple: consumer demand for sustainable products, combined with the desire of retailers and other buyers to bolster their green credentials, would allow voluntary sustainability certification to harness market forces to benefit both the environment and the people whose livelihoods depend on these natural resources. Currently Rainforest Alliance Certified products make up approximately 11% of global coffee production, 7% of cocoa production and 22% of tea production, with the certification programme reaching over 400,000 coffee farmers, 890,000 cocoa farmers and 950,000 tea farmers.

The chapter begins by describing in detail the new Rainforest Alliance Sustainable Agriculture Standard implemented from 2021, developed in accordance with ISEAL's Standard-Setting Code of Good Practice. It then goes on to discuss three ongoing challenges that have affected the Rainforest Alliance's ability to reconcile farming, environmental and social sustainability through certification alone. The key challenges are:

- the disproportionate financial burden that certification places on the shoulders of farmers (particularly smallholder farmers)
- the limitations of the traditional high-stakes binary pass/fail approach to achieving certification, which does not sufficiently support remediation and behaviour change to achieve intended results
- the limited scope of farm-level certification, which is not able to address larger systemic issues that affect wider communities, such as governance systems, attitudes and market dynamics

The chapter describes the ways Rainforest Alliance has sought to address these challenges. The new standard e.g. has shifted to a due-diligence or 'assess and address' approach that encourages farmers to address potential problems proactively. To address the challenge that many sustainability issues are either culturally or geographically broader than the scale at which certification works, Rainforest Alliance has developed landscape-level projects and community-based partnerships. It has also introduced the Shared Responsibility concept (with its two main components, the Sustainability Differential and Sustainability Investment) to allow a more equitable division of costs and rewards across the supply chain, rather than relying on markets to self-regulate and somehow arrive at a fair price for certified products. In the long term there may be a need to move to a new model of 'true' pricing, in which all social and environmental externalities are properly accounted for in a product's market price.

Chapter 12 discusses the growing importance of forest certification as a governance structure for sustainable management of forest resources. This chapter focuses on two well-established standards:

- the Forest Stewardship Council (FSC): an international membership organisation founded in 1993 in Canada by representatives from various forest industry and environmental organisations, aiming for globally responsible forest management
- the Programme for the Endorsement of Forest Certification (PEFC): founded in 1999 in Europe as an alternative certification scheme for smaller private forest owners

Forest certification has faced a number of challenges, including acceptance and take-up by forest owners, credibility amongst consumers and others, and the degree to which it has delivered measurable benefits on the ground. The chapter reviews available research on the degree to which certification reduces deforestation rates and improves biodiversity. It also looks at studies of its impact on forest owners and the degree to which certification costs and potentially lower yields from restrictions on certain practices are offset by price premiums and greater market opportunities, as well as an improved corporate image and improved stakeholder relations. The chapter also looks at the degree to which forest certification has positively influenced workers' rights, health and safety, and forest companies' relationships with local communities. It includes a case study on the motivations and experiences of small private forest owners as well as potential improvements such as group certification to spread costs.

Part 3 Measuring compliance and improving performance

Chapters in Part 3 look at ways of measuring and improving compliance with standards. As highlighted in a number of chapters, new developments in technology will play an increasingly important role in monitoring and verifying safety, quality and sustainability in agri-food supply chains. In particular, they can significantly improve both traceability and transparency. A key challenge is the creation of robust, smart, digital chains of custody i.e. systems capable of tracking and tracing every stage in the development of a product from raw materials to the finished retail article, and in reverse when needed, to create a transparent and unbroken trail of accountability. These digital chains of custody have the potential to deliver more efficient assessment and mitigation of risks and, if required, effective management of product recalls. Digital chains of custody can optimise the use of resources (and reduce emissions) and improve the effectiveness of validation and verification of product claims regarding safety, quality and other environmental and social criteria (such as carbon footprinting).

Chapter 13 provides an overview of the principles, components and potential applications of different technologies to achieve the goal of smart, agile agri-food supply chains, as well as current challenges (such as how to optimise interoperability within and between organisations) and how these can be addressed. Technologies reviewed include blockchain and other distributed ledger technology (DLT) solutions for food traceability, Internet of Things (IoT) technologies, the role of big data and Artificial Intelligence (AI) and the use of Digital Twins (DTs).

Chapter 14 revisits themes from Chapter 1 in exploring how to adapt good agricultural practices (GAP) to new technologies, including advances in controlled environment agriculture (CEA) such as vertical farming. This can be particularly important in the case of leafy vegetables sold as 'ready to eat' at the point of purchase. As the chapter shows, this requires reimagining of GAP in the context of CEAs rather than just translating thinking about field-based food safety aspects to controlled environment conditions. This approach to defining GAP includes updating approaches to risk analysis to recognise the distinctive safety issues in CEA. It also requires setting out a robust food safety hurdle system via clearly-identified control measures with transparent and validated performance criteria to create bespoke CEA food management safety systems. The chapter also explores the potential role of technologies such as IoT, AI and digital twins to promote and deliver GAP.

As Chapter 15 points out, the agricultural sector has a unique role in climate change mitigation. One aspect is the contribution of the sector to the global greenhouse gas (GHG) emissions while the other is the potential to offset emissions through sequestration. However, a key bottleneck is effective methodologies for quantification of GHG emissions and the determination of the level of soil organic carbon for individual farms and specific fields. One approach is on-farm measurement of GHG emissions. The chapter discusses the principles, strengths and weaknesses of different measurement techniques at the plot scale (flux chamber measurement), the field scale (eddy covariance (EC) measurements) and the landscape scale (remote sensing). The EC technique e.g. allows long-term, direct and continuous measurements of turbulent fluxes over an ecosystem, like a grassland or a field, to quantify gases such as carbon dioxide. Remote sensing approaches can potentially use also precision imagery to derive different ground-based variables or fluxes though, while the technology is established for biomass estimates using the normalised difference vegetation index (NDVI), the estimation of GHG fluxes is still under development. The chapter also reviews animal-based GHG emissions measurement techniques, looking at direct gas measurement techniques, spot sampling techniques and indirect measurement techniques.

However, on-farm measurement is costly, labour intensive and often requires expert knowledge. Modelling emissions is an obvious alternative but modelling also requires specific input data from measurements to avoid bias derived from data relevant to one location but not another. Additionally, modelling might not be able to provide the same level of accuracy for GHGEs as with on-farm measurements. The chapter reviews the major types of livestock, crop and grassland models, how they work and their relative strengths and weaknesses.

As the chapter points out, current solutions are still lacking (either too costly, not accurate enough, or both). Future developments will focus partly on the technological developments to find cheaper solutions for accurate measurements and developing systems that combine the advantages of low-cost modelling and remote sensing with the accuracy of on-site measurements. In the short-term modelling will most likely used for estimations. Further development of the models depends on more data of a sufficient temporal and spatial resolution and representing a wide range of different environmental conditions are required to provide a robust database. In the long run, remote sensing will be probably a likely solution for farm-specific measurements of emissions from croplands and grasslands as well as monitoring of grazing animals. One potential approach is inversion measurements (or top-down measurements) using remote sensing combined with modelling (bottom-up approaches).

The final chapter in the book reflects on the wide range of themes covered in the preceding chapters. As it points out, there is now a plethora of both compulsory and voluntary standards that impact food supply chains. As suggested e.g. in the theory of change outlined in Chapter 8, there is a need now to move to greater synergy and possible institutionalisation (e.g. using the hybrid governance models discussed in Chapter 2).

Perhaps more importantly, the chapter highlights that measuring compliance against a set of prescribed normative standards and, as a result, being able to certify that a product meets these standards also does not necessarily embed performance improvement (a theme also discussed in Chapter 5). Continuous improvement requires an organisation not only to comply with prescribed requirements but also to implement management systems that support continuous improvement of processes and operational efficiency. Delivering real improvement requires that characteristics such as integrity, respect, trust, honesty and a genuine desire for change are embedded within organisational practices. Standards need to promote agri-food supply chain resilience, i.e., when subject to a specific shock event, there needs to be the resources and knowhow available to underpin the ability of an organisation or the wider food supply chain to not just 'bounce back' to a steady state, but to 'bounce forward' to a new status that is shaped by the evolving geopolitical and socio-economic environment in which they operate. Standards need to promote agility, and in particular adaptive capacity within organisations in order to build resilience in the often uncertain or changing world in which they operate.

Part 1

Setting standards

Chapter 1

The international context: the role of Codex Alimentarius in setting safety and quality standards for global agri-food supply chains

Tom Heilandt and Amanda Hielm, Codex Alimentarius Commission, Italy

1 Introduction

In January 2023, the 64 ministers present at the 15th Berlin Agriculture Ministers' Conference reiterated their support for the work of the Codex Alimentarius Commission (CAC) or Codex. Codex has 188 member nations, one member organisation (the European Union), and over 230 observers (including other intergovernmental organisations (IGOs) and non-governmental organisations (NGOs) with interest in the work of Codex). This year marks the 60th anniversary of Codex (CODEX@60) which allows for reflection on how fit Codex is for the challenges of the world of today and beyond. Some of these challenges, explored in this chapter, will need intensive intergovernmental debate.

We live in a troubled, fast-changing world where food safety remains an issue, food insecurity is on the rise, and working towards sustainability and promoting healthy diets is vital. Working together in science-based fora like Codex can provide some of the solutions. All consumers are served by Codex, from the youngest, who are still learning how to choose their food, to the schoolchildren who can read labels and should be beginning to learn

http://dx.doi.org/10.19103/AS.2024.0133.01

the fundamentals of nutrition, to the teens who may discover what it means to struggle with nutrition for the first time, to the adults who may have to deal with their own nutrition as well as that of their children and aging parents, and finally to those who want to stay active and healthy for the longest time possible. They are all Codex customers. They are, on the one hand, the most knowledgeable and, on the other, the most perplexed consumers ever, living in a world that is changing quickly and where many people have access to a lot of resources, yet few things are certain or steady.

Setting standards means creating order in chaos. Standards are something to build on and themselves are built on science. Codex is the mechanism that brings the world together to create such standards for food. While the ultimate goal of all food standards work is to protect the consumer, Codex's work also benefits all other actors in the food system. This starts with the producer, by giving indications on what is acceptable and can thus be traded on international markets to helping reduce food waste and loss and thus contributing to food security and sustainability. Governments also benefit through the activities of Codex, both in providing a central forum to discuss cross-border food safety issues and in the development of a reliable, international framework of commonly agreed, ready-to-use international food standards to facilitate global trade. The rest of this Introduction looks at some of the key challenges faced by the global food system.

1.1 We all need food – the food security gap

We need food that is safe, sufficient, and nutritious to live full and healthy lives. Beyond keeping us alive, food is an important element of our culture and society. In our increasingly urbanised and globalised world, fewer and fewer of us produce or even prepare our own food. The food on our plate may have arrived from the other side of the world. Over the last century, the amount of food traded internationally has grown exponentially, and a larger quantity and variety of food than ever before traverses the globe today. This has increased overall food security but, as recent events have shown, these gains are fragile and, if crises accumulate, food insecurity rises. Today there are still too many people who are food and nutrition insecure, and the gap is growing.

1.2 Food is fragile – the food safety gap

Food is a fragile commodity: it can be affected by contamination by microbes, heavy metals, or toxins through soils, production methods, or poor hygiene along the production chain. Though food safety and quality have significantly improved in many parts of the world, the global burden of food-borne diseases remains unacceptably high, and more work needs to be done. The World

Health Organizsation (WHO) estimates that 600 million people fall ill yearly and 420 000 die because of contaminated food. About 30% of these deaths are children under the age of 5.[1]

Food can be tampered with intentionally via food fraud – either for economic reasons or with the intention to cause harm. Insecurity about food, its availability, price, and safety can destabilise a society quickly[2]. From the 1990s research has shown that consumers have became increasingly concerned that the food they eat could be harmful to health[3]. These trends have continued until today with more information being available to consumers everywhere[4].

1.3 Too much safe, good food can kill you - the nutrition gap

Too much food can be as dangerous as not having enough, and a badly composed diet lacking essential nutrients can cause health problems. Information on which nutrients are good for us, how much, and which food components are not is, at times, conflicting and changes over time. To provide a healthy diet, we need to know what is in the food, but we also need to care about what this information means and educating ourselves about good nutrition. Education, labelling, and governments working together with the food industry are vital to make progress in this area.

1.4 Food production impacts the planet - the sustainability gap

To feed a growing population, set to reach 9.6 billion by the year 2050, countries will have to learn to grow what the population needs in a sustainable way and reduce food loss and waste. Hunger and poverty may lead countries to over-exploit the fragile resources on which livelihoods depend. To both meet people's needs for food and other environmental, social and economic objectives, the natural resource base and ecosystems must be managed sustainably. Increased water scarcity and conflicts over access to resources all pose challenges to environmental sustainability and food security. Climate change too is affecting our planet and can make it harder to grow food in some areas and cause safety concerns in others. Without considering how food is produced sustainably, food cannot be safe unless the concept of food safety also takes into account the safety of the environment and the needs of future generations.

Consumers food choices are important in reducing the impact of food production and trade on the environment. However, consumers currently

1 WHO (2015); WHO estimates of the global burden of foodborne diseases.
2 https://en.wikipedia.org/wiki/List_of_food_riots.
3 Margareta Wandel, British Food Journal, ISSN: 0007-070X, Article publication date: 1 August 1994.
4 European perspectives of food safety, Diána Bánáti, First published: 10 February 2014, https://doi.org/10.1002/jsfa.6611.

have difficulty defining what 'sustainability' means for them, and issues such as price, sensory quality, and individual health concerns are more important in influencing what they buy. To change this, policymakers need to communicate knowledge about sustainability in a transparent, evidence-based, and controlled way and to guide consumers by designing science-based and well-regulated sustainability labelling.[5]

There will be necessary debates in the future on if and how Codex, as the world food standard-setting body, can address sustainability issues. The Codex Committee on Food Labelling has recently started a debate on creating principles for sustainability labelling claims[6]. Food production has evolved over millennia and current environmental and other pressures may increase the speed of change. In another 60 years, food systems may be unrecognisable but it is likely that international food standards such as Codex standards will still be needed.

2 The development of food standards

Ensuring that food is safe to eat, and that consumers are not deceived by dishonest practices, have been among the important tasks entrusted to governments since antiquity[7]. Evidence from the earliest historical writings indicates that governing authorities have always been concerned with codifying rules to protect consumers from dishonest practices in the sale of food. Assyrian tablets described how to determine the correct weights and measures for food grains. Ancient Egyptian scrolls prescribed the labelling to be applied to certain foods. In ancient Athens, beer and wines were inspected for purity and soundness. The Romans had a well-organised state food control system to protect consumers against fraud and bad produce. Food additives, substances that are added to food to maintain or improve the safety, freshness, taste, texture, or appearance of food, have been in use for centuries for preservation, such as salt in dried fish or sugar in marmalade. Similarly, the resins added to Greek wine acted as mild preservatives, and in classical times, the pitch linings of amphorae contributed substances that helped to preserve the wine carried in them.

In Europe during the Middle Ages, individual countries passed laws concerning the quality and safety of produce such as eggs, sausages, cheese, beer, wine, and bread – and some of these ancient statutes still exist today. An early example was the thirteenth-century Assize of Bread and Ale enacted in England by Henry III to regulate the quality, weight, and price of bread and beer.

5 Consumers' perceptions on food-related sustainability: A systematic review. M. van Bussel, A. Kuijsten, M. Mars, P. van 't Veer. Journal of Cleaner Production Volume 341, 20 March 2022, 130904.
6 FAO/WHO (2023); Report of the fourty-seventh session of the Codex Committee on Food Labelling, paras. 164–173.
7 FAO/WHO (2018); Understanding Codex.

In the Austro-Hungarian Empire between 1897 and 1911, e.g. a collection of standards and product descriptions for a wide variety of foods was developed as the Codex Alimentarius Austriacus. Although lacking legal force, it was used as a reference by the courts to determine standards of identity/authenticity for specific foods.

At the start of the twentieth century, food safety regulation was still limited in scope, particularly in regulating the growing international trade in food production. In those days, the use of unsafe food additives, especially preservatives, was common practice in many parts of the world. As refrigeration, or cold chains, were significantly less developed, and many homes did not have refrigerators, the foods industry embraced the opportunities that chemistry presented to solve one problem, preservation, while unintentionally creating another.[8] This required research on and regulation of food additives. Many countries independently developed food laws and regulations, and often found different solutions to ensure that food was safe, of sufficient quality, and that consumers had sufficient and accurate information about what was in the products they were purchasing. However, differences between national requirements and specifications often made it difficult to trade food across borders. A major step forward in resolving these differences and developing a common approach came in 1956 when the Food and Agriculture Organisation of the United Nations (FAO) and the World Health Organization (WHO) created the Joint Expert Committee on Contaminants and Food Additives (JECFA) to scientifically examine the safety of food additives and the impact of contaminants (and later also of residues of veterinary drugs) in food.

2.1 The Codex Alimentarius Commission

In November 1961, a FAO conference passed a resolution (12/61) on establishing the Codex Alimentarius Commission[9]. At the joint FAO/WHO conference on food standards in Geneva, 1–5 October 1962, Codex's operational guidelines, first meeting date, and financial statements were established (Alinorm 62/8)[10]. Finally, on 23 May 1963, the 16th World Health Assembly adopted a resolution[11] on establishing Codex, paving the way for the first session of the Commission. CAC1 was held in Rome from 25 June to 3 July 31963, with 120 participants from 30 member countries and 16 international governmental and non-governmental organisations (Alinorm 63/12)[12].

8 Deborah Blum, The Poison Squad, 2018.
9 https://resolutionsportal.fao.org/akn/un/statement/deliberation/fao/1961-11-24/12-1961/eng@/!main.
10 https://www.google.com/url?sa=t&rct=j&q=&esrc=s&source=web&cd=&ved=2ahUKEwi-wdWis_D8AhVpi_0HHSRMBSQQFnoECBIQAQ&url=https%3A%2F%2Fwww.fao.org%2Finput%2Fdownload%2Freport%2F715%2Fal62_08e.pdf&usg=AOvVaw1fR7K_So0OtuOMcdsN1sjQ.
11 https://apps.who.int/iris/bitstream/handle/10665/89162/WHA16.42_eng.pdf?sequence=1&isAllowed=y.
12 https://www.fao.org/fao-who-codexalimentarius/sh-proxy/en/?lnk=1&url=https%253A%252F%252Fworkspace.fao.org%252Fsites%252Fcodex%252FMeetings%252FCX-701-01%252Fal63_12e.pdf.

The mandate of Codex from the Statutes[13] states:

> The Codex Alimentarius Commission shall…be responsible for making proposals to, and shall be consulted by, the Directors-General of the Food and Agriculture Organisation (FAO) and the World Health Organisation (WHO) on all matters pertaining to the implementation of the Joint FAO/WHO Food Standards Programme, the purpose of which is:
>
> - Protecting the health of the consumers and ensuring fair practices in the food trade.
> - Promoting coordination of all food standards work undertaken by international governmental and non-governmental organisations.
> - Determining priorities and initiating and guiding the preparation of draft standards through and with the aid of appropriate organisations.
> - Finalising standards elaborated under (c) above and publishing them in a Codex Alimentarius either as regional or worldwide standards, together with international standards already finalised by other bodies under (b) above, wherever this is practicable.
> - Amending published standards, as appropriate, in the light of developments.

The so-called 'double mandate' of Codex addresses both:

- food safety (i.e. 'protecting the health of consumers'); and
- other food-related issues, such as food quality, food identity, trade descriptions, labelling, claims, import and export inspection, certification systems, etc.

In the case of health and food safety, the principles of risk analysis (with its components of risk assessment, risk management, and risk communication) have been well integrated into all work of Codex food safety committees (Additives, Contaminants, Hygiene, Methods, and Residues (Pesticides and Veterinary Drugs). Scientific risk assessments form the foundation for Codex in taking decisions on what measures to recommend (risk management). This structure opened the way for Codex to become the reference point for food safety in the World Trade Organization (WTO) Agreement on Sanitary and Phytosanitary Measures (SPS-Agreement) published on 1 January 1995.[14] Established in 1995 (and replacing the earlier General Agreement on Tariffs and Trade agreed in 1948), the WTO brings together national governments to agree a common framework to promote global trade by reducing tariffs and other barriers.

In the second area, ensuring fair practices in the food trade, a lot depends on identifying best practices and what can be considered 'fair' in the food

13 Codex Procedural Manual.
14 https://www.wto.org/english/tratop_e/sps_e/spsagr_e.htm.

trade. This is particularly relevant to claims made about food, which should be substantiated and not mislead the consumer. While it has been suggested that investing effort in establishing standards for complex foods or defining quality classes for fresh fruits and vegetables has not been beneficial, other participants have countered that such standards are necessary. An illustration would be the failed attempt to establish a Codex standard for processed cheese due to a lack of global agreement[15]. Because of the complex situation this creates, and the number of different subjects it covers, the WTO agreement on Technical Barriers to Trade (TBT-Agreement, also published on 1 January 1995)[16] makes no attempt to define specific standard-setting organisations but still asks members to base their measures on international standards. Codex standards like the Standard for Sardines have been used in this context to resolve trade conflicts. Discussions regarding 'fair' trade practices are planned in the near future with regard to the possible development of guidelines for sustainability labelling.

At the heart of the Codex mandate are the core values of collaboration, inclusiveness, consensus building, and transparency. Governmental and non-governmental, public, and private organisations alike play a vital role in ensuring Codex texts are of the highest quality and based on sound science. Codex would have little authority in the field of international standard setting if it did not welcome and acknowledge the valuable contributions made by observers. Expert technical bodies, industry, and consumer associations contribute to the standard-setting process in a spirit of openness, collaboration, and transparency. Intergovernmental organisations (IGOs) and international non-governmental organisations (NGOs) can apply for observer status in Codex to attend and put forward their views at every stage of the standard-setting process. The Codex Alimentarius Commission currently has 235 Codex Observers (60 IGOs, 159 NGOs, 16 UN). The close collaboration between Codex and numerous organisations during the work process has greatly aided in the coordination of all global efforts on food standards.

Since its foundation, the Codex system has evolved in an open, transparent, and inclusive way to meet emerging challenges. The Codex Alimentarius international food standards, guidelines, and codes of practice (described here as the Codex standards) contribute to the safety, quality, and fairness of international food trade. As a result, consumers can trust the safety and quality of the food products they buy, and importers can trust that the food they order will be in accordance with their specifications. Codex standards are recommended to governments and trade for application. They are not a

15 FAO/WHO (2017), Report of the 40th Session of the Codex Alimentarius Commission, paras 124–129.
16 https://www.wto.org/english/docs_e/legal_e/17-tbt_e.htm.

substitute for, or alternative to, national legislation. Every country's law and administrative procedures contain provisions with which it is essential to comply. However, the reference made to science-based Codex food safety standards in the World Trade Organization's SPS Agreement means that Codex has far-reaching implications for resolving trade disputes. WTO members that wish to apply stricter food safety measures than those set by Codex may be required to justify these measures scientifically.

3 The Codex organisational structure

Codex develops standards in a cooperative way. As a result, it has become the place where governments from around the world convene with other interested parties to mutually address disputes about food safety and quality. The Codex agenda is completely driven by its members while the Commission considers, where appropriate, the relevant policies, strategies, and guidelines of FAO and WHO, and cooperates in line with its mandate with other intergovernmental organisations. With increased globalisation and increases in the volumes of food traded internationally, the Commission must be capable of responding in a timely manner to the impacts of emerging trends and challenges on consumer health protection or fair practices in the food trade, and to the extent that steps to address those impacts are amenable to standardisation.

Codex prides itself on being open, transparent, and inclusive. All information about Codex work is publicly available for free on the Codex website. The work is member-driven, and it aims to reach consensus on decisions even in the most challenging situations. The Commission itself is the ultimate decision body, meeting annually in Rome or Geneva, and usually attended by more than 600 delegates from over 100 countries.

The Commission is supported by an Executive Committee consisting of elected members and regional coordinators and by an international secretariat. The technical work is done in horizontal and vertical committees or task forces, each of which is hosted and chaired by a member country. The structure differentiates between two different kinds of committees:

- horizontal committees (dealing with thematic issues relating to all foods such as labelling, hygiene, additives, contaminants, or methods of analysis); and
- vertical committees (dealing with a specific food group).

Task forces are formed to deal with specific problems and are dissolved once they have achieved their work. Presently one task force is actively dealing with the challenges of food-borne antimicrobial resistance.

In addition, the FAO/WHO coordinating committees for each of the six Codex regions make sure that specific regional aspects are addressed by Codex. In some cases, these committees develop standards for regional products. The six regional FAO/WHO Coordinating Committees provide vital geographical coverage to identify the challenges and needs particular to each region in food standards and food control. These Committees have recently embarked on a process of revitalisation, aimed at boosting their role as the preeminent fora for the discussion of food safety and quality issues at the regional level.

The Commission adopts food safety standards using the principles of risk analysis and bases its work on the scientific advice provided by joint FAO/WHO expert bodies:

- JECFA (Joint FAO/WHO Expert Committee on Food Additives) – food additives, contaminants, and veterinary drugs;
- JMPR (FAO/WHO Joint Meeting on Pestcide Residues) – pesticides;
- JEMRA (Joint FAO/WHO Expert Meetings on Microbiological Risk Assessment) – microbiological risks; and
- JEMNU (Joint FAO/WHO Expert Meetings on Nutrition) – nutrition.

Codex standards may also address issues related to food quality and labelling.

3.1 The Codex standard-setting process

The Codex standard setting process depicted in has eight steps consisting of the following:

- An initial reflection on which new work is needed (discussion paper);
- Agreement of new work (step 1);
- Drafting (step 2);
- Circulation of the draft for comments to all members and observers (step 3);
- Discussion in the technical committee or task force (step 4);
- Intermediary adoption by the Commission (step 5);
- Circulation for comments (step 6);
- Discussion in the technical committee or task force (step 7); and
- Adoption as a Codex standard (step 8).

The Commission reviews a project document presented by a member and decides whether the standard should be developed as proposed. If new work is approved, the Codex Secretariat arranges for the preparation of a proposed draft standard and circulates it to member governments and observer organisations for two rounds of comments. The text also goes to

Codex committees dealing with issues such as labelling, hygiene, additives, contaminants, or methods of analysis for endorsement of any special advice in these areas. Once adopted by the Commission, a Codex standard is added to the Codex Alimentarius and published on the website.

The process is also designed to allow sufficient input from all stakeholders to achieve a consensus. Steps 6 and 7 may be omitted and the overall process can be finished within 1 year or take longer, depending on the complexity of the issue. The process also allows for input received from the FAO/WHO scientific advice provided by standing committees (JECFA – food additives, contaminants, and veterinary drugs; JMPR – pesticides; JEMRA – microbiological risks; and JEMNU – nutrition). The FAO and WHO expert bodies for risk assessment establish the scientific basis for Codex standards. These expert bodies are independent of the Commission and its subsidiary bodies, such that their work contributes significantly to the scientific credibility of the Commission's own work. Codex strictly adheres to its established principles of risk analysis that ensures the functional separation/independence of science-based risk assessment from the practical realities of risk management.

3.2 Based on science, adopted by consensus

The debate regarding working by consensus within Codex has a long history.[17] It intensified in the late 1990s following a number of decisions taken by a vote in the Commission that was passed by only a narrow majority, most notably the 1995 and 1997 decisions to adjourn debate on maximum residue levels (MRLs) for bovine somatotropins, and the 1997 decision to adopt the revision of the Standard for Natural Mineral Waters (CXS 108-1981).

At that time, the Guidelines for Codex Committees, in the Procedural Manual, already stated that 'the chairperson should always try to arrive at a consensus and should not ask the Committee to proceed to voting if agreement on the Committee's decision can be secured by consensus.' However, the Commission deemed it necessary to strengthen the importance of consensus further and amended the Rules of Procedure (Rule XII Elaboration and Adoption of Standards) to stress that voting should take place only after all efforts to reach agreement by consensus on the adoption or amendment of standards have been exhausted.

In 2000, the Commission also adopted the Guidelines to Chairpersons of Codex Committees and ad hoc Intergovernmental Task Forces and Measures to facilitate consensus (contained in the Procedural Manual). These highlighted the need to ensure that Codex texts were scientifically sound, had been thoroughly discussed by the committees, and were not forwarded to the Commission for

17 Arietta and Wallace (2000);

adoption until consensus has been achieved at the technical level. These texts also provide for:

- the organisation of informal meetings of concerned parties;
- the redefinition of the scope of the texts under consideration to resolve issues on which consensus cannot be reached;
- the formulation of compromises with a view to advancing texts in the Step procedure; and
- the involvement and participation of developing countries.

The concept of consensus was also considered by the 2002 Evaluation[18] of Codex. Its findings stressed that active consensus building was vital to the legitimacy of Codex and recommended that Codex develop guidelines for committees on the meaning of consensus and on decision-making in the absence of consensus. The Evaluation also highlighted the advantages of consensus building and of intersessional facilitation to consider the reduced capacity of delegates to consult with their governments during committee sessions.

Further discussions on consensus were held in 2009 in the Committee on General Principles following proposals to create a Codex definition of consensus. At the time it was decided that a definition might restrict the freedom of Chairpersons to manage the debate and that it was more important to work with Codex Chairpersons to give them the tools to actively contribute to consensus building. The possibility of using facilitators was also introduced to assist Chairpersons in their work. Given that votes arise rarely in Codex (the last two being 2012 and 2022 over the dispute over MRLs for growth promoters used in meat production), it can be argued that the consensus-building system in Codex works well. If a conflict arises and no consensus can be reached it can and should eventually be resolved by vote without putting into question the overall objective of the Commission to strive for consensus.

3.3 Codex achievements

Consumers expect that their governments will take legislative and regulatory action to ensure that only safe food of acceptable quality is sold and that the risk of foodborne health hazards is minimised. Codex has contributed significantly to securing the place of food on political agendas worldwide and, by pooling knowledge, expertise, and best practice, has contributed to higher and more consistent standards to ensure food safety. Governments and food business operators are highly conscious of the consequences should they fail to heed

18 FAO/WHO (2002).

consumers' concerns when it comes to the food they eat. The right to food for everyone is universally accepted, but it is as important that people have the right to expect their food to be safe, of good quality and suitable for consumption. This makes continued Codex work a necessity. Food-borne illnesses are at best unpleasant; at worst, they can be fatal. Outbreaks of foodborne illness can damage trade and tourism and can lead to loss of earnings, unemployment, and litigation. Poor-quality food can destroy the commercial credibility of suppliers, both nationally and internationally, while food spoilage is wasteful and costly, and can adversely affect trade and consumer confidence. Food rejections in trade lead to further food waste, which is difficult to tolerate in a world where many still suffer from hunger.

Codex addresses these challenges by serving as the most important global forum for food safety and quality. By encouraging them to work together, Codex creates invaluable links between different food safety authorities. However, the most important Codex output is the Codex standards. In the 1960s Codex took the initial approach of setting standards for individual foods. This turned out to be impractical, both in keeping standards up-to-date and in their effective implementation. This approach gave way to the practice of creating group standards. As an example, one general standard for fruit juices and nectars replaced the many previous standards for individual fruit juices. This has been done for other commodities and is applied when creating new standards such as those for spices. The benefits of this approach include broader coverage and allowing for innovation in the development of new food products within a food group.

Since 1963, Codex has set commodity standards for all major food commodities as well as 79 Guidelines dealing with general issues such as food import and export inspection and certification systems and 54 Codes of Practice addressing, for example, guidance on preventing contamination (chemical, physical, or microbiological) of food products. In addition, the Commission has created a comprehensive system of thousands of MRLs for food additives, pesticides, and veterinary drugs and maximum limits for important contaminants.

A signature Codex achievement is the widespread adoption of HACCP principles and systems[19]. Food business operators have an interest in ensuring that food products are safe, from farm to fork. One important preventive tool they can use is the Hazard Analysis and Critical Control Point (HACCP) system which identifies specific hazards and measures for their control to ensure the safety of food. Initially developed by the US National Aeronautics and Space Administration (NASA) to provide safe food for astronauts, HACCP was adopted by Codex in 1969 to assess hazards and establish control systems that focus on

19 Codex Alimentarius (2019), Food Hygiene at 50.

the prevention of hazards rather than relying mainly on end-product testing. Any HACCP system that is developed and operationalised must be capable of accommodating change, such as advances in equipment design, processing procedures, or technological developments. HACCP systems are supported by prerequisite programmes such as good agricultural practices (GAP), good hygiene practices (GHP), and good manufacturing practices (GMP). These outline basic operating principles and procedures (e.g. those relating to design and layout of facilities to minimise contamination risks, cleaning and disinfection protocols, etc.).

HACCP can be applied throughout the food chain from primary production to final consumption and its implementation should be guided by scientific evidence of risks to human health. As well as enhancing food safety, implementation of HACCP can provide other significant benefits in terms of developing a robust food safety management system. In addition, the application of HACCP systems is a regulatory requirement in many countries and can aid inspection by regulatory authorities and promote international trade by increasing confidence in food safety.

The successful application of HACCP requires the full commitment and involvement of management and the workforce. It also requires a multidisciplinary approach. This multidisciplinary approach should include, when appropriate, expertise in agronomy, veterinary health, production, microbiology, medicine, public health, food technology, environmental health, chemistry, and engineering, depending on the context. It can be argued that the application of Codex HACCP principles by food business operators is probably the single biggest intervention in food safety that has saved more lives through prevention than any other. It is THE safety belt for food. Codex has recently finalised (2023) an important revision of its signature tool through the Committee on Food Hygiene.[20]

4 The Sustainable Development Goals (SDGs) and the Codex Strategic Plan 2020-2025

Codex has developed a medium-term Strategic Plan 2020–2025[21]. This ties into the United Nations 2030 Agenda for Sustainable Development and its 17 Sustainable Development Goals (SDGs). The focus of Codex on food safety and quality, and on ensuring a level playing field for trade, healthy, and nutritious diets and consumer information, has the potential to improve the world we live in. Codex work has a clear connection to the six following SDGs for safer, better food for everyone, everywhere.

20 https://www.fao.org/fao-who-codexalimentarius/publications/en/.
21 Codex Alimentarius (2019); FAO (2019a, b); WHO (2018); UN (2015).

4.1 SDG1: no poverty

If a country incorporates global food safety standards such as Codex standards into national legislation, then local traders can access new markets and grow their businesses and create employment. Having better access to global markets has the potential to help alleviate poverty in agriculture sectors in many countries.

4.2 SDG2: zero hunger

The first target of this goal is (by 2030) to 'end hunger and ensure access by all people, in particular the poor and people in vulnerable situations, including infants, to safe, nutritious and sufficient food all year round'. Since food security cannot exist without food safety, there is a significant connection between the activities of Codex and the successful achievement of food security. Unsafe food cannot be considered food and is especially dangerous to those already weakened by malnutrition. Codex strives to ensure food safety, thus contributing to food security.

This goal also states the need to 'correct and prevent trade restrictions', which aligns with Codex standards whose function is to harmonise international rules so that food can cross borders safely and without unfair restrictions. This goal also addresses trade and states the need to correct and prevent trade restrictions, another central aspect of the Codex mandate. Codex is positively advocating this goal by contributing to making it possible to bring safe and nutritious food to every home.

4.3 SDG3: good health and well-being

When developing and transitioning economies engage in the development of Codex standards, and then use those standards as the basis for national standards, they are protecting consumer health and well-being both within and outside their national borders. Strengthening food safety systems will enable countries to take measures to reduce the burden of foodborne disease while strengthening food labelling standards will allow consumers to make more informed decisions about the nutritional quality of what they eat.

4.4 SDG8: decent work and economic growth

The targets of this goal are linked primarily to growth, productivity, and employment. The work of Codex has the potential to help countries work towards achieving 'full and productive employment and decent work for all'.

Trade is a vital opportunity for growth in this area and the Codex standards promote fair practices in food trade by eliminating trade restrictions and barriers to trade.

4.5 SDG12: food waste and management of chemicals that could be a public health concern

One of the targets of goal 12 is 'halving per capita global food waste at the retail and consumer levels and reduce food losses along production and supply chains'. Codex's work on accurate and reliable food labelling links directly to this goal. The goal also calls for sound management of chemicals. Chemicals in food are a worldwide health concern and are a leading cause of trade obstacles. The expert advice provided to Codex by FAO and WHO ensures that correct exposure levels are set in Codex standards to protect consumer health.

4.6 SDG17: partnerships for the SDGs

Codex itself is one of the longest-standing and most successful examples of partnership within the United Nations, alongside other bodies such as WOAH (World Organisation for Animal Health – formerly OIE) and the Intergovernmental Panel on Climate Change (IPCC). Codex has, e.g. worked in partnership with WOAH on global issues such as antimicrobial resistance (AMR). As noted earlier, it provides a benchmark for commonly agreed standards in WTO trade disputes. Codex believes in the value of partnerships with all its stakeholders, both members and the over 230 observers from different walks of life. As an example, a longstanding partnership with the International Dairy Federation (IDF) has resulted in global standards, guidelines, and Codes of Practice for milk and milk products. Codex also closely cooperates with the International Standards Organization (ISO) on standardised methods of analysis and sampling and other topics. Codex sees the advent of private standards as an opportunity and cooperates closely with organisations formed by the private sector to create systems that facilitate the application of Codex standards. Examples include the Global Food Safety Initiative (GFSI), set up by leading food manufacturers and retailers to address specific food safety problems encountered by GFSI members, and SSAFE, a non-profit organisation bringing together the private and public sector to ensure the integrity of agri-food supply chains in such areas as food fraud. When Goal 17 describes a 'universal, rules-based, open, non-discriminatory and equitable multilateral trading system', it is pointing directly at the mandate, goals, and daily work of the Codex Alimentarius Commission.

4.7 Codex strategic plan

The Codex Strategic Plan 2020–2025[22] sets out five strategic goals for the next 6 years and defines the vision and mission of the Commission:

- **Codex Vision:** 'Where the world comes together to create food safety and quality standards to protect everyone everywhere.'
- **Codex mission:** 'Protect consumer health and promote fair practices in the food trade by setting international, science-based food safety and quality standards.'

The Plan recognises the important contributions of Codex to achieving the SDGs and using the core values of inclusiveness, collaboration, consensus building, and transparency set out five strategic goals:

- Address current, emerging, and critical issues in a timely manner;
- Develop standards based on science and Codex risk-analysis principles;
- Increase impact through the recognition and use of Codex standards;
- Facilitate the participation of all Codex Members throughout the standard-setting process; and
- Enhance work management systems and practices that support the efficient and effective achievement of all strategic plan goals.

While strategic goals 1, 2, 4, and 5 will be working towards the traditional goals of Codex in setting standards, goal 3 is the task to look further into what happens with the standards once adopted by the Commission and what their value is as an international common good.

5 Codex, the World Trade Organization, and trade harmonisation

One of the founding and guiding principles of Codex is to promote global trade in food[23]. Peace and prosperity depend on food availability which, for many countries, involves on imports to supplement national food production. One of the principal concerns of national governments the world over is the need to ensure that food imported from other countries is safe and poses no threat to the health of consumers or the wider health and safety of their animal and plant populations. However, different national priorities, approaches, and standards create significant potential barriers to trade between countries.

22 Codex Alimentarius (2019).
23 FAO/WTO (2017); Trade and food standards.

The goal of Codex is that, if every country harmonised towards a single set of international food standards, all countries would encounter fewer trade barriers, allowing food products to move more freely among countries. This benefits farmers and their households and has the potential to contribute to reducing hunger and poverty. However, it can be challenging for some countries to integrate Codex standards fully into their national legislation. This may be due to differing levels of capacity to implement standards (e.g. weaknesses in government capacity to implement and monitor standards, lack of sufficient infrastructure to support effective agri-food supply chains, or the challenges faced by agri-food supply chain actors (such as smallholder producers) to implement standards due to constraints in finance, availability of appropriate technologies and expertise. More broadly, differing concepts of food safety, national attitudes, and concepts of sovereign rights can impede the progress of harmonisation and the use of Codex standards. The increasing level of participation by countries in the work of Codex shows that, despite these difficulties, countries value open technical discussion and agree that they are useful as a way to find an international consensus to support the continuous improvement of rules and regulations.

While many trade agreements have implications for food, two WTO agreements are of particular interest with respect to the work of the Codex Alimentarius Commission:

- the Agreement on the Application of Sanitary and Phytosanitary Measures (SPS); and
- the Agreement on Technical Barriers to Trade (TBT).

These are discussed in the following sections.

5.1 The WTO SPS Agreement

The SPS Agreement contains provisions with respect to measures that may be taken to protect animal, plant, and human health. Regarding human health, it acknowledges that governments have the right to take sanitary and phytosanitary measures necessary for the protection of the health of their citizens. However, the Agreement requires them to apply those measures only to the extent required to protect human health. It does not permit member governments to discriminate by applying different requirements to different countries where the same or similar conditions prevail unless there is sufficient scientific justification for doing so.

The SPS Agreement specifically identifies the standards, guidelines, and recommendations established by Codex for food additives, veterinary drug and pesticide residues, contaminants, methods of analysis and sampling, and codes and guidelines of hygienic practice. This means that Codex standards are considered scientifically justified and are accepted as the benchmarks against which national measures and regulations are evaluated. WTO members wishing to set stricter food-safety standards than Codex standards may be required to provide evidence as to the scientific and nondiscriminatory nature of their measures, should a trade dispute arise. If there is a dispute between two members, WTO will encourage them to resolve the matter amicably; only in rare cases are official dispute settlement procedures initiated at the request of members.

Recently, WTO members questioned how the SPS agreement could contribute to sustainability. This will inevitably also have implications for Codex:

> ... the SPS Committee should explore how the implementation and application of the SPS Agreement can support... ***[facilitation of] global food security and more sustainable food systems***, including through sustainable growth and innovation in agricultural production and international trade, and ***through the use of international standards***, guidelines, and recommendations developed by the Codex Alimentarius Commission, the World Organisation for Animal Health and the International Plant Protection Convention as the basis of harmonised SPS measures to protect human, animal or plant life or health. [24]

5.2 The WTO TBT Agreement

The WTO TBT Agreement seeks to ensure that technical regulations and standards, including packaging, product marking and labelling requirements, and analytical procedures for assessing conformity with technical regulations and standards, do not create unnecessary obstacles to trade. The TBT Agreement applies to all regulations and standards and is not specific to food.

On 26 September 2002, in one of its first rulings on the TBT Agreement, the Appellate Body of the WTO upheld a Panel finding in favour of Peru that sardines caught in the Eastern Pacific (from the Sardinops sagax species) could be marketed and labelled as canned sardines in the European Union. The dispute arose when an EC Council Regulation stated that only sardines from the species Sardinia pilchardus Walbaum (mainly found around the Eastern North Atlantic coast) could be marketed as preserved sardines, thereby excluding Peruvian sardines.

The WTO ruling found that the EC Regulation was inconsistent with the TBT Agreement because it was not based on the relevant international standard established by Codex (CXS 94-1981) for preserved sardines and sardine-type products. That Codex Standard sets out what can be written on a food label and characteristics of 21 species from which canned sardine or sardine-type

24 Sanitary and Phytosanitary Declaration for the 12th Ministerial Conference: Responding to Modern SPS Challenges, World Trade Organisation, June 2022.

products can be prepared, including both Sardinops sagax sagax and Sardina pilchardus Walbaum. As a result of this dispute, a mutually agreed solution was reached whereby the EC Regulation was amended to address Peru's concerns and allow trade to continue.

Codex standards for food labelling seek to ensure that consumers have all relevant information about the food they buy and ensure that they are not misled. A potentially useful but also potentially misleading aspect of labelling is related to the sustainability of a food product. Codex is now in the initial stages of discussions on what could be given as useful guidance to countries to ensure that such labels are useful and not misleading.

6 Conclusion and future trends

For six decades, Codex texts have contributed to the safety and quality of the food we eat. The Codex Alimentarius forms a global rule book that everyone in the food chain can follow. The Codex Alimentarius Commission's open and participatory standard-setting procedure – gathering nations to discuss science-based evidence side by side – also serves as a lifeline to build the capacity of countries working to strengthen their own national food safety control systems. Codex has set the standards for all major issues in food trade and most national legislation is based on what Codex proposes.

The General Principles of Food Hygiene (which include the crucial HACCP system), the General Standard for Food Additives and Contaminants, as well as related codes to reduce contamination, are important globally recognised texts on food safety. The Codex Food Labeling System provides nations with all the necessary advice to guarantee that customers are given accurate information about products and are not misled. The Codex Risk Analysis System both for use in Codex itself and for application by national governments offers a comprehensive way for members to protect their populations based on sound science. The risk analysis guidelines for foods made with modern biotechnological techniques (such as genetic modification) and addressing AMR are groundbreaking texts. In the area of nutrition and foods for special uses, Codex has contributed to improving the situation for people intolerant to gluten as well as giving advice on infant and baby foods and on food supplementation. Codex nutrition and labeling standards have also contributed to the implementation of the WHO Global Strategy on Diet, Physical Activity and Health.

Food safety is an important topic that many only think about when they encounter a problem. To change this and keep food safety on the agenda constantly, Codex, FAO, and WHO together with its members have successfully advocated for creating a World Food Safety Day (7 June each year). This is a constant reminder in the calendar that food safety is everyone's business (and there cannot be a food business without safety).

The current Codex strategic plan will guide Codex work from 2020 to 2025. This will be important to ensure that Codex is setting the right standards in the right way and to make sure that the standards developed serve their purpose and are used by members. Speed of standard setting and efficient and effective participation of all members are still challenging issues, as is the financing of the scientific advice underlying Codex and the capacity building for developing countries, together with their effective participation in Codex meetings. The Codex Trust Fund[25] and the Standard and Trade Development Facility,[26] as well as other multilateral and bilateral initiatives, have helped to improve the situation, but there is a long road ahead to reach equality of all.

Codex has faced many challenges to standardisation in the past and will face more in the future. With constant innovations in science and technology, it is now possible to detect minute quantities of, e.g. radiation from Chernobyl (site of a major nuclear disaster in 1986) in blueberries exported from Eastern Europe, and ever smaller residues of pesticides in fruits and vegetables, or the detection of contaminants in food where there is no regulatory level. All these improved detection methods may lead in the future to rejections of imported food and may lead to trade conflicts in the WTO. Codex has developed Guidelines for Rapid Risk Analysis Following Instances of Detection of Contaminants in Food where there is no Regulatory Level (CXG 92-2019) which help to address this particular issue. Other areas for future development include traceability where innovations in tracking technologies, traceability systems, electronic documentation, and labelling technologies have made it easier to track ingredients and products from farm to fork, despite the complexity of globalised agri-food supply chains. This has the potential to tackle ongoing issues such as food fraud. Other opportunities include new partnerships, e.g. with the United Nations Environment Programme (UNEP) to strengthen the way Codex addresses sustainability issues.

Uncertainty remains on where the future challenges are likely to arise for national food safety systems and where governments will need to invest to be prepared. Countries will need to be pro-active and to remain ever watchful, regularly update capacities and stay informed to identify the new challenges but also avoid to potentially over-reacting. By establishing an internal consensus, Codex can minimise this risk.

Over the last 2 years, concerns have been expressed by a number of Codex Members on the use of the Fourth Paragraph of the Statements of Principle (SOP) Concerning the Role of Science in the Codex Decision-Making Process and the Extent to which Other Factors are Taken into Account,[27] which provides

25 https://www.who.int/initiatives/codex-trust-fund.
26 https://standardsfacility.org.
27 Codex Alimentarius (2019), Report of Seventy-Seventh Session of Executive Committee of the Codex Alimentarius Commission.

member countries with the option to 'abstain from acceptance of the relevant standard without necessarily preventing the decision by Codex', when Members 'agree on the necessary level of protection of public health but hold differing views about other considerations'. Subsequent discussions have highlighted the central role of the Chairpersons of Codex Committees and ultimately of the Codex Alimentarius Commission to advise Members not in agreement to clearly indicate the reasons for their non–acceptance of the scientific reasoning underpinning a particular standard as a basis for further discussion.

While there is still no clear direction provided as to whether achieving consensus should be linked to the SOP, the importance of efforts being made throughout the standard-setting process to achieve consensus remains one of the key issues. Providing more guidance on consensus building, not just in relation to the application of the SOP, but to the progression of standards in general would also be beneficial. As mentioned earlier, consensus per se is not defined in Codex, though chairpersons of Codex Committees and Commission procedures are required to make 'every effort to reach agreement on the adoption of standards by consensus'. However, consensus cannot be always reached, and voting occurs. As defined by Arietta and Wallace (2000) 'consensus is a journey and a destination" and, as in every journey, several obstacles can be found on the road before reaching the destination. Voting is just a way to overcome these obstacles although it is not the preferred way.

Since its inception, one of the main features of Codex has been to hold technical meetings in different host countries. This fosters a global community of food safety and quality experts that can learn from each other, including from the systems and procedures operating in a host country. Networking is an important part of Codex meetings: people who have worked together at Codex find it also easier to cooperate when areas of difference arise. In 2018–2019, Codex held physical meetings on 148 days in 20 countries.

In 2020, Codex held only five days of physical meetings, given the global COVID-19 Pandemic. While many physical meetings were cancelled, work did not stop. Codex is a member-driven programme and members were keen to continue the work on international food standards through alternative mechanisms such as virtual meetings. Ongoing research included assessment of whether SARS-CoV-2 can be transmitted through food packaging or other materials in the food supply chain.[28]

After an initial learning and adaptation period, the Codex membership decided to go virtual. In just over 14 days, 147 Codex members endorsed holding a virtual commission which took place successfully between September and November 2020, adopting important Codex standards which included a long-awaited upgrade to the Codex General Principles of Food Hygiene and

28 https://www.efsa.europa.eu/en/topics/efsa-and-covid-19.

the HACCP system. Over 133 member countries, one member organisation (EU), and 54 observer organisations participated with a total of 905 registered participants.

By December 2021, 16 virtual international meetings with thousands of participants had been held. This included a virtual Commission session broadcast from Rome while in-person elections for the new bureau were held in Geneva. An unexpected benefit was that, for the first time in its existence, all Codex members had an equal opportunity to participate virtually in the work of the Commission.

2022 saw 11 virtual meetings and the return to physical meetings, beginning in November with the 45th Session of the Commission. The Codex Committee on Food Hygiene meeting was held physically but with a live broadcast available, i.e. a hybrid format. Since then, most meetings have been held physically with the option for those who cannot be present to attend virtually. As I am writing these lines in May 2023, I am attending the 26th Session of the Codex Committee on Food Import and Export Inspection and Certification Systems. It is the first technical Committee to meet in a hybrid format. Issues with hybrid meetings include costs of setting up and managing virtual meetings, quality of internet connections, challenges managing time zones (including making allowances for those obliged to attend meetings at unsociable hours), and ensuring those participating virtually are able to contribute as fully as those physically present at a meeting, as well as creating opportunities for informal exchanges with peers that can be as important as what is discussed in formal meeting agendas.

Whatever the future holds, Codex will be ready to work closely with members and observers and develop together. As an international tool Codex will uphold its values of transparency, inclusiveness, cooperation, and consensus building and will continue to be science and evidence based and be true to the commitment to improve people's lives when setting international standards.

7 References

Arietta, D. L. and Wallace, L. (2000), *Consensus building fieldbook*. Minneapolis: University of Minnesota Extension and Hubert H. Humphrey Institute of Public Affair.

Codex Alimentarius (2016), Report of the Twentieth Session of the FAO/WHO Coordinating Committee for Asia.

Codex Alimentarius (2018), Codex – A world full of standards.

Codex Alimentarius (2018), Procedural Manual.

Codex Alimentarius (2019), Codex Strategic Plan 2020–2025.

Codex Alimentarius (2019), Food Hygiene at 50.

Codex Alimentarius (2019), Report of Seventy-Seventh Session of Executive Committee of the Codex Alimentarius Commission.

Dressler, L. (2006), *Consensus through conversation*. San Francisco, CA: Berrett-Koehler Publishers, Inc.

Hoffman, S. J., Baral, P., Van Katwyk, S. R. and Poirier, M. J. P. (2022), *International treaties have mostly failed to produce their intended effects*. Edited by Douglas Massey, Princeton University, Princeton, NJ; received December 20, 2021; accepted May 17, 2022.

FAO (2019a), Measuring the SDGs – Improving country data for monitoring SDG achievements and informing policy decisions.

FAO (2019b), Statistical Capacity Assessment for the FAO-relevant SDG Indicators 2018/19.

FAO/WHO (2002), Report of the evaluation of the Codex Alimentarius and other FAO and WHO food standards work.

FAO/WHO (2017), Codex Alimentarius Commission.

FAO/WHO (2018), Understanding Codex 5th edition.

FAO/WHO (2023), Report of the fourty-seventh session of the Codex Committee on Food Labelling.

PNAS Research Article | Political Sciences Open Access International treaties have mostly failed to produce their intended effects.

UN (2015), Transforming our world: The 2030 Agenda for Sustainable Development.

WHO (2007), Assessment of the Risk of Hepatotoxicity with Kava Products.

WHO (2010), Global strategy to reduce the harmful use of alcohol.

WHO (2015), WHO estimates of the global burden of foodborne diseases: foodborne disease burden epidemiology reference group 2007-2015.

WHO (2017), Resource tool on alcohol taxation and pricing policies.

WHO (2018), Progress towards the SDGs: A selection of data from World Health Statistics 2018.

Woolfe, M. and Ditton, T. (2013), Food origin labelling legislation and standards. In P. Brereton (Ed.), *New analytical approaches for verifying the origin of food*. 1st ed. ISBN: 9780857092748.

Zawistowski, J. (2011), *Legislation of functional foods in Asia*, Functional Foods (Second edition). doi:10.1533/9780857092557.1.73

Chapter 2

The role of governments in driving sustainability standards: governance structures and green finance

Louise Manning, University of Lincoln, UK; and Aleksandra Kowalska, Maria Curie-Skłodowska University, Poland

1 Introduction

Agri-food systems encompass all stages of food production from primary production (field, forest or water-based cultivation and harvesting of food) through secondary production (e.g. slaughter, produce sorting and transport) to further stages of processing, packaging, storage and distribution (sometimes called tertiary production) through to retail or food service and, finally, the consumer and domestic activities such as preparing and eating food. Agri-food systems can be described as 'the activities and socio-technical governance mechanisms that frame the way food is grown, produced, manufactured, retailed and eaten' (Rees et al. 2019, p.841). Agri-food systems collate, aggregate, mix, transform, process and transport raw materials and food products which need to consistently meet both business customer (business to business or B2B) and consumer (business to consumer or B2C) requirements in terms of food safety, sensory and nutritional quality.

However, farmers and other agri-food supply chain actors must also increasingly now meet wider sustainability requirements, e.g. in providing food products with 'low environmental impacts which contribute to food and nutrition security and to healthy life for present and future generations'

http://dx.doi.org/10.19103/AS.2024.0133.02

(Gustavsen & Hegnes, 2020, p.1). Food production within agri-food supply chains must now be 'protective and respectful of biodiversity and ecosystems, culturally acceptable, accessible, economically fair and affordable; nutritionally adequate, safe and healthy; while optimising natural and human resources' (FAO, 2012, p.7). Sustainable food production therefore needs to address economic, social and environmental (as well as safety and quality) aspects of food production. In 2015, the UN General Assembly adopted the 2030 Agenda for Sustainable Development delivered through the 17 Sustainable Development Goals or SDGs (Regulation 2019/2088). Agri-food supply chains increasingly need to take account of how they contribute to and support the delivery of these SDGs.

A key concept for businesses is environmental and social governance (ESG) which relates to 'how corporations and investors integrate environmental, social and governance concerns into their business models' (Gillan et al. 2021, p. 2). Camoletto et al. (2022) argue that ESG-related due diligence is a 'central pillar of the transition of the economic system' towards sustainable communities and economies. In addition to environmental issues, social considerations include equality, diversity and inclusivity (EDI), worker and human rights, animal welfare and the impact of corporate decisions on individuals and communities (EC, 2023c). These factors link directly to the SDGs and address the three pillars of sustainability (economic, environmental and social) otherwise described as profit, planet and people.

Agri-food businesses have a key role to play in meeting these broader and more challenging requirements, particularly by adhering to mandatory or voluntary codes of corporate social responsibility (CSR) i.e. 'business practices that address an organisation's various economic, legal, ethical, and philanthropic responsibilities as they pertain to a wide range of stakeholders' (Lee, 2017, p. 211). Scherer and Palazzo (2011) have coined the term political CSR (PCSR) where CSR activities may have a political impact (Frynas & Stephens, 2015), or to describe the responsible use of business and corporate power in the political arena. Indeed, Scherer et al. (2016, p. 276) state that corporates have a role to play as responsible political actors 'engaging in public deliberations, collective decisions, and the provision of public goods or the restriction of public bads [evils] in cases where public authorities are unable or unwilling to fulfil this role [including] the enforcement of social and environmental standards along supply chains or the fight against global warming, corruption, discrimination or inequality.' Business-driven standards extend the range and depth of B2B and B2C quality requirements for food products and the way they are grown and processed. Supply contracts and product specifications increasingly include social and environmental aspects of how the food has been produced or transformed at each stage of the supply chain. In doing so, agri-food governance needs constantly to establish and retain public trust in food (Lang, 2003). However, is a market-based approach sufficient on its own to address the level of environmental and social externalities attributed to agri-food systems? Without suitable oversight and governance, can consumers be protected from greenwashing through unsubstantiated, misleading or potentially fraudulent sustainability claims? Public intervention may be necessary to address negative externalities if the market is not willing or able to address negative environmental or social impacts of food production or where there may be a public benefit from farmers undertaking certain actions which the market is not prepared to pay for (Burrell, 2011).

Externalities are the 'side-effects' of the activities and actions of organisations that can affect people, communities (social) or the wider environment. Positive externalities of agricultural activity (e.g. food, feed/forage, fibre, fuel and other industrial feedstuffs) can also include the aesthetic appeal of the farmed landscape, recreational opportunities, water storage and management, nutrient recycling and fixation, and carbon sequestration. Negative externalities may include, e.g. degradation of natural habitat and resulting loss of genetic diversity and biodiversity. Negative externalities can also include the loss of natural capital, the stock of natural assets (such as soil, air, water, plant, insect and animal life) which provide the foundation for food production and other ecosystem services on which humans depend. Natural capital is one of five types of interrelated assets that also includes human, physical, financial and social capital. These assets can be degraded by poor land use decisions involving agriculture or by poor agricultural practices which may cause, e.g. deforestation, soil erosion, loss of soil health, salination or desertification, as well as diffuse pollution and eutrophication.

This chapter considers the role of supranational policy, and regional and national policy in driving sustainability standards with regards to some of these issues and how they influence the strategies employed within individual organisations and across agri-food supply chains. The chapter also reviews developments in forms of hybrid governance and uses the example of the role of green finance to show how agri-food supply chain strategies are being reshaped.

2 Governance structures to support sustainability: the role of supranational governance

In discussions of governance, it is important to differentiate between the governance landscape, which encompasses multiple initiatives by actors seeking to develop single-issue as well as multi-issue integrated governance structures, and the governance architecture which is developed through established and emergent governance regimes (Schmitt, 2022; Manning et al.

2023). Global governance architectures frame the activities of key UN bodies (Zelli, 2011; Kowalska et al. 2022), for example:

- The World Trade Organisation (WTO);
- The Food and Agriculture Organisation (FAO);
- The Codex Alimentarius Commission (Codex) (a joint initiative between the FAO and another UN body: the World Health Organisation (WHO));
- The UN Framework Convention on Climate Change (UNFCCC) and their annual conferences of parties (COPs); and
- The Intergovernmental Panel on Climate Change (IPCC) which provides the scientific findings informing COP discussions and, in the case of the EU, informs legislation such as the EU sustainability reporting standards (Commission Delegated Regulation (EU) 2023/2772).

These governance architectures directly influence global, regional and national public policy. An example is the supranational recognition of sustainability issues which has led to the subsequent enshrining in law of net zero carbon policies in individual countries (Perino et al. 2022), or regulatory strategies to reduce nutrient pollution (Möhring et al. 2023) or biodiversity loss.

Supranational policymaking via institutions such as the UN has the advantage of creating a common global policy and pooling expertise and resources to address a common problem (Andreozzi & Tamborini, 2019). However, because these policy initiatives typically rely on consensus supranational institutions can be slow to act. The supranational institutions, and the policies created, can be inflexible, technocratic and lacking in accountability to individual sovereign powers, can fail to reflect regional needs and concerns and can lack the ability to enforce agreed changes on recalcitrant countries (Neshkova, 2014). Supranational policy mechanisms include the COP meetings that inform global agreements to address climate action or biodiversity loss (Dogan et al. 2022). For example, the COP 21 Paris Agreement in 2015 (UNCC, nd) sought to limit the global average temperature increase to 1.5°C above pre-industrial levels. Signatory countries agreed to prepare and submit national climate change action plans known as nationally determined contributions (NDCs) to achieve this outcome. These action plans have then fed down to sectors such as farming and food production.

Another example of supranational activity is the European Union (EU) Common Agricultural Policy (CAP). The CAP has drawn together a range of measures, many prescribed in EU legislation, relating to adoption of specific agricultural practices to maintain land in good agricultural and environmental condition with direct financial payments to landowners (Burrell, 2011). Agri-environment schemes (AESs) were also developed that farmers could join on a voluntary basis by adopting (and receiving additional payment for) measures to conserve and enhance the environment. There has been criticism that some AESs have been too prescriptive, inflexible and bureaucratic, inhibiting the ability of farmers to act entrepreneurially or innovatively in developing and implementing their own practices to achieve the same outcomes (Burton et al. 2008).

The European Commission's 2019 European Green Deal is a green growth strategy that has a series of aims (EU Directive 2022/2464):

- To transform the EU into a modern, resource-efficient and competitive economy with no net emissions of greenhouse gases by 2050;
- To protect, conserve and enhance the EU's natural capital;
- To protect the health and well-being of EU citizens from environment-related risks and impacts;
- To decouple economic growth from resource use; and
- To ensure that all regions and EU citizens participate in a socially just transition to a sustainable economic system whereby no person and no place is left behind.

The Green Deal also impacts specifically on farmers and the food system through the 'Farm to Fork Strategy'. Since agriculture and food production have such a significant impact on the environment, ensuring a more sustainable EU food system is a cornerstone of the Green Deal (Kowalska & Bieniek, 2022).

The EU Farm to Fork Strategy aims to accelerate the transition to a sustainable food system that should (European Commission, 2023a):

- Ensure food security, nutrition and public health, making sure that everyone has access to sufficient, safe, nutritious, sustainable food;
- Preserve the affordability of food while generating fairer economic returns, fostering the competitiveness of the EU supply sector and promoting fair trade;
- Have a neutral or positive environmental impact;
- Help to mitigate climate change and adapt to its impacts; and
- Reverse the loss of biodiversity.

This food system change is proposed to be delivered through four policy approaches:

- Sustainable food production;
- Sustainable food processing and distribution;
- Sustainable food consumption; and
- Food loss and waste prevention.

The legislative and governance framework to deliver more sustainable food systems under the Green Deal is currently under development. The 'Fit for 55' package contains numerous legislative proposals to achieve the EU Green Deal.

3 Governance structures to support sustainability: the move to hybrid governance

As mentioned earlier, ecosystem services (ES) are the benefits citizens derive from natural ecosystems. ES include the following (Baskent, 2020):

- Provisioning services (e.g. production of food, fibre and fuel);
- Regulating services (e.g. promoting clean air and better water quality, flood management, carbon storage, pollination services and pest control);
- Supporting services (e.g. nutrient cycling and healthy soils); and
- Recreational services (e.g. tourism, a sense of identity with a landscape, physical and mental well-being).

These services are often described as public goods. A public good is the opposite of a private good in that, being 'public', it is a commodity or service that benefits all members of a given group, community, national or global population. A public good is provided free at the point of delivery. Since they benefit everyone, rather than any one particular group, any costs of protecting, maintaining and enhancing ecosystem services as public goods must typically be 'paid for' from public funds (e.g. taxation). However, governments must set up and administer appropriate frameworks for protecting ecosystems, often enshrined in law, even when the delivery process is sub-contracted by governments to designated agencies or non-governmental organisations (NGOs) (Fernando, 2022). An example of this process of delivering public goods for public money would be the strategies originally developed within the EU Common Agricultural Policy, which have now evolved to be part of the EU Green Deal.

Public goods ideally must be:

- *Non-rivalrous,* i.e. the availability of a good does not reduce as more people receive the benefit; and
- *Non-excludable,* i.e. accessible to all citizens.

In practice, public goods may be neither. A public road, e.g. may be paid for and maintained by government funding and theoretically open to all to use. However, it is not always non-rivalrous if it fills up with traffic which then prevents others from using it. The road is also not non-excludable since those without a car may not be able to make use of it, meaning it is not accessible for everyone. Should the public road then be described as a public good, in comparison to a road that is only available if a toll payment is paid, or should the public road be described as a quasi-public good? (Fernando, 2022), i.e. a form of hybrid public-private good.

This question of hybridity extends to many services paid for by public money where the benefits in practice are not available to all citizens. Virmani (2004, p.5561) describes these quasi-public goods as 'private goods with substantial externalities', i.e. with a mix of public and private benefits, but which cannot be delivered by private mechanisms alone. The general concept of 'public goods' and, more specifically, public payments for public goods, is increasingly being embedded in national and international legislative frameworks addressing issues such as sustainability, biodiversity and ecosystem services, even if some of these services could only be described as being quasi-public goods (Kretsch et al. 2016).

The quasi-public nature of 'public' goods has encouraged an interest in public-private partnerships (PPPs) as a way of funding and managing the achievement of sustainability outcomes (Wang & Ma, 2021). One driver for this focus on PPPs was the 2007 global financial crash which left many national governments indebted, struggling to fund all their commitments and more reliant on support from the private sector to help fund and implement government policy. Some argue that this way of delivering quasi-public goods provides a more efficient process than solely government-driven initiatives (Zhang et al. 2020), though others have questioned the ethics of a situation where public money may end up benefitting private enterprises focussed on their own, narrow commercial objectives (Cheung et al. 2022). Lu et al. (2021) explore the development of quasi-public-private partnerships (QPPPs) as a means to deliver environmental protection programmes. In these circumstances, especially during significant structural transition, the government's role may effectively be a 'lender of last resort', providing the funding and guarantees required to de-risk and incentivise private sector involvement (Raymond, 2010; Lu et al. 2021; James et al. 2022). Examples of such critical infrastructure where QPPPs develop include water supply, energy supply and internet connectivity.

This shift in governance has even affected public health and safety-related areas such as animal disease surveillance and control programmes which have traditionally been seen as exclusively the province of government agencies (Enticott & Little, 2022). Increasingly, due to the rising costs of statutory animal disease control, regulators are developing cost-sharing governance approaches between the public and the private sector (Enticott et al. 2011). Shared responsibility and shared governance have the potential to engender a sense of commitment and accountability in all the actors involved, especially when making collective decisions in areas where some actors have previously been excluded (O'May & Buchan, 1999). PPPs allow hybrid governance with the potential to deliver innovative, flexible and dynamic governance compared to purely regulatory or market controls (Marks 2015; Pies et al. 2020; Kowalska & Manning, 2022). In agri-food systems, public-private governance structures can align the interests of multiple stakeholders (Noto et al. 2022). However public-private governance structures can also result in placing the needs of some stakeholders above others, with some stakeholders seen as passive recipients of policy rather than active co-creators.

Sonnino (2019) describes participatory agri-food governance as where civil society provides non-governmental interventions that build governance across 'institutions, policy sectors and governance scales.' Private food governance is focused on delivering compliance and prescribed performance levels within agri-food supply chains (Fuchs & Kalfagianni, 2010; Thomé et al. 2021). Private aspects of agri-food governance include individual or collective retailer-led food governance systems, and transnational corporation-led food governance systems, often enacted through the development of private third-party standards and certification schemes. These private standards are based on prerequisite programmes such as good agricultural practices (GAP), good manufacturing practices (GMP) and good hygienic practices (GHP) or environmental and social governance (ESG) strategies which are increasingly being extended to encompass sustainable food production (Fuchs & Kalfagianni, 2010). This more complex landscape of agri-food governance brings together multiple stakeholders operating at different levels of a complex socio-technical system. Setting up the governance architecture including industry-based certification schemes, where agri-businesses obtain certification for demonstrating compliance and ensuring robust verification of these governance regimes comes with additional transaction costs (Abebe et al. 2017). This can limit take-up, compliance and effective monitoring, particularly in low-margin supply chains (Manning, 2024). These limitations can be addressed by developing hybrid public-private governance structures which are more focused and agile than 'one size fits all' regulation.

4 The role of green finance in driving effective governance

As noted earlier, a key driver of delivering public goods, which may not have a conventional product related market value to incentivise their delivery, is via alternative policy instruments or policy mechanisms. The Paris Climate Agreement identified the importance of 'green finance' to address climate change or to support decarbonisation towards net zero. Green finance is defined as any structured financial activity created to ensure a better environmental outcome (https://www.weforum.org/agenda/2020/11/what-is-green-finance/).

During COP 26, held in Glasgow (UK) in 2021, sustainable or green finance was a key focus of the discussions (Dogan et al. 2022). Green finance has been seen as a key catalyst for public, private and public/private approaches to improving sustainability. Green finance can provide both public and private financial flows (banking, micro-credit, insurance and investment) to promote the SDGs, green growth and climate mitigation strategies (Desalegn & Tangl, 2022).

The United Kingdom (UK) HM Treasury/Department for Business, Energy & Industrial Strategy Green Finance Taskforce: Terms of Reference (nd) highlight different types of finance such as capital markets, institutional investment, insurance, commercial finance, retail finance and venture capital. The Green Finance Institute (2023) differentiates between:

- *Financing green* through utilising financial instruments to deliver clean and resilient green growth; and
- *Greening finance* by focusing the financial system on social and environmental as well as economic aspects of sustainability.

AESs are an example of publicly-funded green finance products designed to deliver more sustainable agri-food production and greater protection of biodiversity and the environment. An example in the United Kingdom (UK) would be the government funding Environmental Land Management Schemes or ELMs.

The Green Finance Institute (2023) argues that private sector finance can be mobilised in multiple ways to promote improved sustainability practices:

- **Via the supply chain** through retailers, food manufacturers and food service companies using premium payments for sustainably certified products or other types of incentive. Private mechanisms for delivering sustainable food production practices can also be delivered via a higher market price being paid for food products which demonstrate environmental and social benefits such as certified organic food, or the LEAF Marque Standard in the UK (Pérez Perales et al. 2019);
- **Via investors** who are willing to invest in a range of ecosystem services; or
- **Through finance providers and banks** supporting the transition to sustainable food production, carbon capture and sequestration and biodiversity recovery through green finance options such as green loans, green funds for investment and discounted loan facilities.

Another area is extending financial due diligence analysis to encompass both environmental risks and impacts on natural capital in assessing credit risk (Parker & Sheedy-Reinhard, 2022). Two key concepts in the development of these policy instruments are insetting and offsetting. Insetting is where organisations adapt their activities to reduce their own and their supply chain's emissions (Green Finance Institute, 2023). Insetting is described as "reducing greenhouse gas emissions or sequestering carbon through an activity linked to the supply chain of a given actor or an activity in its direct sphere of influence" (Acampora et al. 2023, p. 3). Offsetting is where an organisation reduces the impact of their activities by purchasing offsets from other entities reducing their greenhouse gas emissions. The latter trade the reduction they have as a 'credit' with organisations which are less able to reduce emissions.

An example in this area is the approach to green finance in the EU. The EU Taxonomy Regulation (EU, 2020/852) came into force on 12 July 2020 and established six environmental objectives that economic activity needs to meet to be determined as environmentally sustainable. These are as follows:

- Climate change mitigation;
- Climate change adaptation;
- Sustainable use and protection of water and marine resources;
- Pollution prevention and control;
- Protection and restoration of biodiversity and ecosystems; and
- The transition to a circular economy.

The EU Taxonomy aims to (European Commission, 2023b):

1. Create a frame of reference for investors and companies;
2. Support companies in their efforts to plan and finance their transition;
3. Protect against greenwashing practices; and
4. Help accelerate financing of those projects that are already sustainable and those needed for transition to net zero.

The European Commission describes green or sustainable finance as making transparent financial or investment decisions that apply ESG principles so that long-term investments are focused on sustainable and decarbonised economic activities and projects (EC, 2023c). A key EU piece of legislation is the 2019 Sustainable Finance Disclosure Regulation (SFDR) (Regulation (EU) 2019/2088) on sustainability-related disclosures in the financial services sector. The SFDR confirms the EU's commitment to the delivery of the SDGs, and the need to comply with the COP Paris Agreement 'making finance flows consistent with a pathway towards low greenhouse gas emissions and climate-resilient development.' The SFDR also requires the EU economy to transition 'to a low-carbon, more sustainable, resource-efficient and circular economy in line with the SDGs' and as a result ensure long-term economic competitiveness for the EU. The SFDR defines a sustainability risk as 'an environmental, social or governance event or condition that, if it occurs, could cause a negative material impact on the value of the investment.' The Regulation also requires:

> where the sustainability risk assessment leads to the conclusion that there are no sustainability risks deemed to be relevant to the financial product, the reasons therefore should be explained. Where the assessment leads to the conclusion that those risks are relevant, the extent to which those sustainability risks might impact the performance of the financial product should be disclosed either in qualitative or quantitative terms. The sustainability risk assessments and related pre-contractual disclosures by financial market participants should feed into pre-contractual disclosures by financial advisers. Financial advisers should disclose how they take sustainability risks into account in the selection process of the financial product that is presented to the end investors before providing the advice, regardless of the sustainability preferences of the end investors.

This approach changes the role of financial institutions and the organisations that they provide financial services to and also the scope and depth of disclosures made by financial advisers with regard to financial products that are offered and the organisations seeking to avail themselves of access to financial products (loans, insurance, etc.).

Building on SFDR, the EU Corporate Sustainability Reporting Directive (CSRD) (Directive 2022/2464) applies to all large companies in the EU employing more than 500 employees and comes into force in 2024. Organisations from third countries with a net turnover of more than EUR 150 million per annum in the EU will also need to comply. The Directive sets out mandatory guidelines for annual financial and non-financial reports. In the area of sustainability, the CSRD states:

> Sustainability reporting standards should also take account of internationally recognised principles and frameworks on responsible business conduct, corporate social responsibility, and sustainable development, including the SDGs, the UN Guiding Principles on Business and Human Rights, the OECD Guidelines for Multinational Enterprises, the OECD Due Diligence Guidance for Responsible Business Conduct and related sectoral guidelines, the Global Compact, the International Labour Organization's (ILO) Tripartite Declaration of Principles concerning Multinational Enterprises and Social Policy, the ISO 26000 standard on social responsibility, and the UN Principles for Responsible Investment.

Governance architectures are being developed through the creation of normative private standards. One example of third-party standards for ESG reporting is the GRI Standards (GRI Standards, nd). GRI 13 has been developed as a Sector Standard for the Agriculture, Aquaculture and Fishing Sectors. The Standards highlight the elements that should be included in an ESG Report. Another example of normative financial standards is the Equator Principles (EP) which are a framework to assess ESG aspects when financing projects (https://equator-principles.com).

The Taxonomy Regulation has informed the Disclosures Delegated Act (C(2021) 4987 final) which defines the key performance indicators (KPIs) that will need to be reported in terms of the EU's six key environmental objectives. The KPIs for credit institutions are based on Regulation (EU) No 575/2013, Title II, Chapter 2, Section 2, including the green asset ratio (Table 1). Regulation (EU) No 575/2013 defines credit risk mitigation as the technique[s] used by an institution to reduce the credit risk associated with risk exposure on their debt book. There are a number of credit risks that financial institutions need to consider, however, an emerging aspect of credit risk mitigation is mitigating the institutional exposure associated with environmental threats such as climate change or greenhouse gas (GHG) emissions. Reducing the credit risk through greening finance is being driven at the outset through regulation, but will be enacted in practice through finance and insurance institutions. This will require a large-scale transition of and refocus by the finance sector.

The term *brown asset* has been used to describe those investments held by financial institutions with the largest carbon-related credit risk. Brown financing is, therefore, the financing of activities or assets which contribute to biodiversity loss, GHG emissions, or climate change. Some financial and insurance institutions will have a high 'brown exposure' which will drive them over time to identify ways to dispose of such stranded assets. This off-loading of assets will happen more quickly if the 'asset' [organisation] does not transition to more sustainable practices through decarbonisation, increased circularity,

Table 1 Green asset ratio KPIs associated with credit institutions, green finance and the green deal

Environmental objectives	Green asset ratio (GAR)
Climate change mitigation	Proportion of loans and advances/debt securities/equity instruments financing taxonomy-aligned economic activities for the objective of climate change mitigation, compared to total loans and advances/ debt securities/equity instruments of non-financial undertakings and all other on-balance sheet assets.
Climate change adaptation	Proportion of loans and advances/debt securities/equity instruments financing taxonomy-aligned economic activities for the objective of climate change mitigation, compared to loans and advances/ debt securities/equity instruments financing economic activities in sectors covered by the taxonomy for the objective of climate change mitigation.
Other environmental activities	Same ratios for each of the other four environmental objectives should be disclosed, once the screening criteria are defined. That is: the sustainable use and protection of water and marine resources; the transition to a circular economy; pollution prevention and control; the protection and restoration of biodiversity and ecosystems.

Adapted from C(2021)4987, 2023.

and more effective use of resources (Fanizza & Cerami, 2023). Alternatively, the *green asset ratio* measures a financial institution's holding of green assets (in line with the definitions in the Taxonomy) as a proportion of their total asset base. Analysing the green asset ratio, especially the transition strategies and the associated environmental objective(s) and the type and nature of assets held will provide insight into the financial institution's risk profile, their journey on a sustainability transition of their activities (greening finance), and their long term sustainability strategy (Brühl, 2023).

Food supply chain businesses, and the degree to which they have embedded sustainable practices, will directly affect their access to investment capital (CapEx) and operating capital (OpEx) from financial institutions and/or market returns in a fast-changing geopolitical and socio-economic environment. Farming businesses too will need to consider their farming business and their enterprise profile and opportunities for accessing public funding for AES as they develop strategies for land use change. The role of supranational organisations, governments and private organisations within governance architectures is important. The impact of legislation on market governance will also lead to realignment in coming years as the regulatory and market drivers for mandatory ESG disclosure becomes embedded. New green financial instruments and green finance frameworks will also need to be developed and a reticence to engage when the sector is transitioning may impact uptake and rate of change (Khan et al. 2022).

5 Conclusion

This chapter has considered the role of governments in driving sustainability standards with a particular focus on using policy instruments such as green finance strategies as a driver of more sustainable practice in land use and agri-food production and the implications for agri-food systems. The links between supranational policy, regional and national regulations and their impact on private organisations have been explored, together with the potential role of hybrid governance architectures in guiding the sector towards a greener future. The framing of shared responsibility for delivery to the green agenda by both governments and the private sector, as well as citizens in terms of their behaviours, will continue to develop with the aim of reducing GHG emissions, and the negative externalities that are currently facing the agri-food system and wider global ecosystem.

6 References

Abebe, G. K., Chalak, A., & Abiad, M. G. (2017). The effect of governance mechanisms on food safety in the supply chain: Evidence from the Lebanese dairy sector. *Journal of the Science of Food and Agriculture*, *97*(9), 2908–2918.

Acampora, A., Ruini, L., Mattia, G., Pratesi, C. A., & Lucchetti, M. C. (2023). Towards carbon neutrality in the agri-food sector: Drivers and barriers. *Resources, Conservation and Recycling*, *189*, 106755.

Andreozzi, L., & Tamborini, R. (2019). Models of supranational policymaking and the reform of the EMU. *Journal of Policy Modeling*, *41*(5), 819–844.

Baskent, E. Z. (2020). A framework for ;characterizing and regulating ecosystem services in a management planning ;context. ;*Forests*, ;*11*(1), 102.

Brühl, V. (2023). The Green Asset Ratio (GAR): A new key performance indicator for credit institutions. *Eurasian Economic Review*, *13*(1), 57-83.

Burrell, A. (2011). 'Good Agricultural Practices' in the agri-food supply chain. *Environmental Law Review*, *13*(4), 251–270.

Burton, R. J., Kuczera, C., & Schwarz, G. (2008). Exploring farmers' cultural resistance to voluntary agri-environmental schemes. *Sociologia ruralis*, *48*(1), 16–37.

C(2021)4987 (2023). Commission Delegated Regulation (EU) …/… supplementing Regulation (EU) 2020/852 of the European Parliament and of the Council by specifying the content and presentation of information to be disclosed by undertakings subject to Articles 19a or 29a of Directive 2013/34/EU concerning environmentally sustainable economic activities, and specifying the methodology to comply with that disclosure obligation. Available at: https://ec.europa.eu/transparency/documents-register/detail?ref=C(2021)4987&lang=en [Accessed 11 August 2023].

Camoletto, S., Corazza, L., Pizzi, S., & Santini, E. (2022). Corporate Social Responsibility due diligence among European companies: The results of an interventionist research project with accountability and political implications. *Corporate Social Responsibility and Environmental Management*, *29*(5), 1122–1133.

Cheung, H., Baumber, A., & Brown, P. J. (2022). Barriers and enablers to sustainable finance: A case study of home loans in an Australian retail bank. *Journal of Cleaner Production*, *334*, 130211.

Commission Delegated Regulation (EU) 2023/2772 of 31 July 2023 supplementing Directive 2013/34/EU of the European Parliament and of the Council as regards sustainability reporting standards. Available at: https://eur-lex.europa.eu/eli/reg_del/2023/2772/oj

Desalegn, G., & Tangl, A. (2022). Enhancing green finance for inclusive green growth: A systematic approach. *Sustainability*, *14*(12), 7416.

Directive (EU) 2022/2464 of the European Parliament and of the Council of 14 December 2022 amending Regulation (EU) No 537/2014, Directive 2004/109/EC, Directive 2006/43/EC and Directive 2013/34/EU, as regards corporate sustainability reporting (Text with EEA relevance)https://eur-lex.europa.eu/legal-content/EN/TXT/?uri=CELEX:32022L2464

Dogan, E., Madaleno, M., Taskin, D., & Tzeremes, P. (2022). Investigating the spillovers and connectedness between green finance and renewable energy sources. *Renewable Energy*, *197*, 709–722.

European Commission (EC, 2023a). Farm to Fork Strategy. Available at: https://food.ec.europa.eu/horizontal-topics/farm-fork-strategy_en

European Commission (EC, 2023b). EU Taxonomy Navigator. Available at: https://ec.europa.eu/sustainable-finance-taxonomy/

European Commission (EC, 2023c). Overview of Sustainable Finance. Available at: https://finance.ec.europa.eu/sustainable-finance/overview-sustainable-finance_en

Enticott, G., Donaldson, A., Lowe, P., Power, M., Proctor, A., & Wilkinson, K. (2011). The changing role of veterinary expertise in the food chain. *Philosophical Transactions of the Royal Society B: Biological Sciences*, *366*(1573), 1955–1965.

Enticott, G., & Little, R. (2022). Playing games with 'good farming': exploring the potential impact of disease control policies on farmers' cattle purchasing practices. *Journal of Rural Studies*, *92*, 371–382.

Fanizza D., & Cerami, L., (2023). A market for brown assets to make finance green. Available at: https://www.imf.org/en/Publications/WP/Issues/2023/01/20/A-Market-for-Brown-Assets-To-Make-Finance-Green-528413 IMF Working Papers.

FAO, (2012). Sustainable Diets and Diversity. Rome, Italy. http://www.fao.org/3/i3004e/i3004e00.htm [Accessed 11 November 2023].

Fernando, J. (2022). What are Public Goods? Definition, How they work and Example. Available at https://www.investopedia.com/terms/p/public-good.asp [Accessed 2 August 2023].

Frynas, J. G., & Stephens, S. (2015). Political corporate social responsibility: Reviewing theories and setting new agendas. *International Journal of Management Reviews*, *17*(4), 483–509. https://doi.org/10.1111/ijmr.12049

Fuchs, D., & Kalfagianni, A. (2010). The causes and consequences of private food governance. *Business and Politics*, *12*(3), 1–34.

Gillan, S. L., Koch, A., & Starks, L. T. (2021). Firms and social responsibility: A review of ESG and CSR research in corporate finance. *Journal of Corporate Finance*, *66*, 101889.

Green Finance Institute (2023). Available at: https://www.greenfinanceinstitute.co.uk

GRI Standards (nd). Available at: https://www.globalreporting.org/how-to-use-the-gri-standards/gri-standards-english-language/

Gustavsen, G. W., & Hegnes, A. W. (2020). Individuals' personality and consumption of organic food. *Journal of Cleaner Production*, *245*, 118772. https://doi.org/10.1016/j.jclepro.2019.118772

HM Treasury/Department for Business, Energy & Industrial Strategy Green Finance Taskforce: Terms of Reference. (nd). Available at: https://assets.publishing.service.gov.uk/government/uploads/system/uploads/attachment_data/file/675361/Green_Finance_Taskforce_-_terms_of_reference.pdf

James, A., Retting, T., Shogren, J. F., Watson, B., & Wills, S. (2022). Sovereign wealth funds in theory and practice. *Annual Review of Resource Economics*, *14*, 621–646.

Khan, K. I., Mata, M. N., Martins, J., Nasir, A., Dantas, R. M., Correia, A. B., & Saghir, U. S. (2022). Impediments of green finance adoption system: Linking economy and environment. *Impediments of Green Finance Adoption System: Linking Economy and Environment*, (2), 217–237.

Kowalska, A., & Bieniek, M. (2022). Meeting the European Green Deal objective of expanding organic farming. *Equilibrium. Quarterly Journal of Economics and Economic Policy*, *17*(3), 607–633.

Kowalska, A., Budzyńska, A., & Białowąs, T. (2022). Food export restrictions during the COVID-19 pandemic: Real and potential effects on food security. *International Journal of Management and Economics*, *58*(3), 1–16.

Kowalska, A., & Manning, L. (2022). Food safety governance and guardianship: the role of the private sector in addressing the EU ethylene oxide incident. *Foods*, *11*(2), 204.

Kretsch, C., Dijk, J., & Van Schleyer, C. (2016). Public goods and ecosystem services. *OpenNESS Ecosystem Services Reference Book. EC FP7 Grant Agreement*, (308428), 1–7.

Lang, T. (2003). Food industrialisation and food power: Implications for food governance. *Development Policy Review, 21*(5–6), 555–568.

Lee, T. H. (2017). The status of corporate social responsibility research in public relations: A content analysis of published articles in eleven scholarly journals from 1980 to 2015. *Public Relations Review, 43*(1), 211–218. https://doi.org/10.1016/j.pubrev.2016.10.001

Lu, H., Mayer, A. L., Zhou, S., & Wellstead, A. (2021). Unveiling the quasi-public-private partnership (QPPP): Evidence from China's environmental service sector. *Journal of Environmental Policy & Planning, 23*(3), 302–316.

Manning, L. (2024). Operationalising integrity within supply chains. In M. Kaptein (ed.), *A Research Handbook on Organisational Integrity*. Cheltenham: Edgar Elgar.

Manning, L., Brewer, S., Craigon, P.J., Frey, J., Gutierrez, A., Jacobs, N., Kanza, S., Munday, S., Sacks, J. & Pearson, S., (2023). Reflexive governance architectures: Considering the ethical implications of autonomous technology adoption in food supply chains. *Trends in Food Science & Technology, 133*, 114–126.

Marks, A. B. (2015). A new governance recipe for food safety regulation. *Loyola University of Chicago Law Journal, 47*, 907.

Möhring, N., Kanter, D., Aziz, T., Castro, I.B., Maggi, F., Schulte-Uebbing, L., Seufert, V., Tang, F.H., Zhang, X. and Leadley, P., (2023). Successful implementation of global targets to reduce nutrient and pesticide pollution requires suitable indicators. *Nature Ecology & Evolution*, 1–4.

Neshkova, M. I. (2014). Salience and complexity in supranational policymaking: The case of subnational interests. *Governance, 27*(1), 9–36.

Noto, G., Coletta, L., & Vainieri, M. (2022). Measuring the performance of collaborative governance in food safety management: An Italian case study. *Public Money & Management, 42*(8), 627–636.

O'May, F., & Buchan, J. (1999). Shared governance: A literature review. *International Journal of Nursing Studies, 36*(4), 281–300.

Parker, C., & Sheedy-Reinhard, L. (2022). Are banks responsible for animal welfare and climate disruption? A critical review of Australian banks' due diligence policies for agribusiness lending. *Transnational Environmental Law*, 1–26.

Pérez Perales, D., Alarcón Valero, F., Drummond, C., & Ortiz, Á. (2019). Towards a sustainable agri-food supply chain model. the case of LEAF. In *Engineering Digital Transformation: Proceedings of the 11th International Conference on Industrial Engineering and Industrial Management* (pp. 333–341). Springer International Publishing.

Perino, G., Jarke-Neuert, J., Schenuit, F., Wickel, M., & Zengerling, C. (2022). Closing the Implementation Gap: Obstacles in Reaching Net-Zero Pledges in the EU and Germany. *Politics and Governance, 10*(3).

Pies, I., Hielscher, S., & Everding, S. (2020). Do hybrids impede sustainability? How semantic reorientations and governance reforms can produce and preserve sustainability in sharing business models. *Journal of Business Research, 115*, 174–185.

Raymond, H. (2010). Sovereign wealth funds as domestic investors of last resort during crises. *International Economics, 123*, 121–159.

Rees, W., Tremma, O., & Manning, L. (2019). Sustainability cues on packaging: The influence of recognition on purchasing behavior. *Journal of Cleaner Production, 235*, 841–853.

Regulation (EU) 2019/2088 of the European Parliament and of the Council of 27 November 2019 on sustainability-related disclosures in the financial services sector (Text with EEA relevance). Available at: https://eur-lex.europa.eu/legal-content/EN/TXT/?uri=celex%3A32019R2088

Regulation (EU) 2020/852 of the European Parliament and of the Council of 18 June 2020 on the establishment of a framework to facilitate sustainable investment, and amending Regulation (EU) 2019/2088 (Text with EEA relevance). Available at: https://eur-lex.europa.eu/legal-content/EN/TXT/?uri=CELEX:32020R0852

Regulation (EU) No 575/2013 of the European Parliament and of the Council of 26 June 2013 on prudential requirements for credit institutions and investment firms and amending Regulation (EU) No 648/2012 Text with EEA relevance. Available at: https://eur-lex.europa.eu/legal-content/EN/TXT/?uri=celex%3A32013R0575

Scherer, A. G., & Palazzo, G. (2011). The new political role of business in a globalized world: A review of a new perspective on CSR and its implications for the firm, governance, and democracy. *Journal of management studies*, *48*(4), 899–931. https://doi.org/10.1111/j.1467-6486.2010.00950.x

Scherer, A. G., Rasche, A., Palazzo, G., & Spicer, A. (2016). Managing for political corporate social responsibility: New challenges and directions for PCSR 2.0. *Journal of Management Studies*, *53*(3), 273–298. https://doi.org/10.1111/ijmr.12049

Schmitt, L. (2022). Mapping global AI governance: ;a nascent regime in a fragmented landscape. ;*AI and Ethics*, ;*2*(2), 303–314.

Sustainable Development Goals (SDGs). Available at: https://sdgs.un.org/goals [Accessed 11 November 2023].

Sustainable Development Goals (nd). Available at: https://sdgs.un.org

Sonnino, R. (2019). The cultural dynamics of urban food governance. *City, Culture and Society*, *16*, 12–17.

Thomé, K. M., Cappellesso, G., Ramos, E. L. A., & de Lima Duarte, S. C. (2021). Food supply chains and short food supply chains: Coexistence conceptual framework. *Journal of Cleaner Production*, *278*, 123207.

United Nations Climate Change (UNCC, nd). The Paris Agreement. Available at: https://unfccc.int/process-and-meetings/the-paris-agreement

Virmani, A. (2004). Need for a Hybrid Proposal. *Economic and Political Weekly*, 5559–5562.

Wang, N., & Ma, M. (2021). Public–private partnership as a tool for sustainable development–What literatures say? *Sustainable Development*, *29*(1), 243–258.

Zelli, F. (2011). The fragmentation of the global climate governance architecture. *Wiley Interdisciplinary Reviews: Climate Change*, *2*(2), 255–270.

Zhang, L., Wu, C., & Zhang, Y. (2020). Experimental study based on game theory on the private, voluntary supply mechanisms of goods for forestry infrastructure from the perspective of quasi-public goods. *Sustainability*, *12*(7), 2808.

Chapter 3

The role and range of voluntary standards and certification systems governing sustainable agricultural practices

Mathieu Lamolle and Gregory Sampson, International Trade Centre (ITC), Switzerland

1 Introduction

The origin of sustainability standards was not the large multistakeholder frameworks we recognise today. At their root, standards evolved from a group of civil society-based movements denouncing environmental and social excesses. These excesses can be traced back to the Industrial Revolution, a time

http://dx.doi.org/10.19103/AS.2024.0133.03

when technology and mass production required more natural resources, and human's relationship to nature was forever altered: 'there was a psychological change: confidence in the ability to use resources and to master nature was heightened' (Britannica, 2021). The human need to dominate nature, along with its harmful consequences, would only grow over time.

Fast-forward to the 1960s, when advances in agriculture and chemistry, coupled with a post-war population boom, marked the advent of large-scale agriculture. As the use of pesticides became more widespread, scientists began to sound the alarm about the harmful effects of dichlorodiphenyltrichloroethane amongst other chemicals. Ecological consciousness blossomed with the publication of Rachel Carson's Silent Spring (1962), which documented the effects of pesticides on plants and animals for a general audience. Infused with the Protestant thinking of the time, the book 'demanded personal action to right the wrongs of society.' (Griswold, 2013). This same type of thinking would power the rise of sustainability standards.

Carson was not the only scientist to worry about the state of the planet. In 1968, the Club of Rome was born. This informal organisation of eminent scientists, economists, humanists and industrialists addressed the complex issues facing humanity and the planet. The think tank is famous for 'The Limits to Growth' (1972), which marks the first time that the word 'sustainable' appeared in its modern form (Caradonna, 2016). The book warned of the perils of continued worldwide expansion, arguing that the planet's resources would probably not support continued economic and population growth beyond 2100.

The concept snowballed from there. Whilst no book in the English language had used either 'sustainable' or 'sustainability' in its title before 1970, there was a mushrooming of books referencing sustainability after 1980 (Caradonna, 2016: 2). Around the same time, the grassroots organic and fair trade movements took off, signalling the advent of voluntary sustainability standards (State Secretariat for Economic Affairs SECO 2021; Gale et al. 2017) (Box 1). The organic movement has been slowly gaining ground over the years as a response to large-scale agriculture (Luttikholt, 2007). In 1972, the movement became coordinated under the auspices of the International Federation of Organic Agriculture Movements. Perhaps even more relevant, the organic tenets proposed by pioneers such as Lady Eve Balfour became codified into standards (IFOAM, 2020).

2 The integration of sustainability into international discourse

The concept of sustainability was also gaining ground at the United Nations (UN). Indeed, in 1972, the United Nations Environment Programme was established after the United Nations Conference on the Human Environment in Stockholm (Fig. 1). 'Significantly, the emergence of sustainability within the milieu of the UN signalled the increasingly international character of the movement' (Caradonna, 2016: 142).

Box 1: Defining sustainability standards

The terms private standards, voluntary sustainability standards and sustainability standards are used interchangeably. They are defined as: 'Specifying requirements that producers, traders, manufacturers, retailers or service providers may be asked to meet, relating to a wide range of sustainability metrics, including respect for basic human rights, worker health and safety, the environmental impacts of production, community relations, land use planning and others'.

Companies and non-profit associations lead these voluntary standards. They complement national public health and safety standards; international standards of the World Trade Organisation (WTO) that address sanitary and phytosanitary measures, technical barriers to trade and government procurement; and international standards approved by the International Organization for Standardization that address quality management, food safety, the environment, health and safety, energy, information technology security and more.

Source: United Nations Forum on Sustainability Standards (2013), p. 3.

In 1983, the World Commission on Environment and Development, known informally as the Brundtland Commission, was set up to create a global framework for sustainable development. The establishment of the Commission culminated with the publication of the report Our Common Future (1987). The report looked at sustainability with a holistic vision that encompassed 'the three Es' of 'environment, economy and equity', rather than just environmental considerations (Caradonna, 2016: 143). That definition of sustainability still guides us today. Around that time, environmental and social responsibility

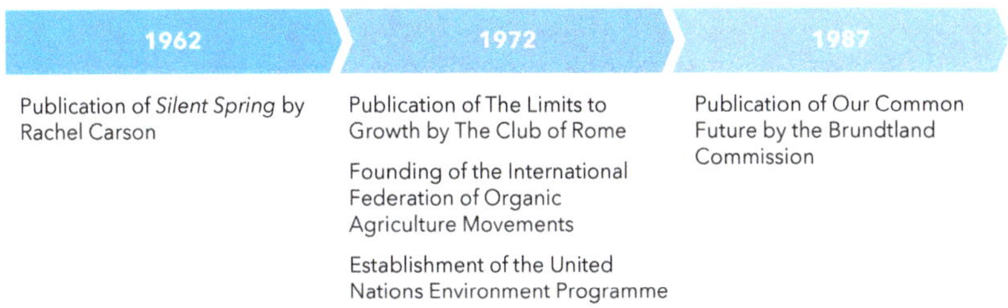

Figure 1 From the 1960s to the 1990s: environmental issues trigger concerns.

were linked through the rise of fair trade movements. Non-governmental organisations (NGOs) seeking more balanced trade relations between North and South started forming. These led to the rise of Fairtrade International in 1997 as an umbrella organisation for national chapters of different local groups (Fairtrade Foundation, n.d.) (Fig. 2).

Two trends marked the 1990s as a turning point for the formalisation and rapid growth of sustainability initiatives. First, international policy discourse adopted the term 'sustainable development' from the Brundtland Commission. Second, social issues became more prominent as globalisation took hold. The 1990s marked the 'era of globalisation', as international value chains began stretching across continents. Major companies fragmented their production processes: products and their spare parts were manufactured, assembled and sold in different countries and regions. This form of cross-border production emerged during the 1990s as a result of trade liberalisation, advances in technology and communication and progress in logistics (Seric and Tong, 2019). Whilst these value chains created new jobs and economic growth, they also contributed to environmental degradation and social inequality.

During the 1990s, activist Jeff Ballinger led a widely publicised NGO movement against Nike for worker abuse and exploitation in developing countries. After pressure from labour activists, campus organisers and anti-globalisation forces, Nike's Chief Executive Officer (CEO) Phil Knight promised reform of organisational policies and processes in 1998. Around the same time, awareness about social conditions in the production of major globally traded commodities such as cocoa grew, as documentaries exposing child labour on plantations were made.

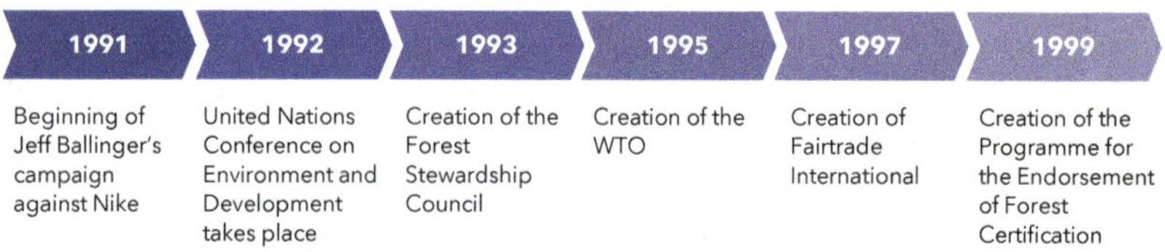

Figure 2 The 1990s: social issues elicit a global response.

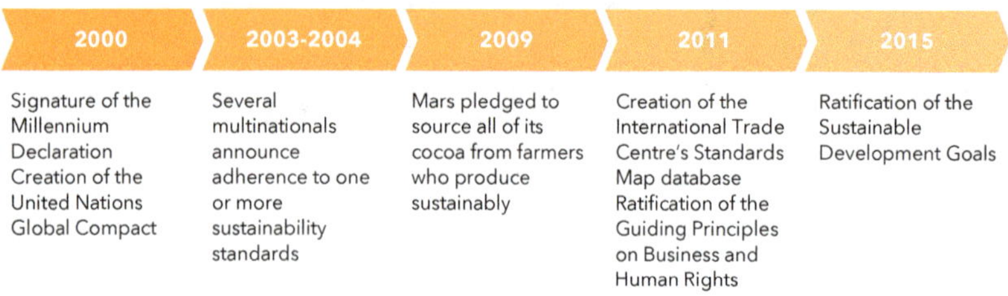

Figure 3 From 2000 to 2015: Companies adopt sustainability standards.

At the international policy level, more than 170 countries met in Rio de Janeiro in 1992 at the United Nations Conference on Environment and Development to focus on the impact of human activities on the environment. An outcome of what became known as the Earth Summit was Agenda 21, which, whilst championing economic growth and free trade, also linked social and economic development with environmental protection (Purvis et al., 2019). This movement marked a pivotal point during which 'sustainability' and 'sustainable development' ceased to be fringe concepts but became mainstream issues endorsed by official bodies such as the United Nations. The 1992 Earth Summit thus represented a turning point in the development of sustainability standards: it was the moment when talk translated into action, providing impetus, e.g. to the founding of the Forest Stewardship Council (FSC) in 1993. In 1995, the World Trade Organisation (WTO) was created, building on the General Agreement on Tariffs and Trade in 1948. The Preamble to the Marrakesh Agreement, which founded the WTO, included direct references to sustainable development and the environment, confirming their roles in official discourse. However, it is worth pointing out that since the establishment of the WTO, its relationship with private sustainability standards has been thorny. A primary goal of the WTO has been to reduce trade barriers and facilitate global trade, which can be in tension with standards that potentially restrict trade because they do not meet sustainability criteria. Fairtrade was founded in 1997 to promote fairer returns for smallholder farmers, raising awareness about the challenges faced by farmers in developing countries. The Programme for the Endorsement of Forest Certification was founded 2 years later. In summary, the 1990s and early 2000s marked a rapid growth of sustainability standards.

The beginning of the twenty-first century saw sharply increased public interest in and concern about climate change, as well as global economic and social inequalities. As a result, a set of global frameworks was launched to promote fundamental values that set certain limits to the globalisation era. After a summit in New York City in 2000, the United Nations General Assembly launched the Millennium Declaration in response to the central challenge of the time: to ensure that globalisation became a positive force for all the world's people (Resolution Adopted by United Nations General Assembly, 2000). National representatives agreed that globalisation offered great opportunities but that costs and benefits were shared very unevenly. More precisely, developing countries and countries with economies in transition faced difficulties in responding to this central challenge. The only way to make globalisation inclusive and equitable was to set a common agenda based on fundamental values (such as freedom, equality, solidarity, tolerance, respect for nature and shared responsibility) to govern international relations. Out of this realisation came the Millennium Development Goals (Box 2). These were set shortly after the 2000 New York Summit to be achieved by 2015.

The Millennium Development Goals served as the North Star for the United Nations, in conjunction with the development agencies, national governments and private sector contributions throughout the 2000s.

Box 2: The Millennium Development Goals

1. Eradicate extreme poverty and hunger.
2. Achieve universal primary education.
3. Promote gender equality and empower women.
4. Reduce child mortality.
5. Improve maternal health.
6. Combat HIV/AIDS, malaria and other diseases.
7. Ensure environmental sustainability.
8. Global partnership for development.

Source: Resolution adapted by General Assembly No. A/RES/55/2, 3 C.F.R. 2 (2000).

At the same time, the UN faced calls for more efficient and effective programmes and the development of public–private partnerships. Out of this notion, 2 months before the launch of the Millennium Development Goals, the UN launched the Global Compact at its headquarters in New York City. This initiative put forward ten principles to guide all business operations and strategies and had two objectives:

- 'Mainstream the United Nations Global Compact ten principles in business activities around the world'.
- 'Catalyse actions in support of broader United Nations goals', such as the Millennium Development Goals, and later, the Sustainable Development Goals.

The coordinated and multistakeholder efforts of the United Nations Global Compact have resonated across the globe, and it has since become the world's largest corporate sustainability initiative. Currently, more than 13 000 companies in over 170 developing and developed countries are Global Compact signatories.

3 The rise of corporate social responsibility and voluntary sustainability standards

As sustainability efforts gained critical mass, companies developed their own corporate social responsibility (CSR) initiatives. By their very nature, these are corporate strategic responses to the particular pressures they face:

- from customers who seek socially and environmentally sustainable products;

- from suppliers and business partners who faced similar customer pressures; and
- from authorities, in response to societal awareness.

These CSR initiatives aimed to protect company reputations (as well as sales and profits) and stave off a rise in regulation. Despite high-level initiatives (such as the Millennium Development Goals) and growing corporate awareness, there was no single accepted definition of CSR in the early 2000s. All definitions share the notion that companies, through their sustainability strategies, should manage their impact on society and the environment. However, there was no agreement or definition on the scope of such a contribution, or what exactly social and environmental sustainability means. The vague and at times differing definitions of CSR included 'a broad range of activities that corporations may engage in, with varying degrees of enthusiasm, to demonstrate that they are addressing important human rights, environmental, and labour issues, many of which have been brought to their attention by activist groups' (Balmaña, 2010).

Whilst the scope of corporate responsibility was not defined, firms were eager to showcase their contributions to broader development goals. Most publicly traded companies, in some form or another, began to issue reports about their CSR and sustainability efforts. In the corporate world, CSR and sustainability strategies have come to mean the same thing: that companies consider limits of natural, human and financial resources, and that they align with and contribute to a globally recognised and voluntary sustainable framework, such as the Millennium Development Goals and the United Nations Global Compact Principles.

This growing attention by the corporate sector was reflected in a 2010 study of more than 700 CEOs of multinational corporations, conducted by UN Global Compact and Accenture. It found that '93% of CEOs believed that sustainability issues will be critical to the future success of their business' and '96% of CEOs believed that sustainability issues should be fully integrated into the strategy and operations of a company (up from 72% in 2007)' (United Nations Global Compact Leaders Summit 2010 – Summary Report, 2010). This strategic importance was equally mirrored in the growth of CSR reports: whilst 464 CSR reports were published in 1998, this number grew to almost 6000 by 2011 (2012 Global Winners & Reporting Trends, 2021).

As companies responded to demands for transparency in contributing to a more sustainable and inclusive world, they aligned with globally recognised frameworks, such as the Millennium Development Goals and the United Nations Global Compact Principles, where such frameworks included high-level objectives. Pressed to show a positive impact, companies looked for more pragmatic tools and began to adopt voluntary sustainability standards to address specific issues in their value chains. Today, these standards are at the heart of many corporate sustainability actions. Often created by industry consortia, NGOs, or multistakeholder initiatives, voluntary sustainability standards go beyond minimum legal standards imposed by legislators. These sustainability standards include a set of criteria defined by the standard owners or stakeholders. To be certified against the standard, an organisation must demonstrate that it fulfills a set of specific criteria. The private, voluntary standards are usually limited to a well-defined sustainability area, value chain, specific sector or product.

The corporate appetite for engagement with these standards is reflected in their rise in the early 2000s. Whilst the first such standards were established decades earlier, they doubled in number from fewer than 200 in 2000 to more than 400 in 2010. At the same time, several global roundtables were established:

- on soy (Round Table on Responsible Soy);
- palm oil (Roundtable on Sustainable Palm Oil); and
- sugar (Bonsucro).

These kick-started a new generation of standards focussing on global food supply chains, backed by considerable industry engagement. Companies were soon eager to publicise their sustainability efforts. A few brief examples are:

- In 2003, Nespresso announced its AAA Sustainable Quality Programme in partnership with the Rainforest Alliance;
- In 2004, Kraft and Procter and Gamble began selling Rainforest Alliance-certified coffee in their stores, representing the first widely sold coffee products to feature that seal of 'assurance';
- In 2006, Walmart, the world's largest retailer, committed not to sell any wild-caught fresh and frozen fish in the United States and Canada unless they met Marine Stewardship Council (MSC) criteria; and
- In 2009, Mars pledged to source all of its cocoa from farmers certified by Fairtrade and the Rainforest Alliance (now merged with UTZ Certified) (ISEAL Alliance, n.d.).

As the use and breadth of the internet gained momentum in the second half of the decade, voluntary standards moved into the virtual realm. The founding of the International Trade Centre's (ITC) online Standards Map database represented one example of the growing presence of sustainability standards on the internet.[1] Simultaneously, supported by the rapid adoption and increased credibility and maturity of the standards system, sustainability standards began

1 https://www.standardsmap.org/en/home

to spread into a new range of sectors and sustainability themes such as tourism, mining, aquaculture, water and jewellery.

By 2010, there was a growing realisation that specific sustainability concerns were interconnected. From the chemical levels in fertiliser to the income received at the end of each month by farmers, these all affect producers equally. In response, many standards expanded their scope of coverage. For example, 4C (The Common Code for the Coffee Community) added a climate module (covering mitigation and adaptation measures) to its existing coffee production guidelines (Deutsche Gesellschaft für Internationale Zusammenarbeit, n.d.). With more than twice as many sustainability standards as a decade before, standards continued to increase their geographic footprint. In 2013, the FSC reached over 180 million hectares of forest managed according to its principles, 450% more than 10 years prior. That year, 32 of the 100 largest global beauty products companies referred to biodiversity in their sustainability reporting (Convention on Biological Diversity, 2013). Certain sustainability standards had an unprecedented impact on consumers. In the United Kingdom (UK), in 2014, for example, surveys reported that 78% of the sample population recognised the Fairtrade symbol (Fairtrade Foundation, n.d.).

However, whilst sustainability standards continued to develop, sustainability problems continued to increase with repeated reports about corporate wrongdoing. Reports revealed, for example, poor working conditions in offshore garment and footwear production. In response, the UN ratified its Guiding Principles on Business and Human Rights in 2011, focussing on human rights affected by business activities (Office of the United Nations High Commissioner for Human Rights, 2021). Such wrongdoing added to the growing notion that tackling sustainability issues in supply chains required collaboration across all supply chain tiers. Scalable impact could only be reached by engaging multiple stakeholders in a common purpose and commitment. This led to corporate coalitions and engagement platforms such as the Consumer Goods Forum, led by more than 400 CEOs from retailers, manufacturers and service providers.

4 Towards a 'sustainability deal' across value chains

The sustainable development goals (SDGs) reflect the results of a global multistakeholder approach to delivering sustainability at scale. Ratified in September 2015 during a historic UN summit, the 17 goals entered into force in January 2016 and outlined a pathway towards sustainable progress by 2030. Compared to the Millennium Development Goals, the SDGs differed in three key areas (see https://sdg-tracker.org):

- First, they apply to all countries; target achievements are nationally owned and led;
- Second, they specifically establish trade as a way to achieve sustainability goals (through, for instance, SDG 12: Ensure sustainable consumption and production patterns); and
- Third, the SDGs contain more sub-goals and associated targets, allowing for detailed contributions by a range of stakeholders, focused actions and clear progress measurements or KPIs.

The 17 goals are highly interlinked. Their achievement requires deep public–private collaboration.

Multinationals from all sectors and geographic locations soon embraced this multistakeholder approach and the ease of measurement and alignment. A quick Google search reveals, for instance, that Airbus, Credit Suisse, Danone, Hitachi, Nissan, Orange, Repsol and Vodafone have started to measure their SDG contributions in detail and have published data-driven reports. Several companies have also quickly pledged to advance sustainability goals in the wake of the SDG launch. Clothing manufacture C&A, for example, announced in 2015 that 100% of the cotton it uses must be organic and 'more sustainable' by 2020 (C&A, 2015). This last example is telling, as what we call the new Sustainability Deal involves public institutions, businesses and voluntary standards that encompass the entire value chain. C&A, a profit-driven corporation, has publicly pledged to achieve a sustainability goal that contributes to the SDGs set forth by a public institution, the UN. This multistakeholder set-up, covering the value chain and its actors, is achieved through corporate action, public alignment and the use of voluntary standards. To certify that its cotton is indeed organic and more sustainable, C&A exclusively buys it from sources supported by voluntary standards such as Better Cotton[2]. Better Cotton not only develops voluntary standards but also provides a clear 'chain of custody' supporting improved transparency and traceability across supply chains.

Sustainability standards have also been aligning their sustainability goals with the SDGs. With the data-orientated design of the SDGs, organisations can measure the extent to which a sustainability standard (and, by extension, a given user of the standard, such as a smallholder cocoa farmer) aligns with and contributes to different SDGs. Governments and financiers are turning to voluntary standards as they measure SDG target achievement and begin to understand the strategic potential of standards. Figure 4 illustrates the linkages between sustainability standards and the SDGs (Bissinger et al., 2020).

2 https://bettercotton.org

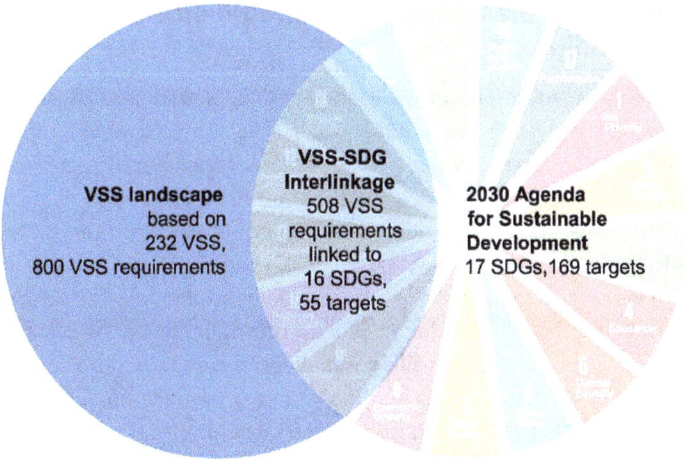

Figure 4 Common objectives: sustainability standards and sustainable development goals. ***Source:*** Adapted from ITC report Linking Voluntary Standards to Sustainable Development Goals. *Note:* Note that the focus of the analysis in Linking Voluntary Standards to Sustainable Development Goals was based on the 126 core SDG targets of the 169 targets in the 2030 Agenda for Sustainable Development.

Companies today find that their roles are evolving. Using these voluntary standards supports due diligence requirements from governments and meets greater expectations from consumers and stakeholders. The ITC has worked with standard-setting organisations for 10 years, listing them in its global, neutral and free Standards Map online portal. At the same time, the ITC has become the convener of a community of sustainability experts, including technical development agencies, international agencies, industry associations, business support institutions, civil society organisations and governments that promote sustainability practices and work to align sustainability frameworks. Looking ahead, policymakers may increasingly recognise and use voluntary standards, signalling a shift away from the 'voluntary' aspect of such sustainability standards. The European Commission, for example, is exploring ways to introduce mandatory due diligence through its Green Deal to give greater weight behind the achievement of sustainability targets (European Parliament, 2020). As part of a mandated due diligence approach and in seeking to tackle the problem of global deforestation and forest degradation, ongoing discussions are focussing on the role that private sustainability standards requirements could play in a 'smart mix' or combination of strategies of supply chain interventions.

The next section of the chapter considers the interconnection of sustainability standards in more detail.

5 Connecting the dots for sustainable international value chains

When it comes to international trade and value chains, these national, international, and private sector initiatives are welcome. However, there is little coordination or understanding of how these rapidly developing policy solutions and multistakeholder initiatives can work together across a value chain. In other words, there is a lack of a connective structure to ensure interoperability or compatibility between national, multilateral, regional, and private industry perspectives on sustainability. Nor is there one internationally recognised body to create a 'level playing field' for all initiatives. Indeed, given the private sector-led nature of many of the prevailing initiatives, this may well be an unrealistic and unworkable aim. More importantly, micro-, small- and medium-sized enterprises (MSMEs) producing and operating along the value chain, which make up the majority of the global industry and trade, have not received much attention or incentives to become sustainable. Their voice is critical to making such policies practical. However, despite these concerns, much has been done and tools have been developed to address sustainability. The most well-recognised tools take the form of voluntary standards and related codes of conduct.

Voluntary sustainability standards, company codes of conduct and other related measures have existed for years. Consumer-facing standards and labels, including Fairtrade, Rainforest Alliance and Organic, are understandably somewhat better known than business-to-business voluntary standards such as GLOBALG.A.P. or SA8000, which do not have their labels on consumer products.[3] Sustainability standards can serve as practical tools to monitor companies' due diligence and their compliance with other social and environmental measures. Financial institutions can also use them to help ensure the sustainability of upstream investments, especially as green finance is being increasingly promoted.

Standards organisations, related NGOs and companies that have developed their codes of conduct have years of experience in developing and implementing environmental, social and labour standards across the world in various sectors. Their collective experience offers insights on how to make the transition to sustainability work for all. Standards are not a 'magic wand' or a panacea for all our sustainability challenges. Rather, these insights serve as a guide and a guardrail to ensure that environmental and social conditions in international value chains do not worsen and that current progress is not

3 It is important to note that private sustainability standards fall outside of the purview of the WTO and its Agreement on Technical Barriers to Trade, Agreement on Sanitary and Phytosanitary Measures, and Agreement on Government Procurement. As we will see, sustainability standards are developed by private actors and NGOs and are voluntary in nature, as opposed to mandatory public regulations. However, even this distinction can get blurry as certain standards, such as organic agriculture standards, are developed by public institutions but nevertheless remain voluntary in scope.

reversed. But can we be more ambitious? Can sustainability standards make innovative and meaningful contributions to help create a new Sustainability Deal? If so, what would this look like, and how would standards help reshape the future global economy? These questions are now explored more fully.

6 Mapping sustainability standards: the International Trade Centre Standards Map

Initiatives to promote social and environmental concerns have emerged over the last several decades. This has occurred in parallel with increasing societal awareness of natural resource constraints, poor social and labour conditions in producing countries, and the general imbalance of power between developed and developing countries. Sustainability standards have proliferated rapidly since the 1990s, initiated mostly by NGOs (Fig. 5). NGO-led initiatives focus differently on social, environmental or ethical topics, depending on their mission. It has become confusing to distinguish and differentiate between initiatives that seem to address similar issues but implement their programmes in practice in very different ways. One example is an initiative requiring a simple membership commitment and signature of a common charter versus another that requires an independent third-party assessment to support product certification.

To assist users of sustainability standards, ITC's Trade for Sustainable Development (T4SD) programme has developed the Standards Map. It is the world's only global repository of such initiatives and standards, including references to codes of conduct, audit protocols, certification programmes, sustainability reporting frameworks, sustainable finance indicators and

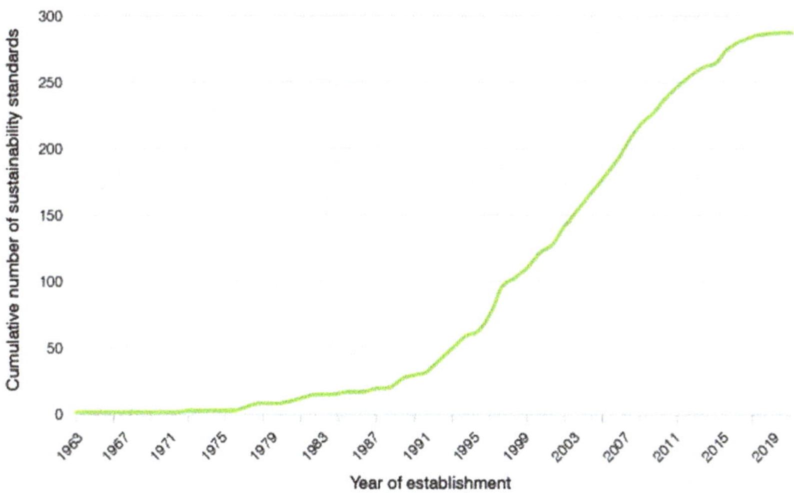

Figure 5 Development of sustainability standards over time. ***Source:*** International Trade Centre, Standards Map (2021).

international standards. In response to the complicated landscape of sustainability initiatives, the Standards Map focusses on making such information more accessible and transparent for everyone, as a global public good (Box 3). The Map provides a publicly available inventory covering over 300 sustainability initiatives applicable to more than 80 sectors and 190 countries, making it easier to compare standards against more than 1600 criteria and assess levels of compliance.

The main challenge in 'mapping' these voluntary standards comes from the fact that they are very different in terms of scope, governance and implementation. Examples include the use of codes of conduct versus certification programmes or demonstrating due diligence through sets of best practices. They can be managed by very different types of organisations (e.g. private or public sector, for-profit or not-for-profit). The standards can be managed through multistakeholder boards and committees, or they can keep the decision-making power in the hands of a single organisation and a few individuals (e.g. a private company with its own corporate sustainability programme versus a civil society multistakeholder-driven programme).

Box 3: Standards Map

Standards Map is a global public tool to navigate the diverse landscape of voluntary sustainability standards. The number of sustainability initiatives, standards, codes of conduct and audit protocols has skyrocketed in the last decade, making it a challenge for companies, consumers and policymakers to navigate the ever-changing landscape of sustainability standards.

Standards Map provides an inventory of public domain standards covering over 300 sustainability initiatives, applicable to more than 80 sectors and 190 countries. Producers, cooperatives and companies across the world wanting to demonstrate that they are operating towards high sustainability standards use the tool. Since its launch in 2011, Standards Map has provided comprehensive, verified information on voluntary standards and similar initiatives covering environmental protection, food quality and safety, labour rights and economic development. This global public good allows users to:

- Identify and review standards by product, sector, area or focus;
- Compare standards side by side with more than 1650 criteria;
- Monitor production trends amongst standards; and
- See where they are in their compliance processes and assess their performance against a range of standard requirements.

Source: www.standardsmap.org

Beyond the organisational set-up, sustainability standards may focus on vastly different topics and operate at various socio-technical levels (e.g. gender empowerment in business management versus biodiversity conservation in the rice sector in South-East Asia). Standards can apply to a single product (coffee, sugar, palm oil, soy and cocoa), an entire industry (textiles, electronics), or even across any sector of the economy (e.g. the International Labour Organisation's Labour Standards).

Given the variety of approaches to sustainability standards, ITC has been developing a comprehensive database of sustainability standards to highlight their differences and similarities. The criteria included in the database are meant to be used by every organisation that deals with sustainability standards:

- Companies that need to compare and contrast sustainability standards used by their suppliers or that want to choose an appropriate sustainability standard based on their market position.
- Business support organisations that need to support their members in search of adequate sustainability standards in their respective sectors and markets.
- Governments, policymakers and trade negotiators seeking to include sustainability provisions in public regulations and trade agreements, and connect such provisions to the language of the private sector, and
- Researchers, academics and civil society organisations involved in the field of sustainability.

The Standards Map database of sustainability standards aims to offer a flexible framework to organise, cluster and analyse standards based on common attributes, missions, objectives, organisational structures, etc.

Standards have evolved significantly over the last half-century. As the next section will show, standards are changing the rules of the game, creating greater transparency along supply chains. Not only are they moving from 'limiting damage' and 'sustaining' towards 'creating positive impact', but they also contribute to an ecosystem that has the potential to create more systemic positive change through integrated action and coordination amongst all actors of international value chains.

7 New approaches: traceability and transparency through technology

As international value chains have evolved, so have consumer, regulator and investor expectations. Today, these groups expect supply chains to be traceable and sustainable. Standards and supply chain actors have joined forces with technology providers to trace products to their origins and provide data to substantiate sustainability claims down to the farmer level. Supply chains are being digitalised through mobile and internet technologies. Procurement professionals and standard setters increasingly use artificial intelligence, the Internet of Things (IoT), and blockchain technologies to track products at each stage of the supply chain.

For instance, Rainforest Alliance has its own traceability system called MultiTrace. This allows it to track consignments from individual farms to final certification. Supply chain actors record their transactions (e.g. from producers to buyers, from traders to manufacturers or blenders) and their sales, purchases, conversions, blends and manufacturing activities within the platform every quarter. Companies must conduct quarterly spot checks by verifying a sample of incoming volumes, matched with traceability documentation for products labelled with the Rainforest Alliance seal or claimed as being Rainforest Alliance Certified. The system allows for organised data collection on consignments for certified products. This helps producers and companies meet chain-of-custody requirements. Data can also be used for internal management, external communication and stakeholder reporting, such as progress towards the SDGs.[4]

In another example, Tony's Chocolonely, a Dutch chocolate confectionary company, uses blockchain technology to track its supply chain back to farmers. Its main objective is to eradicate slave and child labour. It can also trace the farmers' individual contributions of cocoa beans. The technology improves traceability and transparency in value chains and also facilitates the system for suppliers to receive premiums they can use to pay for children's education, invest in the farms, improve access to water and other community benefits.[5]

The main challenge lies not in building the technology, but in collecting timely, accurate, complete and reliable data. Goods from supply chains using primarily micro, small and MSMEs face particular challenges. Products are often aggregated between the farm level and shipment (e.g. by mixing produce from individual farms into bulk quantities that are more cost-effective to transport and process), which means individual consignments can be lost or 'misplaced' along long, fragmented supply chains. Smallholders and others in the supply chain may struggle to keep up with data recording requirements. This complicates traceability. Furthermore, MSMEs, especially agribusinesses, are often located in regions with poor infrastructure, including roads, transportation and bad or no telecommunication network connections. Data collectors are turning to IoT and devices such as tablets to collect data and consolidate this information in a single platform.[6]

4 https://www.rainforest-alliance.org/business/responsible-sourcing/understand-end-to-end-traceability-for-tea/
5 https://tonyschocolonely.com/us/en/our-mission/tonys-impact
6 From the interview with Frans Pannekoek, Tony's Chocolonely 'This Company Is Using Blockchain Technology to Eradicate Slavery In The Chocolate Industry' (Forbes) https://www.forbes.com/sites/chelseadavis/2019/03/31/this-company-is-using-blockchain-technology-to-eradicate-slavery-in-the-chocolate-industry/?sh=6faef5a01407

8 New approaches: reducing carbon footprints

Major businesses have been committing to reduce their carbon emissions or even achieve carbon neutrality in response to climate change. Maersk, the world's largest shipping company, set an ambitious target in 2018 to achieve net-zero carbon dioxide (CO_2) emissions from its operations by 2050.[7] Unilever has committed to lowering its CO_2 emissions by investing in new technologies such as hydrogen, increasing energy efficiency, switching to renewable energy sources and using sustainable transport, amongst other things.[8]

Voluntary standards have also developed solutions to calculate and reduce the carbon footprint of their certified operations. The ProTerra Foundation certification scheme for sustainable and non-GMO crops launched its Carbon Footprint Calculation Project in 2019 to calculate CO_2 emissions in soy production. The method calculates CO_2 emissions per kilogram of soybean and includes emissions data projected from land use change (when forests have been recently cleared to grow soy, carbon stored in trees is released into the atmosphere as CO_2), agricultural production and further processing, as well as carbon emissions associated with logistics. Buyers of ProTerra-certified soy, therefore, can obtain high-quality data from ProTerra for their own CO_2 calculations and decision-making on sustainable soy sourcing.[9]

Calculating or tracking carbon can be a challenge for all organisations, especially MSMEs. Support from capacity-building organisations or the development of simplified tools is one solution. Fairtrade International has developed its own climate change initiative and a new Fairtrade Climate Standard as part of this strategy. The standard aims to enable smallholders and rural communities to gain access to the carbon market by producing Fairtrade carbon credits, which hold a financial value. Small-scale farmer communities create local projects to reduce carbon emissions. Planting shade trees, buying biogas installations and reducing the use of wood cook stoves are some examples. These measures protect the environment where smallholders live and make them more resilient to climate change.[10] Farmers then become eligible to receive Fairtrade carbon credits, which they can sell to buyers who are looking to offset their carbon emissions and their impact on the environment. Farmers receive Fairtrade minimum prices for their products and a premium for the carbon credits they sell to buyers. This enables farmers to cover the costs of implementing carbon reduction projects and to invest in more projects that will help them adapt to climate change.

7 https://www.maersk.com/news/articles/2019/06/26/towards-a-zero-carbon-future
8 https://www.unilever.com/planet-and-society/climate-action/decarbonising-our-business/
9 https://www.proterrafoundation.org/news/carbon-footprint-calculation-project-main-outcomes/
10 https://www.fairtrade.net/standard/climate

9 Sustainability standards and resilience: the case of the COVID-19 pandemic

Coronavirus disease-19 (COVID-19) was a significant challenge for small businesses, especially in developing and least developed countries. The pandemic has created numerous financial, commercial and operational risks for these firms. It has strongly affected two-thirds of African companies, with 75% of MSMEs reporting lower sales and 54% struggling to access inputs. One-fifth of MSMEs surveyed globally by ITC said they risked closing permanently (ITC, 2020). The natural question that arises when discussing sustainability standards today is whether compliance has helped MSMEs stay resilient in the face of the financial, commercial and operational risks brought about by the COVID-19 crisis.

An International Institute for Sustainable Development policy brief has explored how well sustainability standards are achieving the objectives they champion whilst helping to build smallholder resilience to external shocks (IISD, 2023). The study found that adopting sustainability standards can improve smallholders' resilience to external shocks such as COVID-19. When prices for certified products are higher or have premiums, producers can obtain higher incomes and, hence, better capacity to invest, adapt and cope with shocks. For instance, Cambodian rice farmers selling organic-certified rice continued to sell at premium prices during the pandemic, making them resilient to potential commercial and financial shocks caused by COVID-19.

Compliance with sustainability standards forges stronger relationships with buyers, which results in more secure market access or faster recovery of market access after disruptions, according to the policy brief. For example, one buyer of Indian cotton obtained special permission from the authorities to keep timely seed deliveries to its sustainability standard-compliant supplier despite the lockdown; he also kept orders intact and paid a premium price for organic cotton. Certified Guatemalan producers, e.g., have also continued to sell goods to international buyers during the pandemic and have been less affected by COVID-19 shocks than their non-certified peers. Certified producers have generally been better able than non-certified producers to cope with COVID-19 protocols related to hygiene and masks, as they had already received training on sustainability standards' criteria about safety, health and labour protocols.[11]

However, whilst obtaining a premium for certified produce is beneficial, standards do little for inherent price volatility. Prices can fluctuate widely during times of crisis, and a premium on highly volatile prices itself becomes very volatile. Long-term contracts, floor prices and other characteristics of some standards can support producer resilience, but there is no certainty

11 https://www.iisd.org/system/files/2021-04/covid-19-certification-farmer-resilience.pdf

that standards will actually boost resilience. Market diversification helps reinforce the resilience of certified farmers, who also have more options to sell their products.

10 Financial markets: managing risk through sustainability standards

Financiers are increasingly constructing portfolios that avoid social and environmental risks. Voluntary standards are being used as a way to show that these risks are being managed. Today, we see more financial institutions providing what is described as sustainable finance. Sustainable finance integrates environmental, social and governance (ESG) criteria into business or investment decisions for the benefit of clients, the environment and society. Sustainable funds, impact investing, green bonds, microfinance and credits for sustainable products are all examples of sustainable finance. Sustainable finance is a strong instrument to encourage sustainable business practices. It sets a price on sustainability-related risks in business operations. In Ghana, e.g. sustainable businesses can get lower interest rates for loans since such firms appear less risky to banks (Waarts and Kiewisch, 2020). Investors can also influence companies in which they invest to develop more sustainable business practices. In 2020, two-thirds of Procter & Gamble investors voted for better disclosure of the company's impact on forest operations after it was accused of contributing to deforestation. Procter & Gamble then committed to increasing its use of FSC-certified wood pulp to at least 75% across its brands by 2025.[12] Investors demand greater disclosure, especially on climate-related risks, so companies will be required to divulge more sustainability-related information.

Central banks – both in developed and developing countries – are beginning to promote sustainable finance. For instance, in 2019, the Bank of Ghana published its Sustainable Banking Principles[13] and encouraged commercial banks in the country to apply environmental and social risk management policies in their financing. As a result, some banks now offer discounts on loans for sustainable businesses and projects. In the European Union, the European Commission released an action plan to finance sustainable growth in 2018. It aims to reorient capital flows towards sustainable investment, mainstream sustainability into risk management, strengthen sustainability disclosure of financiers and foster sustainable corporate governance.[14]

Financiers use various social and environmental criteria to select clients and invest in businesses. They can use criteria developed by other organisations,

12 https://www.ft.com/content/1dd92502-e95b-4c21-be1c-c18a598acf1a
13 https://www.bog.gov.gh/wp-content/uploads/2019/12/Ghana-Sustainable-Banking-Principles-and-Guidelines-Book-1.pdf
14 https://ec.europa.eu/info/publications/sustainable-finance-renewed-strategy_en#action-plan

such as the International Finance Corporation Environmental and Social Management System[15] (covering stakeholder engagement, emergency preparedness and impact on local communities, for example) or the UN Principles for Responsible Banking, as a basis for investment and lending decisions.[16] Some financiers have developed their own ESG requirements or decide to work with companies that have zero carbon footprints or do not use fossil fuels. Others use voluntary standards as a precondition for companies to apply for funding. In summary, standards can be a credible instrument to confirm investees' compliance with social and environmental practices, as they are usually third-party certified. Financiers can also rely on the expertise of standard setters, rather than having to develop their own expertise in these areas. Finally, voluntary standards are useful as a comprehensive tool to monitor a wide range of risks in business operations.

11 Policymakers: developing due diligence frameworks

Policymakers are starting to use voluntary standards in free trade agreements and public procurement policies and to monitor the supply chains of large firms. Standards are seen as a central component of the smart-mix approach in due diligence frameworks. Over the last decade, some governments have stepped up their role in promoting sustainable production and mitigating social and environmental risks. There are many reasons for this. Governments want to reduce the negative impacts of international value chains on sustainability. They want to see more transparency and consistency in mitigation measures by companies. Governments also aim to create a more level playing field for corporate actors by setting up basic rules for firms to mitigate and report on social and environmental risks in supply chains, thus making a basic level of corporate ESG reporting mandatory.

Several countries have adopted laws that oblige companies to trace their supply chains to ensure that no harmful social or environmental activities are associated with production. The UK was one of the pioneers in this area. In 2015, the UK Parliament adopted the Modern Slavery Act,[17] which includes a section on transparency in supply chains and obliges lead firms operating in the United Kingdom to prepare annual modern slavery statements. Companies must provide details of their supply chains, policies to mitigate modern slavery risks, due diligence policies and performance indicators. The company's board of directors must approve the statement, and a director or CEO must

15 https://www.ifc.org/wps/wcm/connect/topics_ext_content/ifc_external_corporate_site/sustainability-at-ifc/publications/publications_handbook_esms-general
16 https://www.unepfi.org/banking/bankingprinciples/
17 https://www.legislation.gov.uk/ukpga/2015/30/contents/enacted

sign off on it. Following the UK example, France adopted the Law on Duty of Care (Loi relative au devoir de vigilance des sociétés mères et des entreprises donneuses d'ordre).[18] It went beyond slavery risks to include other social issues as well as risks related to the environment. The law obliges French companies to demonstrate they have processes in place to address human rights issues, health and safety risks, as well as impacts on the environment.

Germany's parliament adopted the Supply Chain Act in 2021. Businesses with more than 1000 employees must identify and address child labour, forced labour, pollution, land rights and other risks. The act came into force on 1 January 2023, and firms that fail to comply face fines of up to €8 million or 2% of their annual turnover. Companies also risk being excluded from public tenders in the event of non-compliance. The European Union is also taking action. The European Commission has initiated a Sustainable Corporate Governance Initiative,[19] with a directive on due diligence relating to environmental performance and human rights. The directive aims to reduce fragmentation between existing due diligence laws in member states and to set EU-wide binding requirements for businesses to mitigate social and environmental risks in their supply chains.

Building on the momentum of national initiatives, the Organisation for Economic Co-operation and Development (OECD) has played a pivotal role in developing due diligence instruments aimed at promoting responsible business conduct amongst multinational enterprises. These instruments, such as the OECD Guidelines for Multinational Enterprises and specific sectoral due diligence guidance, provide comprehensive frameworks and practical guidance for companies to identify, prevent, and mitigate adverse impacts related to their operations, supply chains and business relationships. By aligning with national efforts and leveraging its global platform, the OECD contributes to a coordinated approach to due diligence, enhancing transparency, accountability, and sustainability across international value chains.

Governments also use free trade agreements to enhance sustainable production in supply chains. More and more environmental and social provisions have been included in such agreements since the 1990s. Since 2010, such agreements have increasingly referred to sustainability standards as a way for exporters to address social and environmental risks in production. In 2019, 19 such agreements were in place, most of them involving the European Union (Fig. 6). Even though these references to sustainability standards remain recommendations without substantive commitments and mechanisms for

18 https://www.legifrance.gouv.fr/loda/id/JORFTEXT000034290626/

19 https://ec.europa.eu/info/law/better-regulation/have-your-say/initiatives/12548-Sustainable-corporate-governance_en

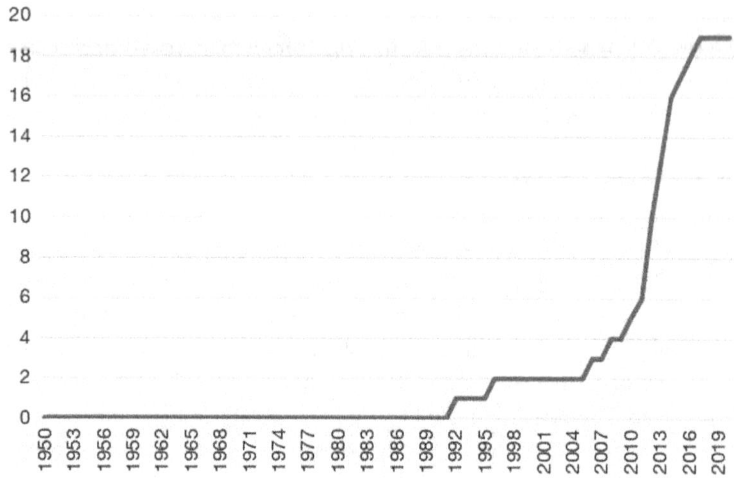

Figure 6 More trade deals refer to sustainability standards**.** ***Source:*** United Nations Conference on Trade and Development, Scaling Up Voluntary Sustainability Standards Through Sustainable Public Procurement and Trade Policy, 2020, p. 40.

measuring performance, standards remain on the radar of governments as a mechanism to check for imports of sustainably certified products.[20]

Sustainable public procurement is another mechanism used by governments to encourage sustainable production. Public tenders increasingly refer to sustainability standards to select suppliers or service providers, as these standards guarantee that suppliers or service providers comply with the ESG practices required by public procurers. Some European and Asian countries refer to standards in their sustainable procurement policies because there is no other way to prove companies' compliance with ESG criteria.[21]

12 Smallholder uptake of sustainability standards

In recent years, several standards have developed solutions and strategies to boost the uptake of standards by smallholder communities and groups to enhance their participation in international value chains and improve their market access. Smallholders may be geographically dispersed, making the delivery of inputs, training and support resource-intensive. They may struggle to access markets because they lack the production quality and volumes required by buyers, as well as the funds and capacity to implement standards and successfully pass audits. Historically, few voluntary standards have focused on smallholders, even though smallholders are responsible for most food production on the planet.

20 https://unfss.org/wp-content/uploads/2020/10/UNFSS-4th-Report_revised_12Oct2020.pdf
21 https://unfss.org/wp-content/uploads/2020/10/UNFSS-4th-Report_revised_12Oct2020.pdf

Fairtrade conducted a comprehensive consultation with farmers and then updated its standard for small-scale producers. The key change in the new standard is that to apply for Fairtrade certification, farmer cooperatives should include more small family farms.[22] Fairtrade standards also aim to improve the organisation of farmer groups and provide Fairtrade minimum prices as a safety net against fluctuating commodity prices. This helps smallholders obtain better prices and sustain their livelihoods.[23] Similar to Fairtrade, Bonsucro, a certification scheme for sugarcane, has also developed a separate production standard for smallholder farmers, which contains fewer criteria for smallholders and a set of indicators to improve farmer group organisation and management.[24] Bonsucro also helps improve sugarcane production through water stewardship, soil management and better management of chemical inputs.[25]

When it comes to the impact of sustainability standards on smallholder farmers and MSMEs, there is some evidence of positive impact depending on specific sectors and standards. One study that focused on coffee standards concluded that certification is commercially viable only if it improves productivity or provides a price premium large enough to cover all certification-related costs (Bermúdez and Voora, 2022). The poorest and most marginalised smallholders usually cannot obtain certifications without external help. In many cases, they cannot get certified because their products lack the quality demanded by buyers. These smallholders often do not have the land, labour, education or other resources to comply with sector standards or find it worthwhilst to go through the certification process. The study also shows evidence of certification-improved coffee prices for farmers and higher incomes, improved agrochemical use and handling, better water use and reduced use of inorganic chemicals (in the case of organic standards, though this can also lead to low yields in the short term). Standards that strongly focus on good agricultural practices also positively impact productivity. Those that target trading relationships or prices are likely to have no impact on productivity (Elliott, 2018).

Several key factors make certification beneficial to smallholders. Effective producer organisations can spread the benefits of certification fairly amongst smallholder farmers (e.g. price premiums distributed equally amongst farmers). Steady relations with buyers can sustain sales and revenues for smallholder farmers. Training on social and environmental practices promoted by standards can improve farmers' sustainability performance. Another study that focuses on Fairtrade and organic standards found that certified farmers receive a 20–30% higher price on their products compared to non-certified farmers. Certified farmers gain higher profits, which boosts household incomes by

22 https://www.fairtrade.net/standard/spo
23 https://www.fairtrade.org.uk/farmers-and-workers/coffee/
24 http://www.bonsucro.com/tools_resources/bonsucro-production-standard-for-smallholder-farmers/
25 https://www.rural21.com/english/news/detail/article/sustainability-standards-traceability-and-certification.html?no_cache=1

16–22% (results vary depending on product, standard or country) (Meemken, 2020). Standards can certainly increase MSME growth through price premiums, improved market access, better sales and access to finance. However, there are several constraints to meeting standard requirements. Small firms must have the right managerial and technical knowledge, as well as funds to cover the costs of certification and implementation of standards criteria.

Whilst some sustainability standards and their associated organisations have made efforts to improve organisational and supply chain standards, more needs to be done, particularly in terms of the producer-enabling environment (policies, regulation and public infrastructure), if standards are to play a pivotal role in the smart-mix approach and deliver on the SDGs, especially SDG 1 and SDG 2.

13 Consumer attitudes to sustainability standards

Consumers are one of the most important stakeholders in the sustainable trade agenda. Markets with a relatively high level of consumer demand for sustainable products can spur increased adoption of sustainability standards.[26] Whilst the uptake of sustainable consumption varies across consumers in different economies, concern about sustainability has deepened over the past few decades. The Conference Board Global Consumer Confidence Survey, conducted in collaboration with Nielsen in 2017, found that 81% of respondents said it was 'extremely' or 'very' important for companies to adopt programmes to improve the environment. When consumer concerns increase, the issue arises of their awareness about the credibility of sustainability standards and labels. People must be better informed about what the labels actually represent to drive them to consume sustainably. The German Government set up an online platform called Siegelklarheit to disseminate information about sustainability labels. Nadine Stiller, an adviser at the Deutsche Gesellschaft für Internationale Zusammenarbeit, explained what Siegelklarheit seeks to achieve (Box 4). Stiller stated that click rates on the platform have risen during the COVID-19 pandemic.

Box 4 Helping consumers make sustainable choices by providing clear, reliable information about sustainability. (https://www.siegelklarheit.de/)

Siegelklarheit translates to 'clarity of labels'. It is an online platform established around 2015 by the German Federal Ministry for Economic Corporation and Development to assist German consumers

26 https://unfss.org/wp-content/uploads/2020/10/UNFSS-4th-Report_revised_12Oct2020.pdf

in assessing labels in different product groups, especially in verifying their credibility and level of ambition. Now, five other ministries are also involved. The labels are assessed as a 'good choice' or a 'very good' choice.

There is a specific methodology for this assessment called the Sustainability Standards Comparison Tool that forms the basis of Siegelklarheit. There are three broad topics: credibility, social, and environmental aspects. For each of the pillars, there are specific subtopics that make the assessment detailed. Aspects that are critical in terms of sustainability are the so-called minimum criteria. Each label must fulfil those criteria in terms of credibility and one of the other two dimensions to get a rating on Siegelklarheit.[27]

Switzerland has established Labelinfo.ch, the most comprehensive label database in the country. Labelinfo.ch offers information on 135 quality labels and 19 declarations in German and French, informing businesses and consumers about labels and helping them make their consumption and purchasing choices more responsible.

Despite efforts like these, greenwashing[28] has been on the rise, with companies providing information that may be imprecise, unclear, incomparable, unsubstantiated or irrelevant. Greenwashing can damage the reputation of credible standards, labels and claims, causing mistrust and confusion amongst both consumers and information providers in business and government. In 2020, the European Commission analysed green online claims from businesses such as garments, cosmetics and household equipment. It concluded that 'in 42% of cases, the claims were exaggerated, false or deceptive and could potentially qualify as unfair commercial practices under EU rules. "Greenwashing" has increased as consumers increasingly seek to buy environmentally sound products'.[29]

To address greenwashing, the United Nations Environment Programme, in collaboration with ITC, developed guidelines for providing sustainability information (United Nations Environment Programme, 2017) in 2017. These guidelines help identify sustainable products, as not all standards use a label. Many use a written message or claims[30] that could be consumer-facing or business-to-business. The ISEAL Alliance has developed a Sustainability Good Practice Guide to inform sustainable standard organisations and enhance the credibility of claims about sustainability standard systems. As few consumers seek additional information, the ISEAL guide recommends that sufficient

27 Interview with Nadine Stiller.
28 This is the process of conveying a false impression or providing misleading information about how a company's products are more environmentally sound.
29 https://ec.europa.eu/commission/presscorner/detail/en/IP_21_269
30 https://www.isealalliance.org/get-involved/resources/video-credibility-claims-and-labelling

information be provided with the claim or logo. The COVID-19 pandemic has focused attention on critical issues of biodiversity conservation and climate change. The pandemic has heightened awareness of the interconnectedness between human health and ecosystems, emphasising the importance of biodiversity conservation. The crisis has also underscored the need for resilient systems, driving increased attention to addressing climate change and promoting sustainable practices. Citizen engagement and concern about sustainability have increased in Europe, where consumers are demanding that fashion businesses act responsibly and consider their social and environmental impacts. Some European consumers also prefer more durable fashion items as they repair and wear them longer (McKinsey & Company, 2020).

Interestingly, people of different ages are demonstrating different sustainable consumer behaviours. The level of engagement between different age groups is varied. Studies show that Gen Z is the least engaged in environmental issues, in contrast with younger Millennials, who are the most engaged, yet supporting ethical brands matters more to Gen Z (Deloitte, n.d.). As the research in behavioural economics attempts to decode the psyche of new consumers, some experts have suggested new approaches for companies – using social influencers, shaping good habits, leveraging the domino effect, talking to the heart with emotions and favouring practical experiences over ownership (White et al., 2019).

Cost must still be considered when analysing changing consumer preferences. Studies suggest that most consumers choose products with a lower impact on the environment (if the cost is the same), and public claims from sustainability standards such as ecolabels are amongst the most important signals to the market to help consumers in this process (Iraldo et al., 2020,). In 2020, data indicated that sustainability was a key consideration for various consumer groups when making their purchasing decisions. However, cost is still a significant barrier. Consumers may want to buy sustainably, but cannot afford to do so (Deloitte, n.d.).

To ensure that younger consumers continue to make more sustainable choices, well-designed and targeted tools are needed to improve awareness and provide these consumers with credible information. Businesses are likely to engage in more sustainable production to meet rising demand and avoid reputational risks. Social media, given its worldwide uptake, especially amongst younger consumers, should be leveraged to promote sustainable consumption and production.

14 Conclusion

The concept of a 'sustainability' initiative or standard has transformed from something that might once have been regarded as just a trend into a vital element of global trade. Sustainability standards play a significant role in international trade by encouraging fair business practices, helping businesses adapt to upcoming rules, promoting agreement in global trade, and sparking innovation. As the world increasingly emphasises sustainability, these initiatives and standards offer a potential framework for building a more environmentally aware and socially responsible global economy.

Sustainability standards are now crucial in international trade for several reasons. As environmental crises and climate change become more pressing concerns, there is a growing understanding worldwide about the importance of sustainable practices. Consumers, investors and businesses are showing more interest in goods and production methods that are environmentally friendly and socially responsible. Sustainability standards, like the ones discussed in this chapter, give companies a way to ensure that their operations meet environmental and social responsibility criteria. They serve as a signal to consumers, showing that a company is committed to ethical practices, which can improve its reputation and attractiveness in the market. Additionally, sustainability standards can help companies prepare for upcoming regulations. As the world works to address climate change, governments and international organisations are developing stricter rules regarding environmental protection and social responsibility. By adopting sustainability standards, businesses gain valuable knowledge and skills in sustainable practices, making it easier for them to follow future regulations. This proactive approach can help companies stay ahead of legal changes.

15 References

Balmaña, R.M. (2010). *International Corporate Social Responsibility: The Role of Corporations in the Economic Order of the 21st Century*. Alphen aan den Rijn: Kluwer Law International.

Bermúdez, S. and Voora, V. (2022). Global Market Report: Coffee prices and sustainability. IISD.

Bissinger, K., Brandi, C., Cabrera de Leicht, S., Fiorini, M., Schleifer, P., Fernandez de Cordova, S. and Ahmed, N. (2020). Linking voluntary standards to Sustainable Development Goals. Geneva: International Trade Centre. https://www.intracen.org/publication/Sustainable-Development/

Britannica, T. Editors of Encyclopaedia. (21 May 2021). Industrial revolution. Encyclopedia Britannica. https://www.britannica.com/ event/Industrial-Revolution

C&A. (2015). Global sustainability report 2015. https://www.c-and-a.com/uk/en/corporate/fileadmin/user_upload/Assets/3_Newsroom/Media_Center/sustainabilityreport_2015/C_A_Global_Sustainability_Report_2015_Summary_English.pdf

Caradonna, J.L. (2016). *Sustainability: A History* (Reprint ed.) [E-book]. Oxford University Press, p. 138.

Convention on Biological Diversity. (2013). Press release on Biodiversity Barometer. https://www.cbd.int/doc/press/2013/pr-2013-04-16-UEBT-en.pdf

Deloitte. (n.d.). Shifting sands: How consumer behaviour is embracing sustainability. https://www2.deloitte.com/ch/en/pages/ consumer-business/articles/shifting-sands-sustainable-consumer.html

Elliott, K.A. (2018). What are we getting from voluntary sustainability standards for coffee? Center for Global Development Policy Paper 129.

European Parliament. (2020). Towards a mandatory EU system of due diligence for supply chains. https://www.europarl.europa.eu/RegData/etudes/BRIE/2020/659299/EPRS_BRI(2020)659299_EN.pdf

Fairtrade Foundation. (n.d.). The history of fairtrade. https://www.fairtrade.org.uk/what-is-fairtrade/the-impact-of-our-work/

Gale, F., Ascui, F. and Lovell, H. (2017). Sensing reality? New monitoring technologies for global sustainability standards. *Global Environmental Politics*, 17(2), pp. 65–83. https://doi.org/10.1162/glep_a_00401

Griswold, E. (4 September 2013). How "Silent Spring" ignited the environmental movement. *The New York Times*. https://www.nytimes.com/2012/09/23/magazine/how-silent-spring-ignited-the-environmental-movement.html

IfOAM. (2020). Our history & organic 3.0. https://www.ifoam.bio/about-us/our-history-organic-30

Iraldo, F., Griesshammer, R. and Kahlenborn, W. (11 March 2020). The future of ecolabels. *International Journal of Life Cycle Assessment*, 25, pp. 833–839. https://link.springer.com/article/10.1007/s11367-020-01741-9#Sec6

IISD. (2023). How can sustainability standards do more to Ratchet up smallholder farmers' incomes?. https://www.iisd.org/articles/deep-dive/how-can-sustainability-standards-ratchet-up-smallholder-farmers-incomes

ISEAL Alliance. (n.d.). Growth of the sustainability standards movement. https://platform.isealalliance.org/sites/default/files/Growth%20of%20the%20Sustainability%20Standards%20Movement%20timeline%20online.pdf

ITC. (2020). SME competitiveness outlook 2020: COVID-19: The great lockdown and its impact on small business. https://www.intracen.org/uploadedFiles/intracenorg/Content/Publications/ITCSMECO2020.pdf

Luttikholt, L. (2007). Principles of organic agriculture as formulated by the International Federation of Organic Agriculture Movements. *NJAS: Wageningen Journal of Life Sciences*, 54(4), pp. 347–360. https://doi.org/10.1016/s1573-5214(07)80008-x

McKinsey & Company. (17 July 2020). Survey: Consumer sentiment on sustainability in fashion. https://www.mckinsey.com/industries/retail/our-insights/survey-consumer-sentiment-on-sustainability-in-fashion

Meemken, E.-M. (2020). Do smallholder farmers benefit from sustainability standards? A systematic review and meta-analysis. *Global Food Security Journal*, 26.

Office of the United Nations High Commissioner for Human Rights. (2021). OHCHR and business and human rights. https://www.ohchr.org/en/issues/business/pages/businessindex.aspx

Purvis, B., Mao, Y. and Robinson, D. (2019). Three pillars of sustainability: In search of conceptual origins. *Sustainability Science*, 14, pp. 681–695. https://doi.org/10.1007/s11625-018-0627-5

Resolution Adopted by United Nations General Assembly No. A/RES/55/2, 3 C.F.R. 2 (2000).

Seric, A. and Tong, Y.S. (3 August 2019). What are global value chains and why do they matter? Industrial Analytics Platform. https://iap.unido.org/articles/what-are-global-value-chains-and-why-do-they-matter

State Secretariat for Economic Affairs SECO. (March 2021). SECO's engagement in Voluntary Sustainability Standards (VSS) an evolving commitment to drive sustainability in global value chains, Claudia Meier, Gregory Sampson, Cristina Larrea, Bernhard Schlatter, Steffany Bermudez, Tuan Duc Dang and Helga Willer (Eds.), The State of Sustainable Markets 2023: Statistics and Emerging Trends. ITC, Geneva.

United Nations Environment Programme. (2017). Guidelines for providing sustainability information: Global guidance on making effective environmental, social and economic claims, to empower and enable consumer choice. https://wedocs.unep.org/bitstream/handle/20.500.11822/22180/guidelines_product_sust_info.pdf?sequence=1&isAllowed=y

United Nations Global Compact Leaders Summit 2010 – Summary Report (Rep.). (October 2010). Retrieved 5 July 2021 from United Nations Global Compact website: https://d306pr3pise04h.cloudfront.net/docs/summit2010%2F2010_Leaders_Summit_Report.pdf

Waarts, Y. and Kiewisch, M. (2020). Balancing the living income challenge: Towards a multi-actor approach to achieving a living income for cocoa farmers. Wageningen University & Research and Mondelez International.

White, K., Hardisty, D.J. and Habib, R. (July-August 2019). The elusive green consumer. Harvard Business Review, pp. 124–133.

2012 Global Winners & Reporting Trends (Rep.). (April 2021). Retrieved 5 July 2021 from CorporateRegister.com website: https://www.corporateregister.com/crra/help/CRRA-2012-Exec-Summary.pdf

Chapter 4

The role of non-governmental organisations (NGOs) in improving safety, quality and sustainability standards in agri-food supply chains

Pilar Pampin and Clare Mike, LEAF (Linking Environment and Farming), UK

1 Introduction

The agri-food sector is greatly affected by climate change-related issues, such as droughts, floods, fires, loss of biodiversity, desertification and deforestation. This impacts not only the productivity of the land but also the quality of life and incomes of farmers. The agri-food sector is also partly responsible for exacerbating these issues: farming with practices that degenerate landscapes and soil, negatively impacting biodiversity and further aggravating climate change through greenhouse gas emissions (Hoekstra and Mekonnen, 2012; FAO and UNEP, 2020; Herweijer *et al.*, 2020). However, this negative impact can be avoided by implementing sustainable practices with an integrated farm management approach, encouraging biodiversity, regenerating soils, improving water stewardship, and adopting integrated pest management (IPM), among many other environmentally sensitive techniques (Kremen and Miles, 2012; Garibaldi *et al.*, 2017; Rose *et al.*, 2019).

http://dx.doi.org/10.19103/AS.2024.0133.04

Sustainability standards play a role in promoting and supporting the uptake of these practices, promoting agroecological farming, reducing the negative impact on the environment, and, where possible, achieving positive outcomes (Swinnen, 2016; Meemken *et al.*, 2021). There has been an increase in the uptake of private sustainability standards within food, textile and similar supply chains around the world (Tayleur *et al.*, 2017; Lambin and Thorlakson, 2018; Ward and Mishra, 2019). Lambin and Thorlakson (2018) attribute this increase in the uptake of standards (especially voluntary sustainability standards) to the increasing complexity and impact of global supply chains and how this is related to their negative impact on the environment. Non-governmental organisations' (NGOs) involvement in providing visibility of these challenges has been recognised as having great magnitude both globally and locally (Harangozó and Zilahy, 2015). This involvement can take various forms, for example, through the creation of partnerships and collaborations, creating and promoting sustainability standards, and generating and sharing knowledge.

Before exploring the role of NGOs in improving safety, quality and sustainability standards, it is imperative to have a clear idea of what NGOs are and some of their key roles. The literature suggests that an NGO needs to have several characteristics. They need to be separate and unconnected from the government, have charitable status, work towards a certain aim related to promoting communities' wellness, safety and health, protecting and enhancing nature and/or providing some kind of service (World Bank, 1995; Willetts, 2009). Moreover, they are not to be part of any political party or associated with criminal activities (Willetts, 2009). From definitions by other authors and our own experience, Linking Environment and Farming (LEAF) has created the following definition:

'Non-governmental organisations (NGOs) are non-profit organisations that set up and operate independently from local, state, or international governments, giving them a lobbying role. NGOs generally work towards a specific cause or goal and tend to address public, social, or political issues, by combining research and experts to address them. They also can represent more widespread, larger scale/international projects.'

NGOs can drive and enable change while working towards a solution to a social, environmental, political or economic issue. One way of achieving this is through lobbying, gaining leverage over policymakers and industry, influencing policy and shaping it to reach their organisational goals. Sometimes this can be through developing programmes, projects and schemes. NGOs are well-positioned to facilitate collaborative working towards a common goal, by connecting individuals, businesses and governments. As they are independent, they can make constructive critiques and act as a neutral conduit, linking together experts from around the world to address key issues. They can be subject matter experts and hold recognition from society and their peers, making NGOs' voices and opinions important to the general audience. Moreover, they usually have access to a wide range of funding for specific projects and can campaign to raise further funds.

An example of an NGO in the agri-food sector is Fairtrade International. They are an international non-profit organisation whose aim is to empower farmers and workers in developing countries to gain more control over their livelihoods through trade. They connect producers and consumers with the aim of enabling fair and secure relationships between them. They achieve this by promoting tools and services and banning discrimination, forced labour and child labour (Fairtrade, no date). They take pride in being partially owned and managed by producers, giving them representation in the general assembly.

In conclusion, NGOs have many key roles in the space of sustainable agri-food supply chains, relating to farming systems, land management, policies, and standards. The outcomes of the activities of environmental NGOs are closely related to delivering such goals. We will be exploring in this chapter the expected outcomes of environmental NGOs and their roles, illustrated through available literature on the subject, as well as through LEAF's experience and the use of case studies.

Disclaimer: LEAF has over 30 years of experience in social and environmental standards, and this has formed the foundation of this chapter. However, the same principles apply to the role of NGOs in improving quality and safety standards, advocating for fair prices, and enabling access to knowledge and markets (note: this is not an exhaustive list of all contexts and processes where NGOs play a role). Generally, NGOs play a supporting role, enabling organisations, businesses and individuals to fulfil an obligations through different actions and approaches.

2 Outcomes environmental NGOs hope to achieve and how to measure them

Environmental NGOs might target a range of specific outcomes; however, usually, there is a common thread connecting the overarching impact desired through the organisation's work. This impact and the related outcomes are explored in this section, with examples of how they can be achieved through different approaches.

As expected, environmental benefits are one of the most significant outcomes targeted by environmental NGOs (Akkucuk and Sekercioglu, 2016). Some environmental benefits targeted by NGOs can include increased biodiversity, improved habitat areas and the quality of those areas, promotion of food with a lower carbon footprint, and actions to create healthier soils, among other things (Nasiritousi, 2019). Simultaneous and holistic positive change across a range of key sustainability challenges is a goal environmental NGOs seek to achieve. There can be a trade-off in terms of certain agri-practices as they may lead to either negative or complementary effects on specific environmental aspects. For example, applying certain crop protection products could improve yields but could also negatively impact local biodiversity, reducing the diversity of plant species, fauna, insects and pollinators. Conversely, water quality could be improved in line with practices that enhance soil health, i.e. one set of practices that are adopted could deliver multiple environmental benefits. NGOs, in terms of the standards they advocate, aim to achieve the broadest possible positive environmental impact. For example, LEAF aims to achieve this by promoting Integrated Farm Management. Integrated Farm Management (IFM) is a whole farm business approach, which emphasises a holistic view of the farming system. This approach encourages farmers and growers to consider all nine sections of the approach, which hold individual, and collective imperative importance and must be integrated to ensure the delivery of sustainable farming. Further information on Integrated Farm Managemen and LEAF is given in Section 3.

It is important to assess and understand the baseline at which the farm business is operating at the start of applying the standards to be able to measure, change and impact as a result of adopting IFM. This is one purpose of standards when they are associated with standardised metrics, which can show the transition towards more sustainable farming practices and enable real and measurable change, enhancing understanding of sustainable farming practices and promoting harmonised goals. These measurable indicators can be in the form of key performance indicators (KPIs), sustainability metrics and data from farmer feedback.

To deliver environmental benefits, there is a need to increase awareness and change mindsets at the farm level. This can be related to encouraging farmers' engagement and passion for environmental issues, which can sometimes lead to behaviour changes. The phenomenon by which farmers make decisions to uptake sustainable practices was studied by Greiner and Gregg (2011) in Australia. They found that farmers were mostly driven by stewardship ethics and personal and family well-being to adopt conservation practices, rather than by economic or social goals. This is supported by Petty and Haugtvedt (1992), who support that a change in behaviour can be achieved by a change in beliefs (persuasion theory).

Building capacity can drive mindset change, enabled through upskilling, providing support to different actors in supply chains, and sharing research and case studies. Training and workshops encourage knowledge exchange, and this, in turn, can lead to an increase in the uptake of sustainable farming practices and sustainability standards, as a consequence of having a positive impact on awareness (Barreiro-Hurle *et al.*, 2008; Greiner and Gregg, 2011; Aidoo and Fromm, 2015). Case studies are a powerful tool to change mindsets and build capacity, as it has been shown that farmers are more open to the uptake of new practices when the knowledge comes from fellow farmers, farming in similar conditions and having similar challenges (BenYishay and Mobarak, 2018; Thu *et al.*, 2020). For example, LEAF provides training in Integrated Farm Management, which helps farmers to understand how all the different aspects and resources on their farms are interconnected and can influence or enhance each other. Training enhances awareness of the importance of farming with nature and having a positive impact on the environment. LEAF also offers numerous case studies that farmers can access for guidance on adopting Integrated Farm Management and achieving LEAF Marque certification. These case studies provide contexts, along with the challenges and potential solutions their peers encountered and applied to their own systems, together with the outcomes.

Encouraging collaboration is a desirable outcome for NGOs. This includes building mutually beneficial and lasting partnerships, leading to new projects, programmes and initiatives, and convening a range of stakeholders to achieve mutually agreed and beneficial outcomes (Harangozó and Zilahy, 2015). NGOs often lead in improving cooperation and networking between organisations (or farmers) with common goals. A good example of an NGO that leads collaboration in this space is the ISEAL Alliance (https://www.isealalliance.org), which is a global network of sustainability standards. They work with their members by establishing codes of conduct for sustainability systems, delivering training and advice, and enabling instances of collaboration and knowledge sharing between members. For some NGOs, achieving integration of their own causes and aims into government policy is a desirable outcome (Peng *et al.*, 2022). This can be linked to achieving consumer recognition of industry-leading schemes (Lambin and Thorlakson, 2018).

Many environmental NGOs are focused on agri-food supply chains. As discussed, LEAF is a renowned environmental NGO and charity that works to deliver more sustainable food and farming. Whilst the NGO is based in the United Kingdom, it has global reach and impact, which will be explored in a later section of the chapter. Another highly active NGO in this sector is the Rainforest Alliance (https://www.rainforest-alliance.org). This NGO works internationally with the mission to protect natural landscapes, enhance biodiversity and improve the livelihoods of farmers and communities (Rainforest Alliance, no date). These aims are delivered through their own certification processes, advocacy, promotion of landscape management standards, training and supply chain services. A different and interesting example is SUSTAIN (https://www.sustainweb.org), which is an alliance of organisations from the food and farming sector that work in collaboration towards addressing contemporary issues such as the cost-of-living crisis, sustainable farming and fishing, climate and nature, the food economy and ensuring good food for all (SUSTAIN, no date).

Although there are a great many NGOs operating within the agri-food sector, and they might have similar high-level goals, their motivations can vary depending on the nature and objectives of the organisation. There are some NGOs that act as alliances, where their members have an important influence on their views, ideas and the initiatives they promote and deliver. Influence can be presented both directly, through input and feedback, or indirectly, for example, through an opportunity to select the people in decision-making positions. Moreover, it is no less important to consider that members of an alliance may have commercial interests in their own activities and views being recognised as 'sustainable' or 'good' by the NGO. We can compare this to other types of NGOs, where external stakeholder views do not directly impact their direction, nor do they impact what the organisation promotes and the activities it engages in. Therefore, we observe that whilst NGOs can have similar general objectives (e.g. promoting sustainable food and farming systems), they can also have inherently different motivations: members' interests versus charitable drivers (e.g. helping farmers). Notably, one commonality among different examples of NGOs is access to both public and private funding and financial support, allowing further action towards their goals and increased impact.

In summary, most environmental NGOs in the food and farming sector share similar aspirations and work towards the same ultimate outcomes and impact. Each might have different approaches that they adopt to achieve these outcomes, with a different focus, specialising in distinct issues, having global or national reach, and having different structures and motivations. Through achieving these outcomes, NGOs have a role to play in improving sustainability standards and advocating for sustainable farming systems.

3 Linking Environment and Farming (LEAF): overview

LEAF is a charity leading in delivering sustainable food and farming on a global scale. Its mission is 'to inspire and enable more circular approaches to farming and food systems through integrated, regenerative and vibrant nature-based solutions, that deliver productivity and prosperity among farmers, enriches the environment and positively engages young people and wider society.' (LEAF, no date/b). To achieve this mission, LEAF works with farmers, researchers, the food industry, customers, teachers and children, in the various projects, partnerships and programmes it is involved. LEAF manages LEAF Marque, which is the standard against which farming businesses are assessed by third-party organisations (certification bodies) to achieve LEAF Marque certification. LEAF is based in the United Kingdom but has a global reach. In 2020, there were 27 countries with LEAF Marque-certified businesses and 300 000 ha of LEAF

Marque-certified land worldwide (LEAF, 2022b). In 2022, LEAF's partnership with Farmer Time, linking schools and farms to exchange knowledge and experience, had reached communities in Australia, New Zealand, Ireland and Scandinavia (LEAF, 2022a). LEAF is also working with non-UK farmers in various projects and initiatives, facilitating conversations, providing access to support and tools, and helping them in their journey towards more sustainable farming.

3.1 Integrated Farm Management and LEAF Marque

Sitting at the heart of LEAF's charitable objectives and delivery is IFM, a whole-farm, site-specific approach to farming and food production. IFM has nine interconnected sections covering themes such as business planning, soil health, plant health, habitat and biodiversity enhancement, energy and waste management, water management, livestock management and engaging society (LEAF, no date/a). IFM encourages farmers to view their business through a lens that combines economic performance, environmental protection and enhancement and social health.

LEAF supports farmers, growers and external stakeholders with resources, templates, case studies, guidance documents, blogs and podcasts on a wide range of sustainability topics, including resource management, waste, biodiversity, biosecurity, regenerative farming techniques and much more. An example of such a resource is LEAF's Simply Sustainable Series, a series of booklets that address how to sustainably achieve IPM, biosecurity, water management, soil management and biodiversity management, among others. Available through the LEAF website, 'Farmer and Grower Help Centre' and YouTube channel, case studies and videos showcase the experiences of farmers in implementing nature-based solutions and their journey to becoming LEAF Marque certified, along with webinars covering projects and topical themes.

LEAF Marque is a private voluntary environmental assurance standard owned and managed by LEAF. Organisations can be certified to the standard to demonstrate they are supplying sustainably farmed produce in the United Kingdom and globally. The requirements for LEAF Marque certification are based on the principles of IFM, which implies that these principles can be implemented in all types of farm businesses. In 2020, 45% of UK fruit and vegetable produce was LEAF Marque certified (LEAF, 2022b). For more information on the LEAF Marque standard, please visit the LEAF website and search for the most up-to-date version. The LEAF Marque standard is available in English, Spanish, Italian and French. A LEAF Marque-certified farm business will have considered and demonstrated improvements in all nine sections of the Integrated Farm Management.

Benefits of becoming LEAF Marque certified include improved business resilience, enhanced on-farm and local biodiversity, access to new markets and industry and customer recognition of a commitment to environmental sustainability (Reed *et al.*, 2017). LEAF Marque certification can open up access to restricted markets; for example in the United Kingdom, many food retailers require LEAF Marque as a condition of supply.

To become LEAF Marque certified, a process is followed that starts with completing a self-assessment and decision-making tool, the LEAF Sustainable Farming Review (LSFR). The objective of this tool is to capture the practices and resultant outcomes within a farming business, allowing the user to baseline, benchmark, understand areas of strength and weakness and set targets for continuous improvement. The tool covers all sections of the principles of Integrated Farm Management and supports the user to consider their current and future performance standards in a context where supply chain, regulatory and policy requirements, and those of private sector financiers and similar, rely on evidence of compliance with KPIs and demonstration of positive impact. Built into the LSFR are guidance documents, templates and links to partnering websites to support the user in completing the tool appropriately and implementing IFM.

LEAF members have access to a directory of LEAF Endorsed Advisors and Consultants, who have undergone a robust verification process to demonstrate strong knowledge of IFM and sustainable and regenerative farming practices. Advisors assist farms both in the United Kingdom and overseas, often benefiting from local knowledge and application of IFM and the LEAF Marque standard.

3.2 LEAF Demonstration Farms, Innovation Centres, Education and Public Engagement

The LEAF Network comprises over 40 LEAF Demonstration Farms and 17 LEAF Innovation Centres. Innovation Centres are focal points for research and innovation, pioneering new approaches to IFM and determining how it might be applied on-farm at scale. They are often located at universities and research institutions. LEAF Demonstration Farms are working farms that follow an IFM approach, delivering and promoting sustainable farming practices. LEAF Demonstration Farms are accessible to all through direct engagement or open days and provide the opportunity to showcase practical and achievable practices and outcomes, where visitors and hosts can benefit from peer-to-peer learning. Moreover, LEAF Demonstration Farms can help illustrate how an IFM approach can be adapted to suit different farming systems, philosophies, and contexts.

An important pillar of LEAF's activity is LEAF Education and Public Engagement. LEAF Education works with farmers, teachers, children and young people to encourage engagement with food production, farming and the environment in education, equipping future generations with this knowledge.

LEAF Education achieves this by providing farmers with training and support to deliver successful educational experiences. LEAF Education also engages with teachers, delivering training and classroom resources to integrate food and farming into the curriculum. LEAF leads public engagement by annually organising Open Farm Sunday, the primary farming open day in the United Kingdom. Every year on this day, visitors get the chance to visit the farms that are taking part, and learn from the farmers about farming, food and the countryside. At the same time, farmers get the chance to interact with their local communities and create meaningful connections. LEAF Open Farm Sunday offers a range of varied experiences and is suited to all farms, as visits can be adapted to the farming system and physical setting, and are easily accessible to both farmers and the general public who want to participate. In 2022, LEAF Education worked with 35 610 young people, 1570 teachers and 485 farmers (LEAF Education, 2022).

4 The role of NGOs in improving sustainability standards and advocating for sustainable farming systems

4.1 Collaboration

One of the most important roles that NGOs play is collaborating with other NGOs, food supply chain stakeholders, farmers and governments, for example, through partnerships, projects and campaigns, often resulting in informal or formal commitments.

The Sustainable Development Goals (SDGs) are 17 objectives developed by the United Nations to create a framework to achieve a sustainable and peaceful future. These objectives are internationally relevant and are based on the importance and interconnection of social, environmental and economic sustainability aspects. Therefore, they can be used as a guide to drive positive action by state and non-state agents (UN, 2019). SDGs can be successfully achieved with the collaboration of NGOs, the business sector and government (Harangozó and Zilahy, 2015). These authors explain that relationships between businesses and NGOs can be both confrontational or cooperative; however, more and more these relationships are proving collaborative and highly impactful. The rationale behind this is that whilst businesses have financial resources, NGOs have the tools, relationships and knowledge to reduce negative environmental impacts, promote good practices, positively influence public opinion, and share a more global view of environmental and social challenges (Harangozó and Zilahy, 2015, Rodriguez *et al.,* 2016, Lambin and Thorlakson, 2018). NGOs can help businesses consider how economic and environmental sustainability is not only compatible with their business practices but also critical to their future viability. NGOs can also support companies in going beyond policy by setting their own standards that are more specifically adapted to their conditions and needs. Some NGOs, for example, the World Wildlife Fund (WWF), through fundraising or coalitions, are able to invest themselves to develop initiatives and campaigns that will impact industry or societal perception and behaviour, positioning themselves as a priority organisation to work with. Exploring the example of the WWF's Basket Metric more closely, this industry-facing framework is built around indicators and outcomes that aim to reduce the environmental impact of an average shopping basket by 50% by 2030. The metric also serves to provide a roadmap of actions for grocery retailers to achieve their own specific ESG priorities. To date, five of the UK's grocery retailers have signed a pledge to deliver the intended outcomes of the metric, reducing their environmental impact (WWF-UK, 2022).

Harangozó and Zilahy (2015) also delve into the collaborations between governments and NGOs and highlight their importance in achieving the SDGs. Governments can work with NGOs and develop partnerships that support the delivery of their own policy goals, for example, through engaging with existing standards, frameworks, campaigns and initiatives rather than developing their own, which can lead to improved consistency of approach, avoidance of duplication and optimisation of impact.

NGOs can play a role in supporting different stakeholders within supply chains, for example, helping farmers to improve economic, environmental and societal resilience (Fontana and Pisalyaput, 2022). They can act as an intermediary between farmers and those considering ways of encouraging improved standards, acting as an objective voice and mediator. For example, if a supply chain organisation or finance provider wishes to impose requirements as a condition of supply or to qualify for business loans, NGOs can review and highlight potential issues of accessibility and practicability and also recommend solutions or alternatives, improving engagement and saving time, resources and the failure of an imposed system. These organisations can also have an intermediary role in encouraging long-term contracts, deals and partnerships that provide security for investment for farmers, allowing them to have further engagement with less risk (Bitzer and Glasbergen, 2015).

4.2 Lobbying, accountability and credibility

One of an NGO's most commonly acknowledged roles is that of lobbying and stimulating accountability, using leverage over government and business and trying to influence policy and practice to achieve their own goals (Damlamian, 2006). This is often operationalised through developing standards, frameworks, roadmaps or initiatives. NGOs are more agile, with more flexible governance structures, and can change their own direction and that of others more readily than government policy can, as evidenced by the uptake of voluntary third-party sustainability standards by governments and supply chains alike (Lambin and Thorlakson, 2018). NGOs engage with companies and governments through lobbying on specific issues, endeavouring to build support for a cause or to prevent perceived detrimental activity from occurring. As a result, there can be conflict between different NGO campaigns: some may lobby against stricter environmental measures, whilst others lobby for stronger environmental protection. Moreover, NGOs can sometimes rely on each other in some aspects of a specific campaign but also have different views in other aspects. This depends on the topic, context and urgency of the desired outcome, as they represent different stakeholder groups with different values and priorities (Lambin and Thorlakson, 2018). NGOs can apply both positive and negative pressure to cause change, for example, through so-called 'smear campaigns' or through competitive or persuasive pressure. Often, campaigns focus on specific companies or practices, for example, Greenpeace targeting crude oil companies. Hatanaka *et al.* (2005) explain that NGOs can demand that retailers take responsibility for their impact, leading them to adopt environmental third-party certifications. The role of NGOs in lobbying and enforcing accountability is backed by Thorlakson *et al.*'s (2018b) research, which reviewed 450 businesses from the agri-food supply chain to explore their engagement with the SDGs and sustainable practices. They found that businesses having headquarters in a country with high NGO activity, among other factors, led to a higher adoption of sustainable farming practices, this being linked to civil society pressure.

NGOs can also influence supply chains, as they can hold actors within them accountable for the claims they are making (Damlamian, 2006) and also support them with solutions to enable companies to make credible claims. For example, LEAF Marque has credibility as it utilises independent auditors and helps supply chain actors support claims through robust verification and reporting. Both supply chains and NGOs benefit from this kind of interaction as it helps both to reach their long-term goals in many cases (Damlamian, 2006). Intrinsically, this form of accountability is delivered by the independent credibility provided by NGOs to stakeholders and non-NGOs (Damlamian, 2006). In summary, NGOs can give credibility to claims by offering standards, certifications and schemes (Lambin and Thorlakson, 2018). These organisations can act as trusted intermediaries between different levels of the supply chain, supporting relationships between them and unlocking benefits for all.

4.3 Provide subject matter expertise

By providing subject matter expertise, NGOs can support supply chains to improve business resilience and environmental and social standards through advice and assistance with designing, implementing, monitoring, recording and reporting (Nasiritousi, 2019; Escher and Brzustewicz, 2020). NGOs offer not only specialist knowledge but also the tools to achieve positive outcomes. NGOs can provide insights through understanding the wider context, allowing the networking of different actors, all working towards a common goal (Bitzer and Glasbergen, 2015; Peng *et al.*, 2022). For example, LEAF provides clients with baseline and contextual reporting and a connection to expert advisers, helping to drive and deliver measurable change at farm-level and within wider supply chains.

For target setting and evaluation, NGOs can help by providing a roadmap or framework of appropriate sustainability or quality indicators and outcomes, driving continuous improvement (Nasiritousi, 2019). This can help farmers and wider supply chains plan for the future, engage more effectively with stakeholders, and encourage collaboration and knowledge exchange, for example, in farm cluster groups (Bitzer and Glasbergen, 2015). NGOs relate theory and research to practice and understand how to apply principles in a real-life context (Nasiritousi, 2019; Peng *et al.*, 2022). For this reason, NGOs often participate in research and initiatives that demonstrate proof of concept as well as knowledge exchange and the dissemination of learnings and impact. An example of this is the 'Accelerating & Demonstrating the Journey to Net-Zero' project, led by LEAF and funded by the Co-op Foundation, with the objective to understand and share the journey for different farmers working towards Net Zero, integrating practical knowledge, expert advice, and technology. Participating farmers share their experiences, challenges and successes to inspire others to reduce their greenhouse gas emissions and enhance carbon sequestration where possible (see Case study I for more details).

Case study I: Accelerating & Demonstrating the Journey to Net-Zero

'Accelerating & Demonstrating the Journey to Net-Zero' is a 3-year project, in partnership with the Co-op Foundation, where five diverse farming businesses are provided with access to tools, advisors, support and training to aid them on the way to becoming net zero. The aim is to take them on the journey of reducing their greenhouse gas emissions and enhancing carbon sequestration, to get a better understanding of levers for change, and showcase how this journey is different and accessible for everyone.

The farms were selected to be diverse, in enterprises as well as geographically. Together they cover farming systems from upland sheep dairy to field vegetables, to deer and beef farming and avocado and nut trees. Geographically, they are distributed from South England to the Scottish Borders and all the way to Northern Limpopo in South Africa. The main goal behind this diversity is to demonstrate that every farm, independent of its enterprise or location, can embark on the journey to becoming net zero.

Each farming business is partnered with an external advisor, depending on its needs and objectives, who will provide farm-specific advice, with suggestions for improvements to carbon sequestration, energy efficiency, and biodiversity, among other aspects. The advisors will help the farmers build site-specific action plans and assist in implementing monitoring to measure continual improvement. Participants also have access to training that will help them upskill in different areas and gain knowledge in IFM.

All farming businesses are using one or more carbon footprint tools, which was chosen considering their farming systems and with the support of their advisor and LEAF. The aim is to set a baseline, understand the current situation and identify the key areas that need improvement to move towards reducing greenhouse gas emissions, sequestering carbon and getting closer to becoming Net Zero.

At the end of the project, the farmers will become LEAF Demonstration Farms, where they will be able to share their learnings and experience on IFM and their journey to becoming net zero. Throughout the project, farmers will also share on social media, in blogs, videos and posts, their challenges and learnings to encourage peer-to-peer learning and to inspire other farmers who might be farming in similar conditions and do not know where to start on the journey.

4.4 Facilitating knowledge exchange

The role NGOs play as subject matter experts is undoubtedly related to their role in facilitating knowledge exchange. Within this role, NGOs provide specialist training, resources, support and technical expertise. Companies can engage with NGOs to help them, for example, make improvements or deliver on their commitments. Another facet of this role is acting as a facilitators of workshops and events. Hatanaka *et al.* (2005) give the example of NGOs in Ghana providing training and support to producers so that they can achieve desired third-party certification. In this specific case, the training was related to IPM, covering practices to improve the quality and sustainability of their produce.

NGOs are often well-placed to participate or lead in research projects. This may involve, for example, facilitating trials, synthesising information, enabling knowledge exchange, or testing proposed systems, schemes or policy. LEAF, among other agricultural NGOs, benefits from access to a network of progressive farmers who demonstrate the principles of continuous improvement and are willing to participate in research and support knowledge exchange across the sector. As part of a Department for Environment, Food and Rural Affairs (DEFRA) test and trial supporting the roll-out of Environmental Land Management Schemes (ELMS), LEAF led a project to understand how peer-to-peer learning could help farmers deliver more sustainable, regenerative farming practices and outcomes at scale.

NGOs can play a role in connecting farmers with support, and providing access to expert advice, tools and training. NGOs can help relate theory to practice and facilitate continuous improvement, providing guidance on the implementation of best practices and the means for improvement to be implemented. Courses, workshops and training provide know-how and support for an audience to achieve a desired standard, which is highly valued by participants (Greiner and Gregg, 2011; Aidoo and Fromm, 2015). Silva *et al.* (2014) studied coffee producers in Brazil and found that participants' attendance at courses and training had a significant impact on certification achievement, as they generated new knowledge and an opportunity to exchange this with others. Similarly, Barreiro-Hurle *et al.* (2008) explain that receiving frequent, periodic education creates a greater awareness of certification, its value, how to achieve it and an improved willingness to adopt. A noteworthy example is LEAF's collaboration with Corteva Agriscience on an innovative project 'Resilient and Ready,' where farmers are given support, advice and training, to improve their farming business resilience and sustainability (see Case study II for more details).

NGOs within the food sector often host networking events to share knowledge and promote collaborative working. This could include presentations from speakers, sharing case studies and stories of significant work and activity and demonstrating the benefits and value of adopting positive action or purpose. Through networking and knowledge exchange, an audience can learn from those who have already embarked upon or experienced a process and better understand the challenges and solutions, risks and rewards. The use of farmers as a network to transfer knowledge and information to rural communities has been proven to be effective by various studies (Thu *et al.*, 2020). A case study in India (Future for Food, 2021) of a programme named Zero Budget Natural Farming aimed to help 6 million farmers transition towards natural farming methods. Farmer Antiparthi Satyavathi was an early adopter in the transition process and reports that she could see how other farmers started to convert once they had witnessed what she was doing first-hand. Antiparthi explains that farmer-to-farmer dissemination of training and support was very important. Beaman *et al.* (2016) noted that for Malawian farmers to access new technology, there had to be another fellow farmer in their network already using it. A similar phenomenon was found by Nakano *et al.* (2018) among rice farmers in Tanzania, sharing the positive effect of knowledge transfer from a trained 'mentor' farmer to an untrained farmer. Moreover, BenYishay and Mobarak (2018) found that farmers were more likely to adopt a new technology when the advocate farmed in the same conditions as them than when the instruction came from a demonstration farmer or an extension service.

Case study II: Resilient and Ready

Led through a partnership with LEAF and Corteva Agriscience, 'Resilient and Ready' is a 3-year programme that was first implemented in 2018 in light of the evolving landscape of the UK farming industry and policy. Now in its second round, following a successful first cohort of farmers, the programme provides farmers with access to advice, support and mentoring for making changes to their farms. The aim is to investigate the economic sustainability of farming as well as the importance of business resilience, giving farmers the knowledge and confidence to work towards farming more sustainably and building resilience in their business, for the environment and community. Farmers in the programme participate in training and receive expert advice on bespoke areas that sit within IFM. These areas are specific to the farmer's interests, farm type and future direction, providing strength in how to apply knowledge into action in their own system and to improve. Through this, they achieve improvements with issues such as biodiversity, water management, soil health and habitat enhancement. Moreover, the participants have the opportunity to trial new innovative products from Corteva, getting to participate in cutting-edge innovation as well as having the opportunity to generate constructive feedback.

Knowledge exchange is a key driver of change; therefore, at the end of the 3 years, the farmers are given the option to become LEAF Demonstration Farmers, where they are able to independently host events to share their knowledge in sustainable farming practices and IFM with other farmers and stakeholders. To be prepared for this, the farmers host events during the programme and develop their presenting and communication skills.

LEAF, as an NGO, is an excellent partner to collaborate in this programme, as it can connect the farmers with the advice and support they might need and give them access to a network of like-minded farmers to share learnings. At the same time, Corteva can benefit from LEAF's reach, knowledge and experience, whilst having the opportunity to trial their products and help farmers on their journey to becoming more resilient.

4.5 Setting standards

NGOs can establish best practice in sustainable agriculture and set standards for farmers to follow (Meemken *et al.*, 2021). These standards provide farmers with guidance, recognition and support on their journey towards a specific goal (Thorlakson *et al.*, 2018a). Meemken *et al.* (2021) explain how the adoption of sustainability standards is interlinked with farmers paying more attention to improving soil health, biodiversity and water management, working towards becoming more regenerative and resilient businesses. An interesting example is demonstrated in Case study III, 'The Jordans Farm Partnership', where Jordans' oat growers become LEAF Marque certified and continue to improve their practices around IFM, with an additional focus on farmland habitat and enhancing biodiversity. Jordans help their farmers by facilitating access to wildlife advisors, helping to improve aspects such as on-farm monitoring of species and IPM.

Standards can be recommended or requested by governments, NGOs or industry, particularly where such standards are independently verified, such as LEAF Marque. This presents robust and reasonable verification of practices or outcomes to meet ESG (Environmental, Social and Governance) commitments, environmental pledges and other targets (Lambin and Thorlakson, 2018; Thorlakson *et al.*, 2018a, 2018b; Meemken *et al.*, 2021). Often, private voluntary standards are made mandatory or a condition of supply. Bitzer and Glasbergen (2015) investigate the partnerships between NGOs and businesses in global value chains and explain that sustainability standards play a role in creating both market opportunities and practice change. Governments might incorporate sustainability standards into their own policies, or they can be seen as precursors for future public regulation (Lambin and Thorlakson, 2018; Meemken *et al.*, 2021). In the short term, there can be resistance towards adopting a certification or scheme, but there are long-term benefits to be achieved by encouraging stronger environmental frameworks for long-term resilience.

Different NGO-led certifications can compete (for adoption, recognition, etc.), as they represent different stakeholder groups that have different values, views and opinions related to sustainability. This can lead to the promotion of more types of certification to suit everyone's views. Thus, there can be duplication of sustainability certification schemes, which will possibly lead to competition. This can be beneficial as it encourages standard owners to make improvements to differentiate themselves, but it can also be very negative, resulting in a waste of resources and can lead to an economic and time burden for farmers. A negative aspect is that consumers might get confused by the vast number of different certifications with similar standards. Sometimes, the certifications do not compete, but they act in a complementary way, for example when one scheme uses another as a baseline or benchmark (Lambin and Thorlakson, 2018) or a precursor to a higher standard certification.

Case study III: Jordans Farms Partnership

Over the last 5 years, farmers across the United Kingdom who supply oats to Jordans Dorset Ryvita, a UK food brand producing cereal-based food products, have been working with the Wildlife Trust and

LEAF to promote and enhance wildlife on their farms. This initiative, called the Jordans Farm Partnership, aims to support these farmers in improving biodiversity. Farmers have been encouraged to provide at least 10% of their farmland for managing and providing habitats for wildlife to flourish.

All farmers are LEAF Marque certified, which ensures sustainably farmed oats, through the use of the IFM approach. Using this approach has also brought multiple co-benefits, including enhanced soil health, carbon sequestration, and improved water management. For example, as part of this partnership, some farmers have started using cover crops to build soil fertility instead of artificial fertilisers to provide nutrition to grow the oats, in turn improving soil health and structure. There is an emphasis on how different areas of farm management, including landscape and biodiversity management, are interrelated and dependent on each other to co-improve.

Every oat farmer is partnered with their local Wildlife Trust advisor, who provides ecological expertise and advice on how to monitor, manage and improve biodiversity across the farm through site-surveying, practical advice and frameworks. The Jordans Farm Partnership has now resulted in farmers having an average area of 25% devoted to improving biodiversity on farms. Advice is tailored to each specific farm, considering their farming system and local conditions and species. Creating diverse habitats is proving to increase wildlife populations; for example, farmers are taking measures to support nesting grounds and food for birds and protecting insects. Some of the target species promoted on farms are brown hare, lapwing, yellowhammer, bats, bees and pollinators, through creating ponds, planting trees and wildflower plots.

4.6 Incentives

To encourage the transition of farming towards more sustainable and resilient systems, NGOs can incentivise farmers to change their practices to deliver more positive outcomes. Market incentives are a common example of this, in the form of price premiums or access to an alternative market through meeting a qualifying standard (Harangozó and Zilahy, 2015; Nogy, 2017). This is illustrated in the 'La Carta del Mulino' project, a collaboration between the Barilla group, WWF Italy, the University of Bologna, the University of Tuscia and Open Fields. In this project, soft wheat farmers have contracts guaranteeing that, where sustainability objectives are met, a premium price is paid for the crop. Premiums could be linked to protecting and enhancing on-farm habitat and biodiversity, introducing more regenerative practices to promote healthier soils, or minimising or ceasing agrochemical use (WWF-IT, 2021). However, there is an argument against rewarding through price premiums, asserting that there is no guarantee that nature will benefit more from adopting these practices.

Sometimes, incentives are not so obviously economic, but are present in the form of free expert advice, access to events/projects and access to networks. A demonstration of this is the M&S Farming with Nature Project, where M&S growers are offered access to expert advice, tools and support, enabling them to work towards stretching targets, particularly supporting biodiversity on farms and implementing IPM plans and approaches for their farming systems (see case study IV for more detail).

Through voluntary initiatives, NGOs can also play an intermediary role, trying to address issues or make change happen before regulation needs to be introduced (Lambin and Thorlakson, 2018; Peng *et al.*, 2022). It might be that, in the first instance, a bottom-up, farmer-led approach is encouraged and supported by an NGO, empowering farmers to feel more in control of their own business and destiny. If this does not work, market incentives or penalties can be introduced to try and initiate change. Finally, if no action is being taken or change is not happening quickly enough, a top-down approach through policy and regulation can be implemented.

Case study IV: M&S Farming with Nature

Since 2021, M&S, LEAF, FWAG, FERA, GWCT, the Wildlife Trusts and other advisors have been working together with all UK M&S Fresh Produce growers, providing them with support and advice to continue to produce great quality food whilst enhancing their biodiversity and integrating an IPM approach. M&S created this programme to support its farmers to achieve resilience, delivering measurable outcomes through the adoption of environment-friendly farming practices. These farmers have been working to enhance over 9500 ha of wildlife habitat across the United Kingdom.

The farmers have been partnered with local expert advisors from the Wildlife Trusts or FWAG, who have helped them identify and plan actions to enhance their wildlife and biodiversity, for example by building habitat connectivity between the farm and its embedded landscape to promote the movement of species and improving the natural spaces on their farms.

M&S British Fresh Produce growers are also LEAF Marque certified, which ensures that they have applied IFM throughout their farm and have action plans in place. In addition, as part of this programme, they will be participating in extra modules tailored specifically for M&S growers. One of these modules is the M&S Landscape and Nature Module which, as its name suggests, aims to enhance the landscape and habitats for species on farm and off farm, acknowledging that each farm is different and does not take a prescriptive approach, rather a plan that is suitable to each farming system. The other module is the M&S Integrated Pest Management (IPM) module, which aims to promote a sustainable way of achieving plant health and protection whilst reducing the negative impact on the environment and encouraging beneficial insects and pest monitoring.

M&S growers also participate in events with other farmers from the same region, which are led by regional Wildlife Trust and FWAG advisors. These events aim to encourage peer-to-peer learning, networking, and knowledge exchange between farmers, as well as learning new aspects from invited speakers on various subjects that will feed into the M&S modules.

The farmers use the LEAF Sustainable Farming Review (with the added M&S modules), which enables them to report on data around the habitat on-farm, practices used to enhance it, and IPM. This data is used to report on the continuous improvements of the growers and the positive impact this programme has on delivering positive environmental outcomes.

M&S has also trialled a range of innovative approaches on its Indicator and Innovation farms, aiming to provide insights for farmers on biodiversity and soil health.

4.7 Communicating

Communication is a large part of NGOs' activity, generating and supporting change towards more sustainable and resilient farming systems, with a positive impact on the environment (Kong *et al.*, 2002; Peng *et al.*, 2022). NGOs can help to educate consumers, and they can provide links to the end user. This role is particularly important where there is little or no upfront demand from consumers, yet change to a farming system is considered by the NGO to be in their best interest.

An excellent example of an NGO's effort to educate consumers and expand their understanding of food and farming is the work LEAF Education is doing with younger generations, teenagers, teachers, and farmers. More details about LEAF Education's work can be found in Section 3.

5 Challenges in improving sustainability standards for NGOs, farmers, supply chains and government

Finally, we consider the challenges ahead for NGOs seeking to improve sustainability standards in the global agri-food supply chain. These challenges may impact different stakeholders uniquely, for example, producers, supply chains and governments, and may or may not be interconnected.

One of the most significant challenges for NGOs to tackle is the risk of greenwashing (Partzsch *et al.*, 2019). Although there is no set definition of greenwashing, it can be broadly defined as a phenomenon where consumers or stakeholders, in general, are deliberately deceived by the supposed environmental actions or claims of an organisation or business (Vieira de Freitas Netto *et al.*, 2020). This usually relates to the making of robust, credible and substantial claims. Greenwashing can occur when businesses do not have adequate knowledge, evidence or systems to support the claims being made (Partzsch *et al.*, 2019). When an organisation is accused of greenwashing, there is a high risk that other associated activities or organisations are linked to the same accusation. There is concern that not enough is being done to address this issue and that businesses are prioritising their own agendas, ESG commitments and marketing opportunities over driving real and positive change within their sector. However, through the introduction of private voluntary and mandatory regulation, this issue can be addressed, ensuring only evidence-based, reliable claims are being made.

NGOs may also face the challenge of conflict, disagreement or competition with other NGOs holding contrasting opinions. This competition can be between NGOs that work toward specific issues within the same big-picture matter, but they can slightly interfere with each other's work of the other NGOs and compete for funding (Heyes and Martin, 2015). Securing government support can present a major challenge for NGOs (Lambin and Thorlakson, 2018; Peng *et al.*, 2022). It can be difficult to integrate findings into policy, access funding and support and design policy that is suitable across a wide variety and complexity of contexts. For example, in farming, there are different sectors, business models, ownership structures, contract agreements and management approaches (Harangozó and Zilahy, 2015). Economic, knowledge and language barriers can also challenge NGOs' operations; however, this might be something that collaborations and partnerships with other businesses or NGOs can help to overcome. For example, NGOs do not have as much financial leverage as commercial businesses, but if they work together, they can combine the knowledge and experience of the NGOs with the economic resources of the business and work together towards the same goal (Damlamian, 2006).

NGOs need to identify effective incentives to generate change. This can be challenging as different incentives work for different audiences. For example, to encourage farmers to adopt sustainable farming practices, suitable incentives might be invitations to demonstration events, workshops and complementary advice (Greiner and Gregg, 2011). For governments, it can be challenging to identify the point at which legislation needs to be introduced, where voluntary regulation is not achieving the desired outcome (Fontana and Pisalyaput, 2022). Ideally, allowing and enabling NGOs to play their role in improving environmental standards and promoting sustainable farming is preferred. Farmers may also face challenges in terms of the time and financial resources required to meet a standard. Added to this is the potential complexity of compliance with multiple certification schemes and standards, depending on government or supply chain requirements. NGOs can help reduce the burden in several ways, including working with standard holders to prevent duplication, benchmarking standards to allow prioritisation or equivalence to be recognised and supporting farmers on their journey to meet a standard through, for example, access to funding, practical advice and resources (Lambin and Thorlakson, 2018; Peng *et al.*, 2022).

Usually, retailers and brands will define their ESG KPIs and share them publicly. To demonstrate their compliance and as a basis from which to make claims, they could adopt their own certification or standard or ensure compliance with a third-party certification standard to help maintain accountability and report on their achievements and impact (Meemken *et al.*, 2021). Aside from the challenge of selecting appropriate KPIs on which to focus, retailers and brands can find difficulty in persuading their supply chains that meeting a standard and accepting potentially costly certification is of benefit. There may be a requirement for compensation of some sort, whether through premiums, more favourable contracts, access to advice and similar. A neutral NGO voice in this framing can be helpful to advocate for both the retailers and brands and for the supply chain in order to achieve the desired overarching outcome (Bitzer and Glasbergen, 2015; Nasiritousi, 2019).

In summary, there are many direct and indirect challenges for NGOs to navigate or alleviate in order to improve standards in food production. There is little doubt about the positive role that NGOs have in overcoming these challenges to reach desirable outcomes and benefit the fullest range of stakeholders.

6 Conclusion and future trends

The role of NGOs in improving sustainability standards in the agri-food sector is intrinsically linked to the way in which NGOs promote sustainability and protect the environment whilst producing food. NGOs are in a privileged position to enable this to happen, without the barriers that government, businesses and other similar types of organisations might experience. NGOs can facilitate progress towards a common goal through partnerships, enabling accountability, providing credibility, setting standards, communicating and sharing knowledge. This leadership comes with challenges, as every stakeholder has their own specific agenda, and it is an NGO's duty to consider the wider context whilst providing specific advice, resources and support. NGOs must consider this when designing their purpose, objectives, and, where applicable, standards to ensure they are meeting the needs of the agri-food supply chain and wider society.

During the process of reviewing the literature available for this chapter, some research questions were identified that will benefit from further research:

- What is the role of NGOs in providing incentives to adopt sustainability practices and standards, studying the best types of incentives to use as well as considering the bottom-up vs. top-down approach?
- What is the role of NGOs in changing consumer perception in this space? How would NGOs enable this? What to target? And how to measure possible tangible impact in the adoption of sustainability practices and standards?
- How can we encourage collaboration between governments or other agents and NGOs to better utilise the available resources to achieve change?
- How can NGOs have a more meaningful impact on policymakers to prevent greenwashing from occurring?

These questions are all of interest as the role of NGOs becomes more important in ensuring sustainable agri-food supply chain practices.

7 References and further reading

Aidoo, R. and Fromm, I. (2015), 'Willingness to adopt certifications and sustainable production methods among small-scale cocoa farmers in the Ashanti region of Ghana', *Journal of Sustainable Development*, 8(1), 33–43.

Akkucuk, U. and Sekercioglu, C. (2016), 'NGOs for environmental sustainability: the case of KuzeyDoga society', *Fresenius Environmental Bulletin*, 25, 6038–6044.

Barreiro-Hurle, J., Espinosa-Goded, M. and Dupraz, P. (2008), 'Does intensity of change matter? factors affecting adoption in two agi-environmental schemes', *107 EAAE Seminar: Modelling of agricultural and rural development policies*, European Association of Agricultural Economists (EAAE). Jan 2008, Seville, Spain. 15 p.

Beaman, L., BenYishay, A., Fatch, A., Magruder, J. and Mobarak, A.M. (2016), *Making networks work for policy. Evidence from agricultural technology adoption inn Malawi. Impact evaluation report 43.* New Delhi: International Initiative for Impact Evaluation (3ie).

BenYishay, A. and Mobarak, A.M. (2018), 'Social learning and incentives for experimentation and communication', *Review of Economic Studies*, 0, 1–34.

Bitzer, V. and Glasbergen, P. (2015), 'Business-NGO partnerships in global value chains: part of the solution or part of the problem of sustainable change?', *Current Opinion in Environmental Sustainability*, 12, 35–40.

Damlamian, C. (2006), *Corporate-NGO partnerships for sustainable development.* CUREJ: College Undergraduate Research Electronic Journal, University of Pennsylvania.

Escher, I. and Brzustewicz, P. (2020), 'Inter-organizational collaboration on projects supporting sustainable development goals: the company perspective', *Sustainability*, 12(4969), 1–26.

Fairtrade. (no date), *How Fairtrade works* (https://www.fairtrade.net/about/how-fairtrade-works); accessed June 30th, 2023.

FAO and UNEP. (2020), *The state of the World's Forests 2020. Forests, biodiversity and people*. Rome: FAO.

Future for Food. (2021), *Case study of future of food: community managed natural farming: an agroecology movement takes root in India's Andhra Pradesh state*. (https://futureoffood.org/insights/community-managed-natural-farming-india/) accessed May 8, 2022.

Fontana, E. and Pisalyaput, N. (2022), 'Understanding the importance of farmer-NGO collaboration for sustainability and business strategy: Evidence from the coffee supply chain', *Business Strategy and the Environment,* 2022, 1–21.

Garibaldi, L.A., Gemmil-Herren, D'Annolfo, R., Graeub, B.E., Cunningham, S.A. and Breeze, T.D. (2017), 'Farming approaches for greater biodiversity, livelihoods, and food security', *Trends in Ecology & Evolution,* 32(1), 68–80.

Greiner, R. and Gregg, D. (2011), 'Farmers' intrinsic motivations, barriers to the adoption of conservation practices and effectiveness of policy instruments: Empirical evidence form norther Australia', *Land Use Policy,* 28, 257–265.

Harangozó, G. and Zilahy, G. (2015), 'Cooperation between business and non-governmental organizations to promote sustainable development', *Journal of Cleaner Production,* 89, 18–31.

Hatanaka, M., Bain, C. and Busch, L. (2005), 'Third-party certification in the global agrifood system', *Food Policy*, 30, 354–369.

Herweijer, C., Evison, W., Mariam, S., Khatri, A., Albani, M., Semov, A. and Long, E. (2020), *Nature risk rising: why the crisis engulfing nature matters for business and the economy.* Switzerland: World Economic Forum.

Heyes, A. and Martin, S. (2015), 'NGO mission design', *Journal of Economic Behaviour & Organization,* 119, 197–210.

Hoekstra, A.Y. and Mekonnen, M.M. (2012), 'The water footprint of humanity', *Proceedings of the National Academy of Sciences,* 109(9), 3232–3237.

Kong, N., Salzmann, O., Steger, U. and Ionescu-Somers, A. (2002), 'Moving business/industry towards sustainable consumption: the role of NGOs', *European Management Journal*, 20(2), 109–127.

Kremen, C. and Miles, M. (2012), 'Ecosystem services in biologically diversified versus conventional farming systems: benefits, externalities and trade-offs', *Ecology and Society,* 17(4), 40.

Lambin, E.F. and Thorlakson, T. (2018), 'Sustainability standards: interactions between private actors, civil society, and governments', *Annual Review Environment and Resources*, 43, 369–393.

LEAF. (2022a), *LEAF Annual Review* (https://issuu.com/linking-environment-and-farming/docs/leaf_annual_review_2022); accessed June 6, 2023.

LEAF. (2022b), *LEAF Global Impacts* (https://issuu.com/linking-environment-and-farming/docs/leaf_global_impacts_report_2022?fr=sMGNkYjQ5MTUxMzI); accessed June 6, 2023.

LEAF. (no date/a), *Integrated Farm Management* (https://leaf.eco/farming/ifm); accessed June 6, 2023.

LEAF. (no date/b), *What we do* (https://leaf.eco/about-leaf/what-we-do); accessed June 6, 2023.

LEAF Education. (2022), *LEAF Education: Our Work in Action* (https://issuu.com/linking-environment-and-farming/docs/leaf_education_impact_report_2022?fr=sZWM4NTU1NzM5MDQ); accessed June 8, 2023.

Meemken, E., Barret, C.B., Michelson, H.C., Qaim, M., Reardon, T. and Sellare, J. (2021), 'Sustainability standards in global agrifood supply chains', *Nature Food*, 2, 758–765.

Nakano, Y., Tsusaka, T.W., Aida, T. and Pede, V.O. (2018), 'Is farmer-to-farmer extension effective? the impact of training on technology adoption and rice farming productivity in Tanzania', *World Development*, 105, 336–351.

Nasiritousi, N. (2019), 'NGOs and the environment'. In Davies, T. (Ed.), *Routledge handbook of NGOs and international relations*, Routledge Handbooks Online, pp. 329–342.

Nogy, C. (2017), 'Environmental NGO-private sector partnerships: a tool to generate business value and protect the natural world', *Master's Projects and Capstones*, 722, 1–63.

Partzsch, L., Zander, M. and Robinson, H. (2019), 'Cotton certification in Sub-Saharan Africa: -promotion of environmental sustainability or greenwashing?', *Global Environmental Change*, 57, 101924, 1–10.

Peng, S., Fia, F. and Doherty, B. (2022), 'The role of NGOs in sustainable supply chain management: a social movement perspective', *Supply Chain Management: An International Journal*, 27(3), 383–408.

Petty, R.E. and Haugtvedt, C.P. (1992), 'Personality and persuasion: need for cognition moderates the persistence and resistance of attitude changes', *Journal of Personality and Social Psychology*, 63(2), 308–319.

Rainforest Alliance. (no date), *Our approach* (https://www.rainforest-alliance.org/approach/); accessed June 30, 2023.

Reed, M., Lewis, N. and Dwyer, J. (2017), *The effect and impact of LEAF Marque in the delivery of more sustainable farming: a study to understand the added value to farmers*. Gloucester, England: The CCRI.

Rodríguez, J.A., Giménez Thomsen, C., Arenas, D. and Pagell, M. (2016), 'NGOS' initiatives to enhance social sustainability in the supply chain: poverty alleviation through supplier development programs', *Journal of Supply Chain Management*, 53(3), 83–108.

Rose, D.C., Sutherland, W.J., Barnes, A.P., Borthwick, F., Ffoulkes, C., Hall, C., Moorby, J.M., Nicholas-Davies, P., Twinning, S. and Dicks, L.V. (2019), 'Integrated farm management for sustainable agriculture: lessons for knowledge exchange and policy', *Land Use Policy*, 81, 834–842.

Silva, E.C., Castro Jr., L.G., Garcia Costa, C. H. and Teixeira Andrade, F. (2014), 'Determinant factors in adopting socio-environmental certifications in coffee farms', *RESR*, 52(3), 437–448.

SUSTAIN. (no date), *What we do* (https://www.sustainweb.org/what-we-do/); accessed June 30, 2023.

Swinnen, J. (2016), 'Economics and politics of food standards, trade, and development', *Agricultural Economics*, 47(S1), 7–19.

Tayleur, C., Blamford, A., Buchanan, G.M., Butchart, S.H.M., Ducharme, Green, H.R.E., Milder, J.C., Sanderson, F.J., Thomas, D.H.L., Vickery, J. and Phalan, B. (2017), 'Global coverage of agricultural sustainability standards, and their role in conserving biodiversity', *Conservation Letters*, 10(5), 610–618.

Thorlakson, T., Hainmueller, J. and Lambin, E.F. (2018a), 'Improving environmental practices in agricultural supply chains: the role of company-led standards', *Global Environmental Change*, 48, 32–42.

Thorlakson, T., Zegher, J.F. and Lambin, E.F. (2018b), 'Companies' contribution to sustainability through global supply chains', *PNAS,* 115(9), 2072–2077.

Thu, V.H., Tran, D., Goto, D. and Kawata, K. (2020), 'Does experience sharing affect farmers' pro-environmental behaviour? a randomized controlled trial in Vietnam', *World Development,* 136, paper 105062, 1–14.

UN. Independent Group of Scientists appointed by the Secretary-General. (2019), *Global sustainable development report 2019: the future is now – science for achieving sustainable development.* New York: United Nations.

Vieira de Freitas Netto, S., Falcão Sobral, M., Bezerra Ribeiro, A.R. and da Luz Soares, G.R. (2020), 'Concepts and forms of greenwashing: a systematic review', *Environmental Sciences Europe,* 32(19), 1–12.

Ward, A. and Mishra, A. (2019), 'Addressing sustainability issues with voluntary standards and codes: a closer look at cotton production in India'. In Arora, B., Budhwar, P., Jyoti, D. (Eds.), *Business responsibility and sustainability in India, Palgrave studies in Indian management.* Cham: Palgrave Macmillan, pp. 161–193.

Willetts, P. (2009), 'Nongovernmental ōrganization'. In Kütting, G. (Ed.), *Conventions, treaties and other responses to global issues.* Oxford, United Kingdom: Eolss Publishers Co. Ltd.

World Bank. (1995), *Working with NGOs. A practical guide to operational collaboration between the world bank and non-governmental organisations.* Washington, DC: Operations Policy Department, World Bank, p. 13.

WWF-IT. (2021), *BARILLA E WWF INSIEME PER IL PROGETTO "CARTA DEL MULINO"* (https://www.wwf.it/pandanews/wwf-life/progetti-e-iniziative/barilla-e-wwf-insieme-per-il-progetto-carta-del-mulino/); accessed August 3, 2023.

WWF-UK. (2022), *What's in store for the planet: the impact of the UK shopping basket on climate and nature* (https://www.wwf.org.uk/our-reports/whats-store-our-planet-impact-uk-shopping-baskets-climate-nature-2022); accessed August 3, 2023.

Part 2

The expanding scope of agri-food chain standards

Chapter 5

Sustainability standards and certification for agriculture: an overview

Anne Tallontire, University of Leeds, UK

1 Introduction

The number of sustainability standards and certifications applied in agricultural supply chains has grown considerably in the past decade. A recent United Nations report suggests that there are in the region of 450 voluntary sustainability standards (VSSs), most covering agricultural commodities (UNFSS, 2022). This chapter provides an overview of the range of sustainability standards that have emerged, who has developed them, their content and rationale. Concerns have been raised as to whether sustainability standards create barriers to inclusion and in fact limit the benefits of incorporation in so-called modern markets for smallholders. Ultimately, standards need to be considered both in terms of the standards system of which they are only one part and the wider social and institutional context in the country of production. Too often standards are regarded as 'off the shelf' supply chain management tools that ignore the interface with local context and the market power of those who choose to use them.

This chapter starts by presenting a framework to help navigate the array of standards that exist, drawing on reports by international organisations as well as academic sources and critiques from a range of disciplines, largely in the social sciences and business studies. It then explores how retailers

http://dx.doi.org/10.19103/AS.2024.0133.08

and brands select the standards they use, and how this is changing. Recent choices made by brands and retailers accentuate some of the challenges associated with standards for smallholders and further contextualises the mixed picture concerning impacts on smallholders. The chapter provides a summary of the implications of sustainability standards, including how they may contribute to improved practices, but highlights limitations of standards on their own as a tool to hold business to account or indeed to drive a shift to a more sustainable future.

2 What are sustainability standards?

It is useful to start by exploring a definition of standards, they are 'agreed criteria ... by which a product or a service's performance, its technical and physical characteristics, and/or the process, and conditions, under which it has been produced or delivered, can be assessed' (Nadvi and Wältring, 2002, p. 6). Standards have been developed by businesses to facilitate trade, helping to determine price and define contract terms as they convey valuable information between buyer and seller. For example, for centuries, there have been standards for the weight of a sack for commodities such as flour, the permitted level of impurities in that bag of flour or the size of a particular component in a mechanical device. Standards are there to protect the consumer, especially food or product safety standards. They show conformity with a set of expectations meaning that business is transacted more quickly, with fewer questions raised by the buyer. As Oakley and Buckland (2004, pp. 132–133) put it, standards 'are about circumspection, practical guidance and measurement that allow good practice to be understood and repeated'. It is a language and set of practices that business understands and is used to working with.

The growth of global supply chains and the increasing phenomenon of global sourcing, facilitated by liberalisation and de-regulation (Utting, 2005), created a gap between the regulatory environment and business norms of companies sourcing goods from around the world and that of their suppliers (Dicken, 2014). Competitive pressures and the desire to cut costs, drive efficiencies, and diversify the supply base have motivated the development of sourcing strategies but have left businesses exposed to new risks. Businesses with brands to protect recognised that they needed to respond to, and manage, ethical risks associated with poor working conditions and impacts of poor production practices on the environment, not least because of the expectations of stakeholders (Busch and Bain, 2004). As business became more concerned with sustainability, the idea of a 'standard' to convey expectations with respect to production practices, inputs and outputs emerged as an idea that, whilst not completely new, accelerated as a governance tool amongst non-governmental organisations (NGOs) and policy groups particularly from the 1990s. According to the International Social and Environmental Accreditation and Labelling or ISEAL Alliance (2016, p. 5), sustainability standards are:

> A set of criteria defining good social and environmental practices in an industry or product supply chain. They are market-based tools and interventions seeking sustainability and development outcomes ... They fall within the subset of 'voluntary' and 'private' standards but with a clear sustainability goal in their mission.

In this definition, social standards are firmly within the scope of sustainability standards that are defined broadly to encompass social well-being and social justice aspects as well as environmental aspects of sustainability. Various NGOs promote sustainability standards to business as tools not only to manage supply chains but also to contribute to the Sustainable Development Goals (SDGs) (WWF and ISEAL Alliance, 2017). However, as will be shown in Section 3, sustainability standards may engage with different aspects of sustainability or have different emphases and have different origins and systems for accountability.

As Bush et al. (2015) explain, standards have become a tool for governance *in* supply chains (enabling companies to claim a certificate and benchmark their own practices), *of* supply chains (whereby companies set expectations of their suppliers) and *through* supply chains (whereby civil society actors use standards as a means of making businesses accountable). In this chapter, the focus is on standards that are explicitly applied to the supply chain referring to those that are focused on governing of supply chains but also recognising that several standards originate from processes of governing through standards. Standards that are about information sharing or transparency, e.g. Global Reporting Initiative (or GRI) or broader sets of principles that are part of public–private partnerships such as the Global Compact, both of which operate at a broader level than the supply chain, are out of the scope of this chapter. Similarly, environmental management systems that are applied at the single organisation level rather than the supply chain level are not part of this review.

3 A framework to understand sustainability standards

At their most basic, sustainability standards are tools that companies use to manage their supply chains, or more precisely set expectations for their suppliers with regard to environmental and social practices. However, how are the criteria that form the basis of these standards identified and specified? Who sets these standards? In what ways do they differ? Moreover, why are there so many?

In the agri-food sector sustainability standards seem to proliferate. There are some with a high level of consumer recognition, associated with a certificate or label such as Fairtrade. But there are increasing numbers of labels making claims about sustainability to the consumer and also several standards that are less visible to the consumer but are increasingly used by business in their engagements with suppliers. In the early 2000s, policy makers, especially those involved with promoting trade and international market access as a vehicle for development, were concerned about how standards proliferation created barriers for producers in developing countries. This was in terms of both setting expectations for compliance and demonstrating this compliance in the first place and also where producers were selling to multiple buyers this meant a proliferation of audits and paper-trails, sometimes with conflicting requirements (Giovanucci and Ponte, 2005; Seville et al., 2010).

Initiatives for standards convergence were created to share good practice and create forums for determining minimum expectations, or even aspirations. The ISEAL Alliance was established in 2002 with the aim of creating a solid foundation for collaborative working between third-party certification and standards organisations, creating credibility tools and platforms for shared learning. Founder members included Fairtrade International, the Rainforest Alliance and IFOAM, the body representing the global organic movement. Large companies organised through the Consumer Goods Forum set up the Global Social Compliance Programme (GSCP) in 2009 with the 'aim of harmonising existing efforts and deliver a common, consistent and global approach across sectors for the continuous improvement of working and environmental conditions in global supply chains' (https://www.theconsumergoodsforum.com/social-sustainability/sustainable-supply-chain-initiative/faq/). A reference code was developed against which members could benchmark their own standards. However, whilst this reference code still exists, in 2017, this approach was replaced by the Sustainable Supply Chain Initiative (SSCI), which focused on guidance for practice rather than standards. The ISEAL Alliance is still highly influential in the standards debate and practice.

Standard setters 'offer different yet similar and mutually recognized standards that are close substitutes' (Reinecke et al., 2012, p. 805). Sustainability is an issue on which there is agreement about what we need to achieve, but not how to get there, meaning that in standards development, choice and application there are roles played by 'ideologies, values' and different understandings of what is legitimate (ibid, p. 810). Moreover, the sustainability standards landscape is not static but is dynamic with features, say Reinecke and colleagues, of a market in which companies choose standards not necessarily on what is technically the 'best' but what meets their needs at a particular time and in relation to specific pressures. There is conflicting evidence as to whether the market for standards leads to a race to the bottom or to the top (Fransen et al., 2019), with some commentators suggesting this might be related to both sectoral characteristics (e.g. consumer concerns, structure of the chain) and has the governance models adopted (Ponte, 2019).

This part of the chapter seeks to provide insight into the standards landscape, drawing on scholarship and empirical evidence to clarify and provide ways of understanding the diversity of sustainability standards. The resultant framework is proposed to place individual standards on a map or to navigate around a complex standards landscape asking the basic but interconnected questions of what, who, why and how:

- Who established the standard?
- What does the standard cover? Which aspects of sustainability does it focus on? How wide is its scope?
- Why did this standard get developed? What is the underlying orientation?

3.1 Who established the standard?

A key distinguishing factor between different standards is the type of actors or organisations that develop them and govern their operation. The most visible standards in agriculture and land management since the mid-1990s have been those developed by multi-stakeholder organisations, often dubbed VSSs usually associated with third-party certification processes, e.g. Fairtrade International or Rainforest Alliance. VSSs are often associated with the processes of governing *through* standards in which civil society, often in partnership with business, creates a standard that they then lobby for business to adopt and offer independent verification, and usually a consumer facing label or certificate. Whilst VSSs were developed with companies or industry bodies as part of a broad coalition of stakeholders, their origins lie in a critique of both the lack of effective transnational regulation and the limitations of private sector regulation or standards developed by companies themselves, which were seen as just about brand protection (O'Rourke, 2003). VSSs have a focus on changing practices, not just recognising and assuring good practices, aiming to use the market as a mechanism for change, through a process sometimes known as non-state market-driven governance (Auld et al., 2008). This embodies the idea that a standard outlining good practice, verified by a certificate or label aimed at consumers, might be a mechanism for protecting the environment and defending the social and economic rights of producers and workers. Giovanucci and Ponte went so far as to suggest that in the coffee sector 'sustainability standards can play a key role for addressing inequalities, and indeed genuinely fomenting sustainability, in the coffee trade and even more broadly in international trade' (2005, p. 299).

Also technically voluntary, but compulsory for producers wishing to sell to the particular company, there are private company standards. Vermeulen (2015) calls these 'Do it yourself' standards where a company has its own standard that it applies to its suppliers, thereby exercising governance *of* its supply chain. This kind of standard is not usually associated with a consumer-facing label, and the standard often remains a business-to-business (B2B) communication though details of such schemes often appear in annual sustainability reports directed at investors (KPMG, 2008). These standards may be developed in collaboration with civil society, but they are proprietary instruments, sometimes with an associated trademark.

As noted previously, private standards have roots in communications about delivery and payment terms and quality and food safety expectations. Global food retailers have long had their own internal standards, often rooted in food safety legislation and due diligence requirements. Examples include Nature's Choice (Tesco), Filières Qualité (Carrefour), Field-to-Fork (Marks & Spencer) and Filière Controlleé (Auchan) (Henson and Humphrey, 2010). These tended to be checked by internal sourcing or purchasing teams supported by quality assurance. When pressed in the 1990s to show sustainability or ethical credentials, often companies added extra responsibilities to these teams who would need to acquire the requisite skills and knowledge (Barrientos and Smith, 2007).

Within the agricultural commodity and beverage sector, private company standards covering a range of sustainability issues have evolved from a focus on quality assurance. Nespresso's AAA Sustainable Quality programme was launched in 2003. As the name says, initially the programme focused on coffee farmers delivering the requisite quality and a continuous improvement process, and with stronger, more direct ties between the brand and the producers from whom they purchased coffee. By 2011, Nespresso sourced more than 60% of its coffee from around 45 000 farmers who belong to the AAA Programme (Reinecke et al., 2012). Given climate change and other environmental changes affecting yields and susceptibility to pests and disease, companies like Nespresso want to ensure more secure access to supply. In developing their standard and their engagement strategy with their suppliers, they have drawn on the expertise of NGOs such as the Rainforest Alliance and Technoserve and global traders (Alvarez et al., 2010).

Whilst having a proprietary standard is important for many companies, linked to quality expectations and reputation management, some companies recognise that there are benefits in working with other companies across a sector or multiple sectors. Such collective initiatives are sometimes created with the aim of limiting standards proliferation, or failing that, to create a benchmark that conveys an understanding of expected practice across the sector, against which private, single-company, standards can be compared. This was the origin of the Global Coffee Platform (GCP) that has evolved from the work and membership of the 4C Association that established a 'common code for the coffee community' that acted as 'a baseline sustainability reference for the coffee industry' (Ugarte et al., 2017, p. 53).

Such initiatives tend to emerge in sectors where companies may be sourcing from the same pool of suppliers, as in coffee, or the initiative is focused on topics that are considered 'pre-competitive' that may relate to the institutional or infrastructural landscape in a country of origin.[1] In such cases, action by one company to improve supplier practices may benefit other companies so without collective action there is a danger of free riding which might become a disincentive for action by a single company. Many agri-food companies, especially those engaging in such collective initiatives, are increasingly recognising that setting standards for suppliers alone will not deliver the improvements needed for environmental and societal sustainability and for the sustainability of their own operations, both materially and in response to societal expectations. Also, companies recognise that they lack the scientific knowledge and expertise with regard to the nature and scale of sustainability challenges, especially with regard to environmental degradation and seek to collaborate with universities and other sources of knowledge generation (Freidberg, 2017). The GCP now works through 'public-private platforms in coffee producing countries, and through thematic work streams at a global level, focusing on climate smart agriculture, economic viability, gender and youth' (Ugarte et al., 2017, p. 53).

A long-standing grouping of food and agriculture companies with a focus on pre-competitive topics is the Sustainable Agriculture Initiative Platform (known as the SAI Platform, https://saiplatform.org/). Established in 2002, with roots in discussions between Danone, Nestlé and Unilever, its original focus was sustainable agricultural practices in arable in the European Union (EU) and United States (US), given that they were already collaborating through the emerging roundtables on tropical products such as oil palm and soy (SAI Platform, 2022). Membership has grown to over 150 of the largest food and agriculture companies across 30 countries, covering a vast array of production systems. Like many other collective initiatives, its focus is the development and sharing metrics covering both good practices and outcome measures. It has a specific focus on commonalities rather than differentiation through a consumer-facing label or standard, working on the development of sustainability metrics with the aim of driving continuous improvement.

Collective standards initiatives across the cocoa sector have origins in attempts to combat forced labour and child labour spearheaded by the World Cocoa Foundation (Nelson and Phillips, 2018). In addition, facing quality and supply issues because of climate change as in the coffee sector, there have been collective industry efforts to work on more sustainable and climate resilience cocoa production practices. Some collective initiatives have decided to concentrate all their energies on sharing good practice and tackling the root causes of the problems faced by the sector. With roots in a shared code,

1 Note that anti-trust laws prohibit collaborations that may restrain competition, including the exchange of information concerning individual prices, rates, coverages, market practices, claims settlement practices or any other competitive aspect of an individual company's operation.

the Ethical Tea Partnership no longer promotes or audits its own collective standard, it works in partnership with development bodies and governments to support capacity building measures including those associated with climate adaptation (Ethical Tea Partnership, 2021) and the adoption of 'living' wages (Malawi Tea 2020, 2017).

Collective standards have some similarities to VSS established by Multi-stakeholder Initiatives (MSIs) in that they are created by more than one entity and increasingly involve measures to build capacity and respond to processes beyond a single site of production or value chain that affect the achievement of their standards. However, in contrast to VSS that tend to originate in the third sector, collective initiatives tend to be business led, though they may involve NGOs in an advisory capacity. Whilst they have an important function in generating collective action, some have been critiqued for an overly strong focus on production. For example, the collective actions at landscape level in cocoa might 'help sustain the industry under deteriorating conditions' but fail to understand the underlying issues, especially with respect to the vulnerability of smallholders, limiting longer term transformation (Nelson and Phillips, 2018, p. 252). Moreover, where the focus is on metrics to drive continuous improvement in production practices amongst the supplier base, there may be less attention paid to the supply chain relations that may facilitate or constrain change.

A final comment to make regarding the origin of standards is that most purport to be 'global' or 'universal', but most standards derive from companies or stakeholders in the Global North. There are of course long-standing standards from the Global South, often oriented at differentiating a country's exports to northern markets such as the Kenya Flower Council (Said-Allsopp and Tallontire, 2015). Others have emerged in direct competition with so-called global standards, especially where a country dominates production as has been seen with the development of standards for palm oil in Indonesia and soy in Brazil (Hospes, 2014). More recently, sustainability standards have started to emerge in regional markets in the Global South, such as in East Africa, often related to the formalisation of supermarket supply chains (Turley et al., 2022).

3.2 What does the standard cover?

As noted previously, there is significant agreement on the nature of sustainability problems, but less consensus on how to resolve them. We see an increasing number of standards often in the same sector and competing in the same markets, e.g. in coffee and chocolate we see VSSs such as Rainforest Alliance, UTZ and Fairtrade as well as private company and collective standards, and in flowers there are VSS and collective standards at international and national levels. That said, there is increasing consensus on the content of sustainability standards, or at least what might be included as criteria if not how these are monitored or verified through an audit process. Sustainability standards cover at least two of the three pillars of sustainability, that is they cover social and environmental issues, and sometimes economic, though the latter dimension is usually less explicit in most standards. Table 1 provides an overview of general patterns of inclusion and exclusion, which to some extent are informed by global normative frameworks agreed at United Nations level and increasingly by guidance from 'meta standards' organisations such as the ISEAL Alliance or the Global Social Compliance Programme (Reinecke et al., 2012).

Since the early 2000s, there has been increased consensus that standards should cover internationally agreed International Labour Organisation (ILO) Conventions that have the status of core labour rights (Barrientos and Smith, 2007). However, the extent to which they cover the complexity of sustainability is limited at best. Sustainability standards criteria tend to focus on inputs and outputs with respect to environmental issues, lacking 'ecological thinking and literacy' (Milne and Gray, 2013, p. 24) and with limited engagement with environmental conventions such as on biodiversity or climate (Smith et al., 2019).

Table 1 Criteria found in sustainability standards in food and agriculture

Pillar of sustainability	Typically included criteria	Less developed or emerging areas
Social	ILO Fundamental Principles and Rights at Work:[a] 1. Freedom of association and the effective recognition of the right to collective bargaining; 2. The elimination of all forms of forced or compulsory labour; 3. The effective abolition of child labour; 4. The elimination of discrimination in respect to employment and occupation; and 5. A safe and healthy working environment.Other specific working conditions linked to health and safety as relevant to the sector	Process rights such as the right to collective bargaining and freedom of association may exist on paper but less respected in practice
Environmental	Pesticide use Water use	Landscape impacts/ impacts beyond the site of production Climate adaptations Biodiversity
Economics	Price usually associated with quality except for fair trade standards Quality specification	Value addition Supply chain relations Living wages/income

[a]Originally adopted in 1998 and revised in 2022; reproduced from https://www.ilo.org/declaration/lang--en/index.htm%5D
Source: Compiled by author.

However, in recent years, several standards bodies and groups of companies have begun to consider how to work on issues at the landscape level, e.g. water stewardship and spill-over impacts of agricultural operations on regional land use patterns (ISEAL, 2022a,b).

On social issues, there is a focus on codified rights as opposed to processes rights associated with building up social capital (Said-Allsopp and Tallontire, 2015). The economic dimensions of sustainability are rarely covered explicitly or are merged with social criteria, such as minimum wages. In only a few cases are wages expressed as living wages (e.g. the Ethical Trading Initiative's Base Code), though there has been considerable momentum to push for living wages and incomes in a number of sectors in recent years, e.g. driven by the Living Income Community of Practice, https://www.living-income.com/. Only Fairtrade standards cover price issues separate from issues of quality and whilst trading relations are covered in the Fairtrade trader standards, e.g. payment and credit terms and length of contract, they are notoriously difficult to implement (Doherty and Davies, 2013).

Numerous detailed critiques exist of sustainability standards content, e.g. Ponte (2019, p. 214) who argues that sustainability standards 'focus on the manifestations rather than the root of the problem', rarely covering 'systemic and structural' issues such as inequality and supply chain power relations. Improvements tend to occur 'unit-level', and it is difficult to account for 'overall' reductions in, e.g., child labour or damage to the environment.

3.3 Why did the standard get developed?

The reasons for the establishment of sustainability standards, and why they vary in their scope is addressed by a vast literature. The purpose of the sustainability standards is considered here in two ways, firstly their level of stringency, and secondly their engagement with a theory of change.

Based on detailed empirical analysis of the vast array of sustainability standards in the cut flower sector, at national and international levels, Riisgaard (2011) developed a simple typology of standards. Type 1 standards were identified as schemes that weigh size over principles, i.e. they were keen to be inclusive across the sector and create a benchmark standard. These standards tended to be about risk management and with respect to labour rights covered outcome standards that provided basic protections such as health and safety. Examples included GlobalGAP, most retailer standards examined and the Kenya Flower Council's standard. In contrast, there were schemes she dubbed Type 2 standards that were more concerned about maintaining principles rather than growing the membership. Adopters of this type of standard wanted to differentiate themselves in the market. In terms of social standards content, these standards tend to focus more on enabling rights and have a closer relationship with the labour rights movement. In the floriculture sector, examples included Flower Label Programme and Fair Flowers Fair Plants as well as Fairtrade and Rainforest Alliance.

This analysis focusing on 'stringency' provides insight into the overall ethos of a standards body, or what it is trying to achieve. In the past decade, there has been an increased popularity in the use of theory of change as a model to explain in more depth the rationale behind the content of and overall approach to implementing a sustainability standard. Theories of change have roots in international development and aid planning, based on the argument that you have to know where you want to go and the assumptions that you are making about the processes of change in order to better understand and account for impact (Valters, 2015). The ISEAL Alliance advocate making a theory of change explicit, expressed as 'Defining the Intended Change', as part of their work on good practice in impact assessment (ISEAL Alliance, 2014).

There are differences between theories of change adopted amongst members of ISEAL that operate in similar or the same sectors, which may account partly for the persistence of so many sustainability standards. Table 2 compares

Table 2 Comparison between theories of change for fairtrade and UTZ

Standard	Impacts (envisaged)	Vision	Reference
Fairtrade (small producer organisations, simplified)	• Improved household income, assets and standard of living; • Less risk and vulnerability, increased food security • Improved access to basic services Increased environmental sustainability and resilience to climate change • Inter-generational sustainability of rural communities • Increased cooperation and gender equality within communities Increased dignity, confidence, control and choice • Enhanced influence and status of small producers • Fairer and more sustainable trading system	'A world in which all small producers and workers can enjoy secure and sustainable livelihoods, fulfil their potential and decide on their future'	Fairtrade Theory of change 2018 Fairtrade International (2018)
UTZ theory of change	• Improved livelihoods of farmers, farm workers and families • Long-term viability of the sector • Safeguarding natural resources	Sustainable farming is the norm Better crop Better income Better environment Better life	UTZ theory of change 2017, as represented in Ingram et al. (2017)

Source: Compiled by author.

the Fairtrade smallholder theory of change with the UTZ standard[2] with respect to two aspects: first, their vision and secondly, the impacts envisaged. Fairtrade International focuses on decision-making and the security of workers and smallholders at the centre of their vision of sustainability, whereas UTZ focuses on crops and the environment and income, with improved agricultural practices at the farm level being the main driver for increases in income (though there is some role noted for the buyer in capacity building). Whilst both theories of change refer to improved incomes for farmers, the route by which they aim to achieve this outcome differs. With UTZ, the focus is on investment in building up capacity to ensure maintenance of quality and improve yields, which will hopefully lead to increased incomes. The Fairtrade route to impact engages with a broader set of mechanisms linked to rural livelihoods, social dynamics related to gender and inter-generational inequalities and reducing environmental vulnerability. This difference might be attributed to UTZ's roots in good agricultural practices that have been developed within industry, originally as risk management reduction for quality and safety, which were then extended to environment and worker health and safety whereas Fairtrade has origins in the trade justice and international development movement and advocating for producers.

The theory of change focus for some of the collective initiatives can be less clear-cut. SAI Platform's focus on metrics, rather than standards to be met, 'assumes that incentives will come in the form of discovered efficiencies and improved relations with supply chain customers', argues Friedberg (2017, p. 1392). She goes on to say that this needs empirical evidence, as well as more clarity on the theory of change approach being used. She states: 'The fact that metrics-based governance expects suppliers to both report on and somehow improve their performances – often without promise of monetary gain – raises the obvious question of how that is supposed to happen' (Friedberg, 2017, p. 1392).

Exploring sustainability standards through the questions of who develops standards, what they cover and why they were developed can help observers to navigate the huge variety of sustainability standards that have emerged. What standards look like on paper, however, is only part of the story; we also need to consider how they are used in practice. In Section 4, how companies choose to use different kinds of standard in practices is explored and critiqued.

4 Sustainability standard selection by companies

Given the variety of sustainability standards that exist, how do companies choose between them or even decide to use standards in the first place? In the

2 Formerly independent, UTZ is now part of the Rainforest Alliance, https://www.rainforest-alliance.org/utz/.

past 15 years, sustainability standards and related tools have been used not only with respect to high-value products with high consumer visibility but also to staple crops such as maize and those ingredients with multiple uses, so-called *flex crops* such as oil palm and soy. Indeed, part of the business benefits of using sustainability standards that the ISEAL Alliance promotes to companies include their 'contribution to supply chain coordination, followed by supply chain risk management and supply chain transparency and traceability' (Molenaar, 2022, p. 18). Ultimately, companies wish to demonstrate that they are responsible businesses to their shareholders and to ensure their own resilience in the face of environmental challenges that threaten productivity and availability. Increasingly, governments and regulators are also creating drivers for the adoption of sustainability standards in certain sectors such as biofuels. The EU Renewable Energy Directive makes specific reference to the Roundtable on Sustainable Palm Oil (RSPO) and the Roundtable on Responsible Soy as indicators of sustainable sources (Ponte, 2019). The choice by companies about whether to commit to a sustainability agenda, their priorities and the tools to use, as well as how well they implement these elements will vary according to the mix and strength of the external pressures on the business and their internal drivers.

Multi-stakeholder approaches are 'increasingly being seen as the most legitimate private rule-makers' (Cheyns and Riisgaard, 2014, p. 409) and in the eyes of many consumers, these are the 'best' standards. In the early 2000s, concerns were raised about the content of MSI standards and the extent to which the 'right' issues were being prioritised or whether topics were 'cherry picked', privileging the concerns of consumers over those of producers (Giovanucci and Ponte, 2005; Tallontire et al., 2014). However, further rounds of standards' revision in the wake of stakeholder engagement and impact assessments have led to improvements, especially to the MSI standards aligned to bodies such as the ISEAL Alliance.

However, legitimacy or the 'best' sustainability practice is not always the prime motivation for standards adoption. Moreover, third-party or collective standards do not exist in all sectors, and in many markets, there is minimal consumer or civil society pressure that sets expectations on the approaches to use. Indeed, a company's approach to managing the supply chain is not all about communicating or responding to consumerism: '[M]any studies also show that firms are not that directly worried about lessening sales as a result of NGO shaming tactics' (Fransen, 2019, p. 796). In choosing the tools to use, many companies are as concerned about relationships with suppliers as they are with consumers. Where there are environmental or land pressures on agricultural production, sustainability standards may be used to improve practices and secure access to produce by creating more direct and stable relationships with suppliers.

Whilst third-party-certified standards offer benefits to businesses, many of which have adopted them as tools to manage sustainability and to communicate this, critiques of the mainstreaming of VSS highlight that even if such standards are chosen, stringency in the application or the planned benefits resulting from the standard body's theory of change approach depends on how they are used. However, this is an area ripe for more research.

A recent analysis of company choices of standards, based on a range of case studies of multinational companies sourcing agricultural products, identified key factors influencing the choice of 'instrument', including different types of standard, as well as investment and capacity building or alternatively, avoiding sourcing from a location (Rueda et al., 2017). This first highlights the importance of the sustainability issues faced in the production location including the level of risk, especially environmental risk, the visibility of that risk, the availability of technology and knowledge to deal with the issue and the overall regulatory framework in that region. A second consideration is the extent of leverage a specific company has over its suppliers, which affects whether a company adopts its own approach to embedding sustainability standards or works with others. A third factor is the characteristics of the market, including the extent of brand recognition and consumer-facing communications as well as the significance of origin or 'terroir' for the consumer experience.

The resulting decision-tree the authors produce helps explain why chocolate producers have tended to adopt cross-industry collective initiatives as they are sourced from scattered suppliers over whom they have limited leverage. Multi-stakeholder standards were more common where there was a product with limited transformation from producer to consumer, and where there was good traceability capability, such as in the banana supply chain. Companies developed their own standards where there was a concern for quality and a desire to develop closer links with suppliers, such as with Starbucks Café Practices. Based firmly on commercial criteria, this analysis misses out the values of the buying company beyond market success and survival and focuses on environmental decision-making, not social criteria, but helps illustrate the factors that are routinely taken into consideration.

Whether a company uses a VSS with third-party certification, participates in a collective industry initiative or uses its own private standard, the link to the company's values and its over-arching sustainability strategy is an important indicator of its potential to initiate and deliver longer-term change. There may be trade-offs between implementing the principles underpinning the standard and business interests, with operation of purchasing practices and the exercise of supply chain power eroding sustainability benefits for the supplier (Doherty and Davies, 2013; Ponte, 2019). In other circumstances, a standard may be central to supplier engagement and overall company strategy. In her analysis of UK retailers use of Fairtrade standards, Smith (2010) argued that some used the standards in an arms-length way, passing on the standards' requirements to actors further upstream whereas other retailers used Fairtrade as part of their overall supply chain strategy, making 'category shifts' so that all a certain product category rather than just some individual lines were certified, often combining this with capacity development. Similarly, other supermarkets have used a 'sharing' strategy in applying standards, whereas others have been characterised as 'pushing' a standard and associated expectations onto producers (Muller et al., 2012).

However, it is important to remember that the adoption of standards is not static, and companies may shift their approach in response to contextual factors, e.g. consumer demands, or internal company strategy. An increasing trend in some markets is companies dropping VSS, especially Fairtrade, in favour of their own company standards. Since around 2016, a number of companies that had previously made very public affiliations to Fairtrade announced changes to their approaches. The first high-profile company to do this was Cadbury (a brand now owned by Mondelez) who worked out a deal with the Fairtrade Foundation so that the traceability requirements would not be quite so strict (Doherty, 2016). In 2017, the UK supermarket Sainsbury's decided that it would no longer buy its tea for its 'Red Label' line from on Fairtrade terms, though they were keen to still buy from the same farmers, establishing their own Fairly Traded scheme, a private standard. This proved controversial. Fairtrade supporters highlighted the potential for confusion amongst consumers confusion about the claims on the packaging and the underlying practices. Fairly Traded is very similar, but not identical to, to Fairtrade, and developing a bespoke retail standard rather than engaging with Fairtrade as an independent standard led to concerns about the benefits it delivered to producers and the future of the Fairtrade standard (CAFOD, 2018). There were also objections from the farmer co-operatives supplying tea, worried that they would no longer have a say in the use of the 'social premium' from sales (Fairtrade Africa, 2018).

Sainsbury's produced a report on the outcomes of their new approach in which they tried to address some of the concerns from the wider sustainability standard community about the legitimacy of the company's 'do-it-yourself' approach, building on an earlier study that set out their sustainable sourcing strategy (Sainsbury's, 2019). They stressed improvements in transparency in their tea supply chain from East African suppliers, a sector that has been critiqued for poor transparency (see e.g. Dolan, 2008). They also referred to generating a better understanding of challenges faced by small producers and that quality and environmental practices by producers had improved. Bringing standard setting in-house might be perceived as part of organisational learning and development of the sustainability strategy with respect to agricultural sourcing. However to date, they have not rolled this approach out to other commodities.

Nevertheless, this case illustrates some of the decision-making behind the use of internal company standards, a concern to have closer relationships with suppliers and to link standards' use to a company's strategy, rather than relying on policy instruments governed externally. Scrutiny is therefore needed not only for the specific standards but also for a company's approach to sustainability more broadly, and if it is integrated within their overarching business strategy and the extent to which it is linked to organisational transformation and a vision of sustainable development as opposed to company focused sustainability (Schaltegger et al., 2012). More fundamentally, one might ask if a company is motivated by 'the business case' or whether its strategy is linked to a broader understanding of the company's relationship with society or as a corporate citizen, that is its understanding of corporate social responsibility (Garriga and Melé, 2004; Sheehy, 2015).

5 Standards and smallholders

The companies managing supply chains and adopting standards do not necessarily do this with the needs, livelihoods or experiences of smallholders in mind. As examined throughout this chapter, sustainability standards are focused on governance of supply chains (whereby companies set expectations of their suppliers to minimise the risk of being exposed to poor practice and to move towards more sustainable practices) and through supply chains (whereby civil society actors use standards as a means of making businesses accountable) (Bush et al., 2015). Further, sustainability standards were not designed with a specific focus on the needs and capacities of smallholders, with notable exceptions such as Fairtrade's standards for Small Producer Organisations (providing a framework for smallholder empowerment through their organisations) and Bon Sucro which has the same core criteria but has fewer overall requirements for smallholders (Elder et al., 2021). Rarely do the governance structures of standards bodies include the voice of producers, on any scale, even in MSIs (Bennett, 2017). The structures and processes of some sustainability roundtables create barriers for smaller scale actors and those from the Global South (Ponte, 2019, p. 269) and exclusion of smallholders from direct representation is sometimes justified on the basis that the subject matter is too technical (Tallontire et al., 2014).

Smallholders have been involved in global markets for over a century, with ties and relationships developed during the colonial era. Smallholder producers, usually defined as having 2 ha or under, are an important part of the global food system, supplying at least 35% of global food calories (Lowder et al., 2021), though if 'family farms' are included, as much as 70% of global food calories and a third of global crops (Guarin et al., 2022) arise from these micro and small businesses. However, as Guarin et al. (2022) point out, over the past 20 years many supply chains have been managed more directly by buyers (e.g. coffee and cocoa), or new 'tightly structured chains' have been created, such as for fresh vegetables and fruits. As discussed in this chapter and elsewhere in this volume, engagement in so-called 'modern markets' requires conformity to a range of standards. These include sustainability standards and quality and food safety standards, as well as meeting the price and logistical requirements of buyers, many of which require access to documentation and formal management control systems that many, if not most, smallholders do not possess, and where benefits tend to accrue based on economies of scale. Many international development institutions, such as the Food and Agriculture Organisation of the United Nations (FAO), have been concerned as to whether standards contribute to a barrier to entry, creating exclusion for some from such markets, thus limiting trade and sustainable development.

There is an increasing volume of evidence being generated on the sustainability impacts of standards, particularly concerning the VSS associated with third-party certifications rather than private company standards (Thorlakson et al., 2018). However much of the evidence is skewed to certain commodities or regions, e.g. Nelson and Pound's review of impact studies on Fairtrade (2009) highlighted that much of the evidence was on coffee from Latin America. Similarly, a study for the FAO on the inclusion of smallholders in chains requiring standards noted that much of the evidence is concentrated on standards such as Global GAP, Fairtrade and organic standards (Loconto and Dankers, 2014).

It has proved difficult to generalise from studies, but often the results have been less positive than many standard setters or development bodies would like (UNFSS, 2022). To some extent, this is a reminder that inclusion in markets requiring sustainability standards, and by default meeting them, does not necessarily equate to longer term benefits from them for all supply chain actors (Loconto and Dankers, 2014). Moreover, the way that standards are implemented may not have developmental narratives in mind (Nelson and Tallontire, 2014). Others have noted that impact assessment methodologies are not always sensitive or robust, especially given the many variables they have to be taken into consideration (Ruben, 2017) and given considerable study design and sampling challenges (Meemken, 2020). For example, many studies focus on the adoption of established good practices as a proxy for impact (Thorlakson et al., 2018). Similarly, they might highlight the processes that standards-related compliance might engender, which then may lead to development processes such as access to resources, development of skills and the strengthening of farmer (and worker) organisations so that they can better articulate members' needs (Elder et al., 2021). Moreover, most studies lack consideration of how different supply configurations or the role of farmer organisation might affect outcomes (Meemken, 2020).

A feature of many studies of the impacts of standards on smallholders is that environmental and labour practices might be improved, but often costs increase more than income, limiting opportunities for poverty reduction at the household level (Oya et al., 2018). Frequently benefits that are harder to quantify or are less directly comparable may be cited, such as having longer term or more reliable linkages with buyers as compared to the open market. Similarly, standards engagement, when accompanied by training and donor-funded projects to support producers, can generate benefits that go beyond impacts that can be attributed to standards-related compliance, as has been shown in the Tanzanian tea sector by Loconto and Dankers (2014). Increasingly it is being recognised that to understand the impacts of standards, there is a need to take into account the actions of actors in the supply chain and also national-level actors and institutions (Tallontire et al., 2011; Guarin et al., 2022). Often there is much potential in standards engagement, but the implementation process, and support provided can be key.

Indeed, standards compliance can be challenging, but a recent study has revealed unexpected benefits of standards, specifically Fairtrade standards. A study conducted for Fairtrade International by Günther et al. (2022) demonstrated that smallholders with Fairtrade certification were more resilient in the face of the shock of the Covid-19 pandemic. This was attributed to economic criteria in the standard (e.g. minimum price and longer-term contracts) combined with the experience of certification processes which meant that farmer groups had greater organisational capacity that helped with adaptation. The benefits of organisations forming into groups to facilitate both standards compliance and to articulate the voice of small farmers can also come into play in such challenging situations (Nelson et al., 2016). It is increasingly clear that standards by themselves do not bring benefits to smallholders: realising benefits requires a supportive institutional context, both in the country of production and of consumption and along the supply chain (Krauss and Krishnan, 2022). Often the issue is not whether the implementation of standards criteria themselves might lead to positive outcomes, but who is implementing them, how they work with smallholders and listen to their concerns.

6 Conclusion

The origins of sustainability standards lie in the expansion of global sourcing in the food system, offering a form of 'control from a distance' (Gibbon and Ponte, 2005). Over the past decade, for sustainability impacts to be achieved, organisations and the buyers that work for them have recognised the need for greater visibility of the supply chain, particularly making efforts to get closer to the producer as environmental challenges have accelerated. It remains the case that retailers often do have visibility much beyond their first-tier suppliers, especially in supply chains that depend on commodity traders with limited capability to trace the product to the source. As Friedberg quotes from her reading of the Walmart-led Sustainability Consortium's activities: 'If you can't manage what you don't measure, you definitely can't manage what you can't even see' (2017, p. 1398).

There are many expectations placed on sustainability standards in the food and agriculture sector. For some, they are emblematic of good practices, demonstrating companies' contributions to the sustainability, especially with respect to the SDGs and are presented as a tool to enhance trade and hence broad-based economic development (UNFSS, 2022). However, sustainability standards are by definition exclusionary, they set out what is necessary to merit a certificate or label, often in niche destination markets, and compliance may be a requirement to sell produce to a particular buyer that uses sustainability standards to manage its supply chain. This can be problematic for smaller producers and for development, especially where expectations of compliance are not accompanied by support to achieve compliance (Loconto and Dankers, 2014). Moreover, certification with sustainability standards can erode producer incomes and resilience where efficiency pressures and poor supply chain management practices undermine the capacity of producers to address underlying technical and social issues (Oxfam, 2010) and add to the margins of buying companies where risks and compliance costs are pushed onto suppliers (Ponte, 2019).

This chapter demonstrates that there are no easy answers regarding the impacts of standards in the agri-food sector; sustainability standards are applied in highly complex and contested contexts. However, this examination, asking simple questions of who establishes standards, what they cover and the rationale for the standard, together with considerations of how companies choose standards seeks to provide routes into understanding the complexity, going beyond the claims on labels and marketing blurbs. These questions help us to consider what might be a plausible pathway to generating sustainability benefits, of what kind and for whom.

A standard's goodness of fit into the sustainability strategy or indeed underlying values of companies buying produce is emerging as an important factor, together with the understanding of the practices and priorities of producers, seeing standards as part of wider systems, at company, chain and institutional scales. If standards are to deliver benefits, it is crucial to make them more relevant to farmers, so that they can see the benefits financially as well as in terms of protecting their environment and community. However, what may become more important for sustainability than standards setting out criteria for products might be sets of expectations for companies themselves and considerations of how company practices interact with national-level institutions to support social and environmental sustainability. The adoption of sustainability standards on the basis of 'the business case' and the sustainability of the buying company may drive compliance but it may bring fewer lasting benefits than an approach to standards development and adoption that is based on a broader understanding of a company's responsibility to society.

7 Where to look for further information

United Nations Forum on Sustainability Standards (UNFSS)

- https://unfss.org/.

The UNFSS 'helps producers, traders, consumers, standard-setters, certification-bodies, trade diplomats, non-governmental organizations and researchers to talk to each other, find out more about Voluntary Sustainability Standards and influence decision-makers at the intergovernmental level' and aims to provide impartial advice on VSS, including where they may disrupt trade or cause specific challenges to small producers.

Regular overviews of the range of standards in operation and trends can be found in the biennial flagship reports of the United Nation Forum on Sustainability Standards (UNFSS) https://unfss.org/home/flagship-publication/.

UNFSS holds regular webinars on a variety of topics related to sustainability standards.

It also commissions and publishes useful reports such as its October 2020 report linking sustainability standards to the SDGs and their specific targets, https://unfss.org/2020/10/13/linking-voluntary-standards-to-sustainable-development-goals/.

ISEAL Alliance

- https://www.isealalliance.org/.

The ISEAL Alliance is a membership organisation of sustainability standards bodies. ISEAL is concerned with assuring the credibility of the standards of its members and has developed a series of good practice guides, including on standards development and on impact assessment.

Considerable evidence about the effectiveness of sustainability standards is collated by the ISEAL Alliance's Evidensia website, https://www.evidensia.eco/ which is supported by a range of international donors.

The State of Sustainability Initiatives

- https://www.iisd.org/ssi/.

The State of Sustainability Initiatives (SSIs) is long-standing initiative of the International Institute for Sustainable Development in Canada, with roots in a series of multi-stakeholder meetings (2003–6) on sustainable commodities production and trade, known as the Sustainable Commodity Initiative (SCI). SSI has the support of multi-lateral and bilateral development agencies. It undertakes research on VSSs with the aim of supporting 'better environmental and social performance in important commodity sectors'. It produces regular surveys of sustainability standards, their effectiveness and with respect to trends in specific sectors.

Sustainable Food Lab

- https://sustainablefoodlab.org/.

The Sustainable Food Lab is a non-profit organisation established in 2004 with members from the private sector together with scientific institutes, foundations and NGOs. It takes a whole systems approach to deliver programmes that focus on developing solutions through collaborative pre-competitive projects (e.g. on regenerative agriculture and living income), development and testing of tools and frameworks (for example of greenhouse gas emissions calculations to support sustainability practices by smallholders).

Research Network Sustainable Global Supply Chains

- www.sustainablesupplychains.org/.

This a network funded by the German Federal Ministry for Economic Cooperation and Development (BMZ) and hosted by four scientific and research organisations. It collates research papers by leading scholars on global supply chains, with special section on sustainability standards, making them publicly available.

8 References

Alvarez, G., Pilbeam, C. and Wilding, R. (2010). Nestlé Nespresso AAA sustainable quality program: an investigation into the governance dynamics in a multi-stakeholder supply chain network, *Supply Chain Management: An International Journal* 15(2): 165–182.

Auld, G., Bernstein, S. and Cashore, B. (2008). The new corporate social responsibility, *Annual Review of Environment and Resources* 33(1): 413–435.

Barrientos, S. and Smith, S. (2007). Do workers benefit from ethical trade? Assessing codes of labour practice in global production systems, *Third World Quarterly* 28(4): 713–729. https://doi.org/10.1080/01436590701336580.

Bennett, E. A. (2017). Who governs socially-oriented voluntary sustainability standards? Not the producers of certified products, *World Development* 91: 53–69.

Busch, L. and Bain, C. (2004). New! improved? The transformation of the global agrifood system, *Rural Sociology* 69(3): 321–346.

Bush, S. R., Oosterveer, P., Bailey, M. and Mol, A. P. J. (2015). Sustainable governance of chains and netwoords: a review and future outlook, *Journal of Cleaner Production* 107: 8–19.

CAFOD (2018). Sainsbury's tea isn't Fairtrade, Available at: https://blog.cafod.org.uk/2018/03/02/sainsburys-tea-isnt-fairtrade/, accessed 8 December 2022.

Cheyns, E. and Riisgaard, L. (2014). Introduction to the symposium: the exercise of power through multi-stakeholder initiatives for sustainable agriculture and its inclusion and exclusion outcomes, *Agriculture and Human Values* 31(3): 409–423.

Dicken, P. (2014). *Global Shift: Mapping the Changing Contours of the World Economy*, London: Sage.

Doherty, B. (2016). It's not a very merry Christmas for Fairtrade chocolate, Available at: https://theconversation.com/its-not-a-very-merry-christmas-for-fairtrade-chocolate-69761, accessed 8 December 2022.

Doherty, B. and Davies, I. (2013). Where next for fair trade, *Business History* 55(2): 161–189.

Dolan, C. S. (2008). In the mists of development: fairtrade in Kenyan tea fields, *Globalizations* 5(2): 305–318, https://doi.org/10.1080/14747730802057787.

Elder, S., Wilkings, A., Larrea, C., Elamin, N. and Fernandez de Cordoba, S. (2021). *State of Sustainability Initiatives Review: Standards and Poverty Reduction*, International Institute for Sustainable Development (IISD). https://www.iisd.org/publications/ssi-review-standards-poverty-reduction.

Ethical Tea Partnership (2021). Climate change and tea, October 2021, Available at: https://ethicalteapartnership.org/wp-content/uploads/2021/11/Climate_report_web.pdf, accessed October 2021, 23 November 2022.

Fairtrade Africa (2018). Open letter to Sainsbury's from fairtrade producer, Available at: https://www.fairtrade.org.uk/media-centre/news/open-letter-to-sainsburys-from-fairtrade-producers/, accessed 23 November 2022.

Fairtrade International (2018). Journeys to change: fairtrade theory of change, Available at: https://www.fairtrade.org.uk/wp-content/uploads/legacy/doc/Fairtrade_Theory_of_Change%20-%202018.pdf, accessed 23 November 2022.

Fransen, L. (2019). Political consumerism and corporate strategy towards sustainability standard-setting: in or out of sync? In: Boström, M., Micheletti, M. and Oosterveer, P. (Eds). *The Oxford Handbook of Political Consumerism*, Oxford Handbooks, https://doi.org/10.1093/oxfordhb/9780190629038.013.34.

Fransen, L., Kolk, A. and Rivera-Santos, M. (2019). The multiplicity of international corporate social responsibility standards: implications for global value chain governance, Multinational Business Review 27(4): 397–426, https://doi.org/10.1108/MBR-08-2019-0083.

Freidberg, S. (2017). Big food and little data: the slow harvest of corporate food supply chain sustainability initiatives, *Annals of the American Association of Geographers* 107(6): 1389–1406.

Garriga, E. and Melé, D. (2004). Corporate social responsibility theories: mapping the territory, *Journal of Business Ethics* 53(1/2): 51–71.

Gibbon, P. and Ponte, S. (2005). *Trading Down: Africa, Value Chains and the Global Economy*, Philadelphia, PA: Temple University Press.

Giovanucci, D. and Ponte, S. (2005). Standards as a new form of social contract? Sustainability initiatives in the coffee industry, *Food Policy* 30(3): 284–301.

Guarin, A., Nicolini, G., Vorley, B., Blackmore, E. and Kelly, L. (2022). *Taking Stock of Smallholder Inclusion in Modern Value Chains: Ambitions, Reality and Signs of Change*, London: IIED.

Günther, M. K., Afroz, B. and Rathinam, F. X. (2022). Fairtrade certification and producer resilience in times of crises, *Scio Network & Athena Infonomics*, Available at: https://

www.fairtrade.net/library/fairtrade-certification-and-producer-resilience-in-times-of-crises, accessed 23 November 2022.

Henson, S. and Humphrey, J. (2010). Understanding the complexities of private standards in global Agri-food chains as they impact developing countries', *Journal of Development Studies* 46(9): 1628–1646.

Hospes, O. (2014). Marking the success or end of global multi-stakeholder governance? The rise of national sustainability standards in Indonesia and Brazil for palm oil and soy, *Agriculture and Human Values* 31(3): 425–437.

Ingram, V., van Rijn, F., Waarts, Y., Dekkers, M., de Vos, B., Koster, T., Tanoh, R. and Galo, A. (2017). *Towards Sustainable Cocoa in Côte d'Ivoire. The Impacts and Contribution of UTZ Certification Combined with Services Provided by Companies*, Wageningen: Wageningen Economic Research, Report 2018–041, 140, Available at: https://www.rainforest-alliance.org/resource-item/sustainable-cocoa-cote-divoire-impacts-utz-certification/, accessed 2 September 2022.

ISEAL (2014). Assessing the impacts of social and environmental standards systems ISEAL code of good practice, Version 2.0, December, Available at: https://www.isealalliance.org/defining-credible-practice/iseal-codes-good-practice, accessed 22 November 2022.

ISEAL (2016). Researching standards and certification: A guidance note, Available at: https://www.isealalliance.org/sites/default/files/resource/2018-01/Researchers%27_Guidance_Note_Apr2016.pdf, accessed 23 November 2022.

ISEAL (2022a). Driving good water stewardship at district level, Available at: https://www.isealalliance.org/innovations-standards/innovations-projects/driving-good-water-stewardship-district-level, accessed 23 November 2022.

ISEAL (2022b). Blueprint for a sustainable landscape, Available at: https://www.isealalliance.org/innovations-standards/innovations-projects/designing-blueprint-sustainable-landscapes, accessed 23 November 2022.

KPMG International (2008). *International Survey of Corporate Responsibility Reporting 2008. Tech. Report*, KPMG International, Available at: http://www.kpmg.com/LU/en/IssuesAndInsights/Articlespublications/Documents/KPMGInternational-Survey-on-Corporate-Responsibility-Reporting.pdf, accessed 23 November 2022.

Krauss, J. E. and Krishnan, A. (2022). Global decisions versus local realities: sustainability standards, priorities and upgrading dynamics in agricultural global production networks', *Global Networks* 22(1): 65–88.

Loconto, A. and Dankers, C. (2014). *Impact of International Voluntary Standards on Smallholder Market Participation in Developing Countries: A Review of the Literature*, Rome: Food and Agriculture Organization (Agribusiness and food industries series, 3).

Lowder, S. K., Sánchez, M. V. and Bertini, R. (2021). Which farms feed the world and has farmland become more concentrated?, *World Development* 142: 105455.

Malawi Tea 2020 (2017). The sustainable procurement practices, Presentation at Annual Progress Meeting (Malawi Tea Revitalisation Programme) by Rachel Wilshaw (Oxfam) and Will Battle (IDH), Available at: https://ethicalteapartnership.org/wp-content/uploads/2019/06/APM-2017_part-4-Sustainable-Procurement-Practices.pdf, accessed 23 November 2022.

Meemken, E. M. (2020). Do smallholder farmers benefit from sustainability standards? A systematic review and meta-analysis, *Global Food Security* 26: 100373.

Milne, M. J. and Gray, R. (2013). W(h)ither ecology? The triple bottom line, the global reporting initiative, and corporate sustainability reporting, *Journal of Business Ethics* 118(1): 13–29.

Molenaar, J. W. (2022). *The Business Benefits of Adopting Sustainability Standards: A Review of Literature and Evidence in the Last Six Years*, London: ISEAL, Available at: https://www.evidensia.eco/resources/2336/, accessed 23 November 2022.

Muller, C., Vermeulen, W. J. V. and Glasbergen, P. (2012). Pushing or sharing as value-driven strategies for societal change in global supply chains: two case studies in the British-South African fresh fruit supply chain, *Business Strategy and the Environment* 21(2): 127–140.

Nadvi, K. and Wältring, F. (2002). *Making sense of Global Standards*, INEF Report 58/2002, Duisburg: INEF-University of Duisburg, Available at: https://www.academia.edu/78780751/Making_sense_of_global_standards.

Nelson, V. J., Haggar, J., Martin, A., Donovan, J., Borasino, E., Hasyim, W., Mhando, N., Senga, M., Mgumia, J., Quintanar Guadarrama, E., Kendar, Z., Valdez, J. and Morales, D. (2016). *Fairtrade Coffee: A Study to Assess the Impact of Fairtrade for Coffee Smallholders and Producer Organisations in Indonesia, Mexico, Peru and Tanzania*, Chatham: Natural Resources Institute, University of Greenwich, Available at: https://files.fairtrade.net/publications/2016_NRI_CoffeeEvaluation.pdf.

Nelson, V. J. and Phillips, D. (2018). Sector, landscape or rural transformations? Exploring the limits and potential of agricultural sustainability initiatives through a cocoa case study, *Business Strategy and the Environment* 27(2): 252–262.

Nelson, V. J. and Pound, B. (2009). A review of the impact of Fairtrade over the last ten years, Available at: https://www.fairtrademaxhavelaar.ch/fileadmin/CH/Was_ist_Fairtrade_/Wirkungsstudien/2009_Nelson_Pound_A_review_of_the_impact_of_Fairtrade.pdf.

Nelson, V. J. and Tallontire, A. (2014). Battlefields of ideas: changing narratives and power dynamics in private standards in global agricultural value chains, *Agriculture and Human Values* 31(3): 481–497, https://doi.org/10.1007/s10460-014-9512-8.

O'Rourke, D. (2003). Outsourcing regulation: analyzing nongovernmental systems of labor standards and monitoring, *Policy Studies Journal* 31(1): 1–29.

Oakley, R. and Buckland, I. (2004). What if business as usual won't work? In: Henriques, A. and Richardson, J. (Eds) *The Triple Bottom Line, Does It All Add Up?*, London: Earthscan Publications.

Oxfam (2010). Think big go small: adapting business models to incorporate smallholders into supply chains, Available at: https://policy-practice.oxfam.org/resources/think-big-go-small-adapting-business-models-to-incorporate-smallholders-into-su-114051/, accessed 23 November 2022.

Oya, C., Schaefer, F. and Skalidou, D. (2018). The effectiveness of agricultural certification in developing countries: a systematic review, *World Development* 112: 282–312.

Ponte, S. (2019). *Business Power and Sustainability in a World of Global Value Chains*, London: Zed Book Company.

Reinecke, J., Manning, S. and Von Hagen, O. (2012). The emergence of a standards market: multiplicity of sustainability standards in the global coffee industry, *Organization Studies* 33(5–6): 791–814.

Riisgaard, L. (2011). Towards more stringent sustainability standards? Trends in the cut flower industry, *Review of African Political Economy* 38(129): 435–453.

Ruben, R. (2017). Impact assessment of commodity standards: towards inclusive value chains, *Enterprise Development and Microfinance* 28(1–2): 82–97. https://doi.org/10.3362/1755-1986.16-00020.

Rueda, X., Garrett, R. and Lambin, E. F. (2017). Corporate investments in supply chain sustainability: selecting instruments in the agri-food industry, *Journal of Cleaner Production* 142: 2480–2492.

Said-Allsopp, M. and Tallontire, A. (2015). Pathways to empowerment?: dynamics of women's participation in global value chains, *Journal of Cleaner Production* 107: 114–121.

Sainsbury's (2019). Sainsburys fairly-traded update, Available at: https://about.sainsburys.co.uk/sustainability/plan-for-better/our-stories/2019/230819-sainsburys-fairly-traded-update-2019, accessed 1 May 2019.

Schaltegger, S., Lüdeke-Freund, F. L. and Hansen, E. G. (2012). Business cases for sustainability: the role of business model innovation for corporate sustainability, *International Journal of Innovation and Sustainable Development* 6(2): 95–119.

Seville, D., Buxton, A. and Vorley, B. (2010). Under what conditions are value chains effective tools for pro-poor development?, Available at: http://www.value-chains.org/dyn/bds/docs/806/IIED_ValueChainsEffectiveToolsforPro-PoorDevelop.pdf.

Sheehy, B. (2015). Defining CSR: problems and solutions, *Journal of Business Ethics* 131(3): 625–648. https://doi.org/10.1007/s10551-014-2281-x.

Smith, S. (2010). For love or money? Fairtrade business models in the UK supermarket sector, *Journal of Business Ethics* 92(S2): 257–266.

Smith, T., Paavola, J. and Holmes, G. (2019). Corporate reporting and conservation realities: understanding differences in what businesses say and do regarding biodiversity, *Environmental Policy and Governance* 29(1): 3–13.

Sustainable Agriculture Initiative Platform (2022). Transformation through collaboration: annual report 2021, Available at: https://saiplatform.org/wp-content/uploads/2022/05/sai_2022_report_ok_1sept_double_page.pdf.

Tallontire, A. M., Opondo, M. and Nelson, V. (2014). Contingent spaces for smallholder participation in GlobalGAP: insights from K enyan horticulture value chains, *The Geographical Journal* 180(4): 353–364.

Tallontire, A. M., Opondo, M., Nelson, V. and Martin, A. (2011). Beyond the vertical? Using value chains and governance as a framework to analyse private standards initiatives in agri-food chains, *Agriculture and Human Values* 28(3): 427–441.

Thorlakson, T., Hainmueller, J. and Lambin, E. F. (2018). Improving environmental practices in agricultural supply chains: the role of company-led standards, *Global Environmental Change* 48: 32–42.

Turley, L., Bermudez, S., Voora, V., Wilkings, A. and Elder, S. (2022). *Voluntary Sustainability Standards in East Africa*, International Institute for Sustainable Development, Available at: https://www.iisd.org/system/files/2022-09/voluntary-sustainability-standards-east-africa.pdf.

Ugarte, S., D'Hollander, D., Tregurtha, N. and Haase, N. (2017). *SDGs Mean Business: How Credible Standards Can Help Companies Deliver the 2030 Agenda*, Gland: WWF Verlagsgesellschaft World Wide Fund for Nature, Available at: https://www.isealalliance.org/get-involved/resources/wwfiseal-report-sdgs-mean-business-how-credible-standards-can-help-companies, accessed 23 November 2022.

UNFSS (2022). 5th flagship report: voluntary sustainability standards (VSS), sustainability agenda and developing countries: opportunities and challenges, united nations forum on sustainability standards, Available at: https://unfss.org/home/flagship-publication/, accessed 23 November 2022.

Utting, P. (2005). Corporate responsibility and the movement of business, *Development in Practice* 15(3–4): 375–388.

Valters, C. (2015). Theories of change time for a radical approach to learning in development, Available at: https://cdn.odi.org/media/documents/9835.pdf.

Vermeulen, W. J. V. (2015). Self-governance for sustainable global supply chains: can it deliver the impacts needed?, *Business Strategy and the Environment* 24(2): 73–85.

WWF and ISEAL Alliance (2017). SDGs mean business: how credible standards can help companies deliver the 2030 agenda, Available at: https://www.isealalliance.org/get-involved/resources/wwfiseal-report-sdgs-mean-business-how-credible-standards-can-help-companie, accessed 9 November 2022.

Chapter 6

Carbon farming, regenerative agriculture, and environmental sustainability: practices and standards

Louise Manning, University of Lincoln, UK; and Rounaq Nayak, Bournemouth University, UK

1 Introduction

The carbon cycle has always been a crucial element of both ecosystem function and agri-food production. It involves dynamic interactions between sources (processes that release carbon into the atmosphere), sinks (processes that remove carbon from the atmosphere), and reservoirs such as the ocean, atmosphere, geosphere and the biosphere (including soil or forests) that can both store and release carbon (Table 1) (BS EN ISO 14064:1, 2019). Carbon flux refers to the amount of carbon exchanged between carbon stocks, for example, the flux from the atmosphere to the biosphere in which algae, fungi and plants sequester carbon dioxide (CO_2) from the atmosphere, via photosynthesis to convert into organic molecules. These then travel through food chains before being released back into the atmosphere as CO_2 via respiration.

However, the rise in atmospheric CO_2 from around 280 parts per million (ppm) in 1780 (before the industrial revolution) to 424 ppm in 2023 indicates how far human (anthropogenic) activity has disrupted this carbon cycle (NOAA, 2023). A particular problem is the production of greenhouse gases

http://dx.doi.org/10.19103/AS.2024.0133.11

Table 1 Carbon reservoirs within the carbon cycle

Carbon reservoir	Description	Restoration time	Examples
Atmosphere	Atmospheric carbon is the carbon stored in the atmosphere.		The atmospheric carbon is almost completely CO_2 at about 0.04% by volume. CH_4 also present.
Biosphere	Biocarbon: the carbon stored in the biosphere.	Natural ecosystems (decades to millennia); semi-natural ecosystems (years to centuries); agricultural systems (annual to decades).	Living and dead biomass and organic and inorganic carbon in soils in terrestrial and marine ecosystems. Carbon stored in natural ecosystems (peatlands, savannah and forests) and managed natural ecosystems (food, fibre and biofuel).
Geosphere	Geocarbon: the carbon stored in the geosphere.	Geological time.	Fossil fuel reserves (oil, coal, gas), sedimentary rocks e.g. limestone, methane clathrates and marine sediments.
Ocean	Carbon stored or dissolved in the ocean, e.g., via photosynthesis with algae and plankton.		Blue carbon is the carbon stock within the ocean and the coastal ecosystem, e.g. sea grasses, mangroves, salt marshes and wetlands.

Source: Adapted from: Rackley, 2010; Ajani et al., 2013; NOS, nd.

(GHGs). BS EN ISO 14064:2019 (p. 2) defines a GHG as a 'gaseous constituent of the atmosphere, both natural and anthropogenic, that absorbs and emits radiation at specific wavelengths within the spectrum of infrared radiation emitted by the Earth's surface, the atmosphere and clouds.' Atmospheric levels of three GHGs and their contribution to global warming are of particular concern (NOAA, 2023):

- Carbon dioxide (CO_2) (424 ppm in 2023);
- Methane (CH_4) (1.9 ppm in 2023); and
- Nitrous oxide (N_2O) (0.34 ppm in 2023).

The agricultural, forestry and other land (AFOLU) sector is estimated to contribute between 13% and 21% of global greenhouse gas emissions (GHGEs) in the period 2010–2019 (Nabuurs et al., 2022), with 10–14% directly coming from agriculture (Paustian et al., 2016). GHGs from agriculture have different sources:

- Carbon dioxide (CO_2): produced during manufacture of synthetic fertilisers and by the disturbance of soil organic matter (which acts as a carbon sink) due to agricultural practices such as tillage. A wider issue is land use change, e.g. deforestation (since trees store carbon) to create land to grow crops, e.g. for livestock feed (which increases atmospheric CO_2 and other emissions);
- Methane (CH_4): produced by fermentation during the digestion of food by cows and emitted when cows burp. Methane is also produced by manure; and
- Nitrous oxide (N_2O): produced by manure, and through the use of nitrogen-based fertilisers.

The United Nations Food and Agriculture Organisation (UN FAO), for example, has stated that the 2018 global GHGEs due to agricultural activity (both from farming and related land use/land use change) represented 9.3 billion tonnes of carbon dioxide equivalent (CO_2eq). Carbon dioxide equivalent (CO_2e or CO_2eq) is the standard unit used to ensure an equivalent measure for different GHGs based on their global warming potential (GWP). GWP is a measurement of the impact of each GHGE gas in terms of its capacity to absorb heat in the atmosphere (Vallero, 2019; BS EN ISO 14064-1:2019, 2019). The lower residency period of a gas like CH_4 is taken into account by GWP100 or the 100-year GWP calculation. Key GHGs considered in the GWP100 calculation are (IPCC, 2007a,b; Greenhouse Gas Protocol, 2015):

- CO_2 (lifetime 5–200 years depending on carbon cycling; GWP100 of 1);
- CH_4 (lifetime 12 years; GWP100 of 25); and
- N_2O (lifetime 114 years; GWP100 of 298).

Whilst GWP100 is a useful metric, it has been argued that it does not fully account for the difference between stock gas and flow gases. CO_2 is an atmospheric stock gas that contributes to the overall atmospheric stock of GHGs over its lifetime (IPCC, 2007a; Ajani et al., 2013). As a flow gas, methane is short-lived in the atmosphere, and some argue that the impact of CH_4 using GWP100 is over-estimated and suggest the GWP* method would better account for the production and breakdown of CH_4 (Rocha, 2022).

2 Carbon farming

One way agriculture can contribute to reducing GHGEs is what has been called 'carbon farming' (Dumbrell et al., 2016). Carbon farming can be viewed as a whole farm approach that optimises carbon capture (sequestration) by implementing practices that improve the rate of carbon dioxide (CO_2) removal from the atmosphere and storing CO_2 in plant material and/or soil organic matter (CCI, 2023). Other definitions (e.g. In the European Parliament publication 'Organic farming. Making agriculture fit for 2030') expand the definition of carbon farming to include both practices for sequestering atmospheric carbon and/or for reducing the production of GHGEs at the farm level (McDonald et al., 2021). Carbon farming is defined here as:

> *'the management of both land and livestock, all pools of carbon in soils, materials and vegetation, plus fluxes of CO_2, CH_4 and N_2O... [including] carbon removal (sequestration and permanent storage of carbon in soils and biomass), avoided emissions (preventing the loss of already stored carbon) and emissions reductions i.e. reductions of GHGEs below current levels of farm emissions'.*

Decarbonisation of farming activities can be achieved by either reducing the intensity of GHGEs per unit of output or, alternatively, or in combination, offsetting GHGEs associated with the unit of output by equivalent GHGE capture through another activity (Rogelj et al., 2018; Pearson et al., 2023). The aim of carbon farming is to achieve 'net zero' in the balance of carbon emissions/capture whilst maintaining food production, therefore contributing to more resilient food production systems (Pozza and Field, 2020). Practices to optimise soil organic carbon (SOC) can improve overall soil health, which, in turn, not only promotes sequestration but also improves crop production and climate resilience in agriculture. These practices can also enhance the delivery of ecosystem services associated with healthy soils, which both support crop production and deliver broader societal benefits. These services include reducing the risk of flooding through increasing soil water-holding capacity, reducing water-run-off from farms which reduces pollution risk (e.g. leaching of nutrients and crop protection products into watercourses), as well as reducing soil erosion risk, improving water and air quality and promoting biodiversity (including promotion of beneficial species such as pollinators and predators of insect pests) (McDonald et al., 2021; Green, 2022; CCI, 2023). These services are seen as characteristics of a healthy ecosystem namely: 'the dynamic complex of plant, animal and micro-organism communities and their non-living environment, interacting as a functional entity' (BS EN ISO 14008:2020, p. 2). Ensuring agroecosystems function effectively is increasingly seen as the foundation for a more climate-resilient and sustainable agriculture.

As this suggests, a key concept underpinning carbon farming is resilience, which has been defined as the ability of an organisation or system to 'anticipate, prepare for and respond and adapt to incremental change and sudden disruptions to both survive and prosper' (BS 8001:2017, p. 16). Natural ecosystems and agri-food systems are dynamic, and this means that, to be resilient, they need to be able to buffer and adapt to shocks (Manning, 2023). Carbon farming offers the potential to ensure resilience in agri-food systems by both mitigating the contribution of agriculture to climate change and allowing

Table 2 Carbon farming practices

Activities	Impact on farming business and society
Agroforestry (including silvopasture, woodland pasture, maintaining hedgerows, riparian restoration). Energy production (agrovoltaics, biofuel production) Fibre production (hemp, flax, cotton, wool, leather) Resource appraisal and optimisation strategies (e.g. fertiliser, crop protection products). Integrated crop and livestock systems Crop production practices (application of biochar, green manure or mulch to land; biomass planting, compost application, crop residue retention, maintaining crop cover; direct drilling, inter-cropping with perennial pastures, multi-species cover cropping, no-till cropping practices, planting tree belts, retaining/incorporating crop residue/stubble after harvest) Livestock production practices (forage planting, maintaining appropriate grazing systems, feed additives to reduce methane production) Nature-based practices (using natural vegetation to reduce salinity and increase land sequestration capacity)	Agroecosystem improvement Carbon sequestration Increased agricultural yields (food security) Increased soil moisture retention Improved efficiency of input use Improve nutrient storage capacity Improved profitability Improved soil biodiversity and earthworm activity Improved soil organic carbon (SOC) Improved soil structure Improved water quality Nutrient cycling Reduced point source and diffuse pollution Reduced soil erosion by water and wind Resilience Restoration (biodiversity) Restoration (soil)

Source: Adapted from: Lal, 2015; Dumbrell et al., 2016; CCI, 2023.

it to adapt to its impacts (Basso, 2022; Talebian et al., 2023). One way carbon farming can promote resilience is by protecting natural capital, defined as the 'stock of renewable and non-renewable natural resources (e.g. plants, animals, air, water, soils and minerals) that combine to yield a flow of benefits to people' (BS 8001:2017, p. 15).

Table 2 highlights a range of carbon farming practices designed to achieve these objectives by both reducing GHGEs and enhancing carbon sequestration, as well as contributing more broadly to agroecosystem function. They can be grouped into categories such as agroforestry, (bio)energy production, fibre production, resource appraisal and optimisation strategies (including integrated crop and livestock systems), specific crop production practices, livestock production practices and nature-based practices (Dumbrell et al., 2016; Higashi et al., 2023). An example is agroforestry, defined by Paudel et al. (2022), as 'a land management system that integrates trees, agriculture crops and animal farming in order to provide a diverse range of ecosystem services'. Agroforestry systems can have a range of elements including: (Nair et al., 2021)

- structural components: spatial (boundary, strip, hedges and mixed systems) or temporal (overlapping, sequential, simultaneous); and

- functional components: productive (food, fodder, fuel and other) or protective (windbreak, shade, shelterbelts) with environmental (reducing diffuse pollution) and social (animal welfare) benefits and soil conservation and soil restoration).

The multifunctional benefits of agroforestry systems include enhanced carbon sequestration (since trees and hedges can store significant amounts of carbon), provision of a sustainable biofuel source, improved soil health, reduced nutrient loss and soil erosion and improved biodiversity. All of these contribute to a more self-sufficient and resilient farming system with lower emissions and higher carbon sequestration potential (Jose et al., 2019). Another example of a more integrated and renewable approach is agrovoltaics, which combines the integration of photovoltaics (solar PV panels) on land with crop or livestock production (John and Mahto, 2021; Jain et al., 2021).

There are, however, significant barriers to the adoption of carbon farming practices. There is a wealth of research on the mechanisms of soil carbon sequestration and on how particular agronomic practices can enhance sequestration, improve soil health and deliver other ecosystem services (Rumpel, 2022; Horwath, 2022). However, this has not always been translated into specific, practical guidance tailored to individual farms. In addition to a lack of knowledge in some cases, farmers are concerned about the costs of transitioning to new practices, and there remain concerns about how far these practices affect yields (Sharma et al., 2021). A particular concern is how to effectively measure the effectiveness of carbon farming practices. The European Commission (2023) has highlighted that robust and transparent regulatory frameworks need to be in place for certifying carbon removal from atmospheric stocks and carbon sequestration.

3 Measuring carbon balances

A carbon balance can be defined as the net carbon balance (including all gains and losses) on an individual farm. Calculating a carbon balance requires a measurement, reporting and verification (MRV) system capable of measuring both losses (through GHGEs) and gains (through carbon sequestration). Measuring the range of GHGEs and quantifying levels of carbon sequestration in a farm setting requires a range of different and complex instrumental methods. This means that using on-farm instrumental measurement techniques to determine a carbon balance is too complex, costly and labour-intensive for routine use (see Ch 15 by Matthias Kuhnert in this volume). Instead, instrumental techniques are primarily used for research purposes to better understand the processes involved and, in particular, to provide data for developing models capable of representing these processes and predicting gains and losses in different farming settings.

Although it cannot replicate the level of accuracy of instrumental measurement, a more practical approach to measuring a carbon balance is the use of models. Current commercial carbon calculator tools available for use by individual farms are therefore based on models. Modelling first requires a clear definition of the farm and its activities. While the operational area of the farm is well-defined, products that are imported and exported from the farm need to be considered. GHGEs can be separated into different categories:

- Imported products;
- Emissions produced on the farm; and
- Products that capture carbon from the atmosphere (e.g. harvested crops) which are then exported from the farm.

Imported goods, e.g. fertiliser or seedlings, have a carbon footprint from their production. Exported goods like harvested crops and manure will have either a positive or negative impact on the farm's carbon balance. On-field application of manure can increase the SOC content, but long storage will result in emissions, and the sale of manure and resultant crops will reduce the carbon balance. Exported products will be also transported, stored and packed, which is difficult to predict at the farm stage. Therefore, farm-specific emissions are often calculated as emissions up to the farmgate, which includes all emissions until the goods leave the farm, but no more.

Modelling requires inputting farm-specific data to allow a model to calculate losses (through GHGEs) and gains (through carbon sequestration) to produce a carbon balance for that farm. For practical reasons, carbon calculation tools ask farmers only to input some of the data required (typically the kind of data farmers already use to run their operations). Typical data required includes:

- cropping data: crop types, acreage, inputs and yields;
- livestock data: herd or flock size, type of feed and manure management; and
- Fuel, energy and water use.

Other data may be derived from other sources (e.g. climate and weather data from meteorological databases). Where data are not available, models assume a value which may be based on generic data from relevant, available studies (e.g. field trials conducted by academic institutions) (Kuhnert et al., 2022). Examples of model-based carbon calculation tools include:

- AgreCalc (https://www.agrecalc.com);
- Cool Farm Tool (https://coolfarmtool.org/);

- Farm Carbon Calculator (https://calculator.farmcarbontoolkit.org.uk/); and
- Sandy (https://www.trinityagtech.com/carbon-audit).

An example of a US carbon calculation tool is COMET-Farm (http://comet-farm.com/) (Paustian et al., 2017). These tools have all evolved separately, have different assumptions and model structures, and therefore, can report different results. Some tools focus on the carbon footprints of particular products, whilst others prioritise an overall carbon balance, which can, for example, be used for carbon trading purposes.

A key issue is how to gauge and compare the robustness of these tools (Paustian et al., 2019; Smith et al., 2020). While several MRV protocols exist, a standardised protocol with clear guidelines is still lacking (Kuhnert et al., 2022). Some of the existing protocols are discussed here. In 2006, the Intergovernmental Panel on Climate Change (IPCC) introduced a tier system to categorise model complexity and accuracy, which was then updated in 2019 (https://www.ipcc.ch/site/assets/uploads/2019/12/19R_V0_01_Overview.pdf). The system has three categories (Whittaker et al., 2013):

- Tier 1 indicates the lowest level of complexity.
- Tier 2 indicates medium complexity.
- Tier 3 indicates complex process-based models.

Tier 1 models describe a linear relation between an activity (e.g. fertiliser application on the field) and related estimated GHGEs. This provides a simplified, generic approach to estimate GHGEs, based on a range of scientific data establishing a general relationship between these variables. However, accuracy may be reduced because the model and the data on which it is based do not reflect particular local conditions. In contrast, Tier 2 models use country-specific values that provide a more accurate estimate for farms in the target country. Tier 3 models use country-specific data combined with more complex, process-based models and detailed inventory measurements. Broadly speaking, increased complexity in the modelling is associated with potentially more accuracy, but usually is also linked with an increased data requirement. Tiers 2 and 3 are sometimes characterised as 'higher tier' methods and are generally considered more accurate, providing that sufficient data is available at farm level. These trade-offs can, for example, be seen in the Cool Farm Tool (Hillier et al., 2011) where some data are input by the farmer, some may be derived from an underlying database but, for some values, default data are used. The SOC component of the Cool Farm Tool, for example, uses a refined Tier 1 approach based on IPCC 2019 Guidelines. AgreCalc also uses a Tier 1 approach. With the Trinity AgTech Sandy system, carbon sequestration is calculated by a blended model using soil test analysis and modelling to Tier 2 IPCC 2019 Standards.

The COMET-Farm Tool uses a Tier 3 approach (Paustian et al., 2017). As with other carbon calculation tools, standard farm-specific information (e.g. field management activities, number of animals) are added by the farmer (Oldfield et al., 2022). This tool also allows users to mark the target area (farm location) by using an interactive map, which all required environmental data is then extracted from, from example, using meteorological databases to provide model input data. GHGEs estimates for crops are calculated using the process-based DayCent dynamic model which follows official United States Department of Agriculture (USDA) GHG inventory guidelines for entity-scale reporting (Eve et al., 2014).

The United Kingdom (UK) British Standards Institute (BSI) has a relevant standard: PAS 2050:2011 Specification for the assessment of the life cycle greenhouse gas emissions of goods and services. PAS 2050 was developed to ensure the consistent application of a methodology to quantify product carbon footprints. It is important to note here that the methodology is related to a specific product, produced in a prescribed way. A more recent addition is: PAS 2060:2014 Specification for the demonstration of carbon neutrality. Other internationally recognised carbon footprint standards include:

- ISO 14067:2018 Greenhouse gases – carbon footprint of products – Requirements and guidelines for quantification; and
- Greenhouse Gas (GHG) Protocol Product Life Cycle Accounting and Reporting Standard (2011).

Some carbon foot printing calculators for farming practices are certified to ISO 14067 and can integrate with PAS 2050 and the GHG Protocol e.g. Sandy and Agrecalc which are certified to PAS 2050.

A recent UK review has compared the performance of a number of carbon calculation tools, particularly the degree of divergence in calculation of farm GHGEs between different tools (DEFRA, 2023; see also overview in: https://www.cla.org.uk/news/why-do-farm-carbon-calculators-give-such-different-results/). The study reviewed use of the following calculators across a range of different farm types:

- AgreCalc;
- Cool Farm Tool;
- Farm Carbon Calculator (Farm Carbon Toolkit);
- Sandy;
- The Farm Carbon Calculator (Solagro); and
- Carbon Footprint Tool (Eggbase Ltd.).

The study concluded that all the tools produced useful output to support on-farm GHGEs reduction. However, significant differences in measured emissions (differences between highest and lowest emissions in percentage terms) were found across different farm types, for example:

- poultry farms: up to 350%;
- dairy farms: up to 143%;
- cereals: up to 175% (measurement of carbon stock changes); and
- general cropping: up to 116% (measurement of carbon stock changes).

The study examined areas of divergence such as methods for accounting for crop residues, calculating enteric emissions (with differences between calculators in accounting for livestock numbers, activity and feed intake), including embedded GHGEs in fertilisers (i.e. those resulting from their manufacture) and how to account for manure application quantities and management practices). More generally, calculators used different system boundaries and emissions factors (which estimate average emissions for a given activity). Since most are based on Tier 1 models, calculators also struggled to account for the effects of potential mitigation practices in enabling farmers to make decisions about which mitigation measures to focus on.

Initiatives to harmonise methodologies (e.g. in setting appropriate system boundaries) include the Science-Based Targets Initiative (SBTi) Forest, Land and Agriculture (FLAG) Target-setting guidelines for emissions reduction (https://sciencebasedtargets.org/sectors/forest-land-and-agriculture) and the ongoing work of the Greenhouse Gas Emissions Protocol (https://ghgprotocol.org/). Other potential improvements to improve robustness and consistency of results include asking farmers to provide more detailed (primary) data where the greatest GHGEs are likely (e.g. enteric emissions) and agreeing a common set of robust databases for calculating embedded GHGEs consistently.

4 Product carbon footprints and life cycle assessment

Product carbon standards are often based on establishing a product carbon footprint and conducting a life cycle assessment (LCA). The carbon footprint of a product is described as the 'sum of GHG emissions and GHG removal in a product system expressed as CO_2e and based on an LCA using the single impact category of climate change' (ISO 14067:2018). A product carbon footprint is only valid for the specific production system that has been assessed and may vary significantly for another business, even if it produces the same product. BS EN ISO 14044:2006 (p. 1) defines a life cycle as the 'consecutive and interlinked stages, from raw material acquisition or generation [of the product] from natural resources to final disposition'. An LCA has been stated as the 'compilation and evaluation of the inputs, outputs and the potential environmental and/or social impacts of a product or service system through its life cycle' (BS 8001:2017). LCA methodology comes under the United Kingdom (UK) British Standards Institute (BSI) Standard PAS 2050:2011 Specification for the assessment of the life cycle greenhouse gas emissions of goods and services. Product carbon footprints can be (Carbon Trust, 2023):

- a *cradle-to-gate inventory* i.e. a partial LCA from material acquisition to the boundary of the organisation's gate; or
- *a cradle-to-grave inventory* which is an LCA with the boundary from material acquisition to the end of product life.

As suggested, some LCA models therefore limit their scope up to the farm gate, allowing farms to assess their carbon footprint (de Vries et al., 2020).

LCA tools simulate emissions based on indicators, generic factors and simple equations. Similar to carbon calculator farm tools, there are global or local factors or empirical equations that are developed based on local datasets in the background of the methodology that translate management data into emissions. Environmental aspects are considered, but in a simplistic way. LCA models depend on the quality of the data used in their development. System conditions, system boundaries, the constraints on the datasets that are available, and other assumptions that are made, influence how a LCA tool is applied and how representative the results are for actual practices, especially where production systems within a sector vary widely. This means too that, for LCA tools developed in a particular country, there is a requirement to prove their applicability in other geographic areas. Examples of country-specific LCA models for potential use on farms are:

- the ANCA tool (The Netherlands) (de Vries et al., 2020);
- CAP2ER (France);
- Carbon Audit (Ireland); and
- Bovid-CO2 (Spain) (O'Brien et al., 2020).

An overview of models focussed on livestock emissions is provided by Del Prado et al. (2017).

However, the lack of harmonisation of LCA tools means that the results derived can vary and care should be taken when making direct comparisons between the outputs of different tools. This is a challenge when considering the development of carbon markets and carbon trading.

5 Carbon trading

A carbon credit has been defined as 'any tradeable certificate or permit presenting the right to emit one tonne of carbon dioxide or the mass of another greenhouse gas with a carbon dioxide equivalent (tCO_2e equivalent) to one tonne of carbon dioxide' (ISO/TR 14069:2013, p. 3). A carbon credit is an asset that is tradeable between parties. One party (e.g. a farmer) guarantees to undertake a practice to reduce GHGEs to offset the emissions of another party. An organisation can purchase a carbon credit representing 1 tonne of CO_2eq to offset the 1 tonne of CO_2eq that they are responsible for emitting. This is known as offsetting, which has been defined by PAS 2060: 2014 (p. 3) as a 'discrete reduction in GHGEs not arising from the defined subject (the entity being analysed), made available in the form of a carbon credit .... and used to counteract emissions from the defined subject'. This can be compared with insetting, which is associated with an organisation aiming to counterbalance its GHGEs within its own operations rather than seeking another entity to offset their GHGEs which would be offsetting (Green Finance Institute, 2023).

Carbon codes play a crucial role in certifying and quantifying the carbon sequestered through changes in farming practices. By complying with specific carbon codes, carbon farmers can generate tradable carbon credits, which can then be bought by companies or organisations seeking to offset their own carbon emissions (Sechi et al., 2021). Dales (1968) introduced the concept of a 'cap and trade' system, allowing governments to issue a certain number of permits (carbon credits). Under this type of system, the number of permits issued is lower than the amount of emissions industry is currently producing, thereby creating a market for trading of carbon emission permits. Organisations that are unable to comply with the number of permits allocated to them are either incentivised to reduce emissions or trade with other organisations that have carbon emissions below the permitted allowance. While the 1997 Kyoto protocol established mechanisms for carbon trading, with the first commitment for trading being made for the period between 2008 and 2012, the first voluntary trade was carried out in the late 1980s (Hepburn, 2007).

As of 2023, China is the world's largest emissions trading market and covers around 4800 million metric tons of CO_2eq, with the potential for covering approximately 7800 million metric tons (Tiseo, 2023). Since 2012, multiple carbon codes and carbon trading schemes have been implemented. Developments such as the Paris Agreement (United Nations Climate Change, 2016) have supported expanding the mechanisms introduced by the Kyoto Protocol. Two UK examples of carbon codes supporting carbon trading are the

- Woodland Carbon Code (WCC) (https://woodlandcarboncode.org.uk); and
- Peatland Code (PCC) (https://www.iucn-uk-peatlandprogramme.org/peatland-code/introduction-peatland-code).

Under the WCC, a reduction in GHGEs is achieved by planting and growing trees, while under the PCC, the reduction is achieved by restoring depleted peatland (Wallace, 2021). The WCC links elements of sustainable forestry management with carbon accounting methods. All WCC projects or schemes must be initially validated by an independent validation/verification body accredited by the UK Accreditation Service (UKAS), which assesses compliance with the requirements of the WCC (Woodland Carbon Code, 2022). UKAS accredits a range of schemes, including emissions trading schemes (ETS), requiring that emissions reports must be verified by an organisation accredited to ISO/IEC 14065: 2020 (UKAS, 2023). Organisations accredited to ISO14065 are expected to conform to standards such as PAS 2050 for carbon footprinting. The WCC carbon sequestration scheme and UKAS is also trialling accreditation under PAS 2060 (UKAS, 2023). In the case of the PCC, the validation/verification body must also have UKAS accreditation to ISO 14060-3 or ISO/IEC 14065: 2020 (Peatland Carbon Code, 2023).

The UK Land Carbon Registry (2023) lists and provides public access to transactions associated with all WCC and PCC projects, levels of carbon sequestration and how these have been converted, e.g. to woodland carbon units (WCU) equivalent to 1 tCO_2eq. These credits can then be used in offsetting or carbon neutrality claims conforming with PAS 2060 (Woodland Carbon Code, 2022). Projects under the WCC scheme are reviewed 5 years after the establishment of new woodland, where pending issuance units can be converted to WCU, and then at least every 10 years after the start date (Woodland Carbon Code, 2022). The woodland-related carbon credits are issued based on the degree of carbon sequestration related to the baseline audit in the biomass within the woodland area and the soil (carbon pools) and the additional sequestration achieved within the project boundary, taking into account potential carbon leakage (e.g. deforestation or intensification of agricultural activities elsewhere). As of June 2023, there were 1931 projects in the UK registered with the Land Carbon Registry and an area of 72.6 thousand hectares with 23.8 million tCO_2eq, of which 9.4 million tonnes has currently been validated and 2.0 million tonnes verified at 5 years (Woodland Carbon Code, 2023). One option for farmers to be part of a carbon trading scheme is to switch from conventional farming practices to regenerative practices and then demonstrate and quantify the impact on the carbon balance on their farm. This is discussed in more detail below.

6 Regenerative agriculture

6.1 Overview

One suggested approach to reducing GHGEs is by adopting what has become known as 'regenerative agriculture'. There is growing interest and investment in regenerative agriculture amongst major multinational corporations (MNCs) in the agri-food supply chain. The term regenerative agriculture is not new and has been used for over half a century. Starting with Gabel in the 1970s, to Robert Rodale and the Rodale institute in the US, the term fell out of fashion in the 1990s in favour of a focus on organic farming and agroecology, until a focus on regenerative practices increased again from around 2015 (Giller et al., 2021). More recently, regenerative agriculture has been promoted strongly by civil society, non-governmental organisations (NGOs) and MNCs within the food supply chain (Giller et al., 2021). As with organic agricultural practices, regenerative agriculture is associated with low external input-based agriculture i.e. limiting artificial nutrient use and use of crop protection products. Regenerative agriculture focuses strongly on promoting biological systems and integrating farming practices such as nitrogen use and nitrogen cycling, crop rotations, integrated pest management (IPM) and effective weed management (Francis et al., 1986), and decarbonising through reducing GHGEs and increasing carbon sequestration.

Rodale developed the concept of 'organic regenerative' farming in the 1970s as a system which moved beyond reducing reliance on chemicals to a focus on regenerating and using natural resources as the basis for more self-sufficient, sustainable farming systems (https://rodaleinstitute.org/). Rodale set out seven principles to achieve this:

- Soil health
  - Nutrient retention in the soil;
  - Improve soil structure;
  - More perennial and other plants with vigorous root systems; and
  - More surface cover from plants.
- (Bio)Diversity
  - Increase in diversity of plant species.
- Input (resource) management
  - Eliminating chemical fertilisers and pesticides; and
  - Disrupting patterns of weed and pest infestation.

The theoretical foundation for this regenerative approach lies in the term 'agroecology', which has already been used in this chapter and is defined as an integrated approach applying ecological (as well as social) principles to the design and management of food and agricultural systems (https://www.fao.org

/agroecology/home/en/). Agriculture, in this context, is viewed as an adapted form of an ecosystem in which an understanding of fundamental biogeochemical flows/cycles and functional interactions between organisms within this type of ecosystem can used to optimise its ability to deliver a range of ecosystem services, including food production. Examples of practices promoting agroecosystem function include optimising the physical, chemical and biological health of soils to promote crop growth and health (reducing the need for external inputs such as fertilisers and pesticides), promoting biodiversity (e.g. to encourage beneficial species such as pollinators or predators of crop pests) and ensuring diversity in agricultural landscapes (to optimise functional interactions between species) (Gliessman, 2007; Nicholls et al., 2016; Wezel et al., 2020).

An overview of the range of practices associated with regenerative agriculture has e.g. been provided by the European Academies Science Advisory Council (EASAC, 2022) (Table 3). The practices have been grouped by practice: soil health, biodiversity and input (resource) management and diversification through carbon farming.

Table 3 Summary of the range of regenerative agricultural practices

Practice: soil health and input (resource) management
Cover crops
Crop rotations
No/zero, minimum or reduced tillage
Organic amendments (incl. biochar) rather than artificial fertiliser
Retaining crop residues on soil surface
Practice: (bio)diversity
Avoid insecticides, fungicides and herbicides (input (resource) management)
Buffer strips (e.g. hedgerows, trees/shrubs)
Field margins/borders
Flower strips
Herbal leys
Intercropping, relay cropping
Landscape mosaics in time and space
Natural/semi-natural habitats
Practice: diversification through carbon farming
Agroforestry, e.g. silvo-pasture
Conversion of arable land to grassland
Crop–livestock systems
Perennial crops
Woodland (e.g. for timber or biofuel)

Source: Adapted from: EASAC (2022).

6.2 Issues of definition

Issues of definition are important. To quote one review: 'Despite widespread interest in regenerative agriculture, no legal or regulatory definition of the term "regenerative agriculture" exists nor has a widely accepted definition emerged in common usage'. (Newton et al., 2020), a situation which has continued with competing definitions stemming from different origins and lineages (Brunyee and Semple, 2022; Hurley et al., 2023). The study by Newton et al. identifies two contrasting approaches to defining regenerative agriculture:

- Outcome-based definitions, for example, optimising soil health, increasing biodiversity and improving carbon/nutrient/water cycles; and
- Process-based definitions, for example, reducing/eliminating tillage, use of cover crops and rotations and implementing integrated crop–livestock systems.

This and other research also highlight a further lack of consensus within these two approaches on what 'outcomes' or 'processes' should come within a definition of regenerative agriculture and its associated practices and different practitioner websites focus on different aspects (Tables 4 and 5).

A recent, and potentially influential, initiative to define regenerative agriculture has been made by the Sustainable Agriculture Initiative Platform (SAI Platform) (2023). The SAI Platform represents over 30 leading food processing and other agri-food businesses, including Cargill, Coca-Cola, Danone, Diageo, Kellogg's, Kraft Heinz, Nestlé and Unilever. The SAI Platform has defined regenerative agriculture as: 'an outcome-based farming approach that protects and improves soil health, biodiversity, climate and water resources while supporting farming business development'. It has outlined a framework with four impact areas: soil health, water, biodiversity and climate. The framework sets out material criteria that determine how these four impact areas are affected, and key outcomes which optimise impact area function (e.g. soil health, water resources, biodiversity and mitigation of climate change) and a set of key principles and practices to achieve these outcomes (Fig. 1). This framework is designed to allow farmers in different contexts to assess the specific environmental and production risks they face, identify outcomes to mitigate those risks, prioritise farm management actions and establish metrics for measuring performance.

Given the relatively recent resurgence of interest in regenerative agriculture and, in particular, the lack of consensus on an agreed definition and scope, there are currently no internationally accepted certification schemes for regenerative agriculture. One US example is the Regenerative Organic Alliance, which has established a certification scheme for regenerative agriculture: Regenerative Organic Certified™ (ROC™). ROC is supported by the Rodale Institute, widely

Table 4 Outcome-based definitions of regenerative agriculture: is there a consensus?

Regenerative agricultural practice	Newton et al. (2020) (%)	Schreefel et al. (2020) (%)
Outcome: soil health		
Improve soil health	86	61
Increase carbon sequestration	64	46
Outcome: biodiversity		
Increase biodiversity	46	
Promote ecosystem services	32	25
Additional outcomes		
Improve water quality	46	28
Improve community wellbeing	41	
Improve food nutritional quality	32	46
Increase farm profitability	27	45
Protect/increase yields	23	
Protect/improve productivity	23	

Source: Percentage of 25 practitioner websites analysed in: Newton et al. (2020); percentage of 28 peer-reviewed articles reviewed in: Schreefel et al. (2020).

Table 5 Process-based definitions of regenerative agriculture: is there a consensus?

Regenerative agricultural practice	Newton et al. (2020) (%)	Schreefel et al. (2020) (%)
Practice: soil health		
Reduce tillage (no/minimal/conservation tillage)	41	25
Use cover crops	36	
Use crop rotations	32	21
Use compost, mulch, green manure or crop residues (organic amendments)	27	18
Protect/cover the soil	18	
Practice: biodiversity		
Use no or few external inputs	32	43
Use no synthetic fertilisers	23	
Use ecological principles	14	
Use crop plant diversity (including intercropping)	14	
Use organic methods	14	
Practice: diversification through carbon farming		
Integrated crop–livestock farming	41	36
Incorporate perennials and trees	27	18

Source: Percentage of 25 practitioner websites analysed in: Newton et al. (2020); percentage of 28 peer-reviewed articles reviewed in: Schreefel et al. (2020).

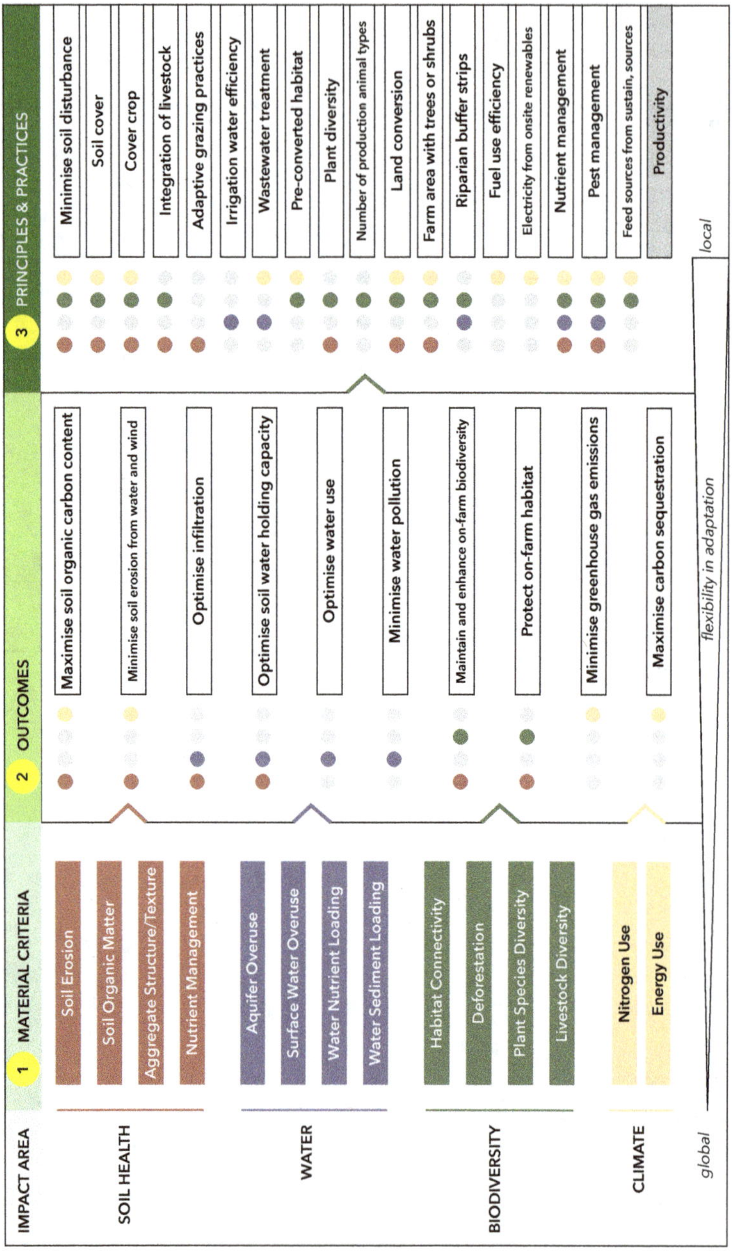

Figure 1 SAI Platform's Regenerating global framework for regeneration agriculture (© SAI Platform, 2023; reproduced with permission)

seen as having pioneered the concept of regenerative organic agriculture in 1970s. The scheme includes a pillar specifically covering regenerative organic agricultural practices: Soil Health and a Land Management pillar with the following elements (https://regenorganic.org/resources/):

- Base requirements;
- Regenerative practices;
- Compost, manure and fertilisers;
- Facilities;
- Use of prohibited substances; and
- Measurement.

Table 6 provides a detailed list of certification requirements for this standard. It is important here to differentiate between the certification of a product in terms of it being produced to regenerative practice standards on farm (or delivering specific outcomes due to the adoption of regenerative practices) and the certification of a farm producing a range of commodities and demonstrating that their outputs have been 'grown' regeneratively. Questions arise such as how much of the ingredients within a food product need to be certified as 'regenerative' for the product to be labelled with claims associated with regenerative practices. Another question is what practices have an impact in terms of carbon farming specifically? These practices have been highlighted in bold in the table.

6.3 The future of regenerative agriculture?

It is widely accepted that agriculture is at a crossroads and needs to transition, especially with another two billion people predicted to be on the planet by 2050. In the short term, the flexibility of the framing of regenerative agricultural practices may be a strength, in that it allows a wider range of actors to choose the particular set of regenerative practices that suit the needs of the farmed environment and farming system. In the long term, if not effectively implemented and monitored, this very diversity could dilute its impact. Research is key to establishing a clear foundation of authoritative knowledge on how particular regenerative agricultural practices work, how to measure their effects and optimise their implementation. A summary of the challenges and opportunities associated with regenerative farming practices has been drawn together in Table 7.

The next section considers how supply chain standards for environmentally sustainable and regenerative agricultural practices are being developed. A number of approaches have already been considered in this chapter. To achieve disruptive changes in land management and agricultural practices on the scale required to deliver climate action and biodiversity action, scaling up of multiple approaches is essential.

Table 6 Regenerative Organic CertifiedTM (ROCTM): Soil Health and Land Management Pillar

1 Base requirements	**Practice description**
1.1 Existing certifications	Proof of existing organic certification.
1.2 **Regenerative Organic System Plan (ROSP)**	Required items include tillage action plan; soil lab test and in-field soil test results; record of native flora and fauna on farm and key performance indicators.
1.3 Water	Operations conserve and restore natural bodies of water, wetland riparian areas, and associated habitats. Irrigation rights are legal.
1.4 **Deforestation**	Operations have not cleared primary, untouched forest or old-growth secondary forests nor converted wetlands, peatlands or protected grasslands into agricultural production.
1.5 Extractive practices	Fracking, mining and other extractive practices including initial exploration shall not be conducted.
2 Regenerative practices	**Practice description**
2.1 **Vegetative cover**	Operations shall aim to cover land with living vegetative cover year-round. Otherwise, maintenance of either dead/rolled/lightly incorporated crop residues or mulch is required when field preparation or planting is not underway.
2.2 **Crop rotations**	Operations shall demonstrate use of crop rotations or perennial systems. Annual crop rotations should include a green manure.
2.3 **Minimal soil disturbance**	The goal of ROC is to reduce soil disturbance as much as possible. To achieve this goal, all tillage events shall be documented with an action plan to reduce intensity of tillage over time. Soil disturbance shall only occur when necessary to accomplish one or more of these objectives: • Incorporate crop residues and/or green manures into soil to feed microorganisms • Control weeds • Prepare seed bed/planting • Break up compacted soil • Develop drainage Shallow cultivation tools must be used whenever possible.
2.4 Rotational grazing	Intensive grazing in which animals are used in high concentrations for brief periods of time. Pastures divided into paddocks with animals moved regularly
2.5 Soilless practices	Aquaponics, hydroponics and other soilless practices are not eligible.
2.6 Control of invasive species	Operation does not allow hunting. Fishing or gathering or rare or endangered species, nor do they cause harm to the species' habitat.

2.7 Regenerative practices	Operations engage in additional regenerative practices (beyond statutory/certification organic requirements). Practices are documented with timestamped photos, preferably linked to GPS coordinates. Examples include: • Forest stand improvement • Agroforestry (including silvopasture, riparian buffers and shelterbelts) • Herbaceous wind barriers and field borders • Perennial planting • Reduction of off-farm inputs and recycling of on-farm biomass • Anaerobic digestion • Mulching • Forage and biomass planting • Integrated crop–livestock systems • Pollinator habitats, insectary strips or wildlife habitat • Water conservation and wetland restoration
3 Compost, manure and fertilisers	**Practice Description**
3.1 General	Aim for self-sufficiency in compost and manures. If importing soil amendments to replace nutrients exported off-farm, they should be from regionally-available wastes and by-products.
3.2 Crop nutrient demand	Imported fertilisers are discouraged unless crop nutrient demand dictates. A comprehensive nutrient management plan or equivalent must be provided when nutrient use exceeds recommended rate.
4 Facilities	**Practice Description**
4.1 Wastewater	Operation does not directly discharge untreated wastewater generated on farm into natural waterways or soil.
4.2 Waste	Operation does not illegally dump, bury or burn waste
5 Use of prohibited substances	**Practice Description**
5.1 Synthetic chemicals	Synthetic pesticides are not permitted; restricted use of organically-approved pesticides that are toxic to pollinators.
5.2 Quick-acting fertiliser	Operation does not use quick-acting soluble fertilisers.
5.3 Genetically-modified inputs and cloning	Operation does not use any genetically-modified additives or processing aids, including emerging technologies such as gene editing.

(*Continued*)

Tables 6 *(Continued)*

6 Measurement	**Practice Description**
6.1 Soil health lab test	Tests should be carried out in accordance with procedures set out by an accredited organisation with expertise in analysing soil health; use ROC Soil Sampling Guidelines.
6.2 Soil health in-field test	Producers follow ROC Soil Health In-Field Test Instructions.
6.3 Computer models	Operators utilise computer-based modelling tools (e.g. Cool Farm Tool) to determine annual GHG emissions and sequestrations. Operators document practices in computer-based models, review annually with auditor and submit documentation to a certifying body.

7 Supply chain standards for environmental sustainability

Key players in the development of integrated farm management standards and adoption of environmentally sustainable farming methods can be considered at the national scale, for example, the UK's LEAFMarque (https://leaf.eco/) which is applied in many countries and at a global scale through standards such as GLOBALG.A.P. The latter, in particular, has a potentially very significant role to play in defining sustainable farming practices in the future and is therefore discussed in more detail below. There is also a specific chapter on LEAFMarque in the book for further information.

GLOBALG.A.P. is a group of farm assurance certification schemes advancing safer and more sustainable farming practices (https://www.globalgap.org/). In just over 20 years, GLOBALG.A.P. has grown to provide some of the most respected and internationally recognised standards supporting the global trade of farmed products, particularly fruits and vegetables, flowers and aquaculture products. The origins of GLOBALG.A.P. go back to 1997 when a group of leading European retailers established EUREPGAP: the Euro-Retailer Produce Work Group Good Agricultural Practice Initiative. Their goal was to harmonise their own internal quality assurance (QA) standards and procedures for suppliers to create a common, comprehensive and independent set of standards for good agricultural practice (GAP) in fresh produce production. FAO defines GAP as a 'collection of principles to apply for on-farm production and post-production processes, resulting in safe and healthy food and non-food agricultural products, while taking into account economic, social and environmental responsibility' (https://www.fao.org/3/i6677e/i6677e.pdf).
In 2001, a dedicated company, FoodPlus GmbH, was set up to manage the EurepGAP standard, with the first individual farm certificate awarded to a UK producer and the first group certificate awarded to banana growers working for Fyffes Group in Belize and Costa Rica. In 2007, EurepGAP became

Table 7 Regenerative agriculture: challenges and opportunities

Challenges	Opportunities
Lack of a clear definition and agreed set of practices increases the risk of 'greenwashing'	Interest in regenerative agriculture reflects a general recognition that 'conventional' agricultural practices are unsustainable and need to change and there should be flexible ways to achieve the outcomes.
The wide range of potential practices and the many ways they can be combined in different contexts can complicate analysis of their effects and benefits. Their value will be enterprise, farm and location specific.	Many key actors in the agri-food supply chain are promoting a shift towards regenerative practices: including governments, major food processors and retailers, financial institutions, NGOs and a growing number of farmer groups/ networks
There is significant debate about how beneficial regenerative agricultural practices are in areas such as improving yields and their contribution to carbon sequestration.	There is a growing body of research on how particular regenerative practices work, how to measure their effects and how to deploy them to best effect in particular contexts. The permanency of carbon sequestration is an important area for research.
Regenerative practices face ongoing challenges such as: developing the skills to manage more complex farming systems, improving the efficacy/ reliability of non-chemical methods of control compared to synthetic alternatives, continued reliance of no-till farmers on herbicides such as glyphosate and closing nutrient loops.	Research is complemented by a growing number of farms successfully implementing regenerative agricultural practices
Yield reductions can impact on food availability within the local and global food market.	Gaining greater understanding of how to feed a growing population and deliver to climate and biodiversity goals is essential. This will be driven by research and also practice-based innovation.
Carbon farming is a term being used, but it has yet to gain significant traction in the agricultural sector. The validity and ability to verify the performance measures developed will be key to gaining wider societal engagement with the trade-offs that come with carbon farming.	Developing regenerative agricultural practices that are valid and can be verified is essential to prevent greenwashing, but also to develop carbon markets and 'willingness-to-pay' programmes focused on consumers.

GLOBALG.A.P. in recognition of its growing international reach. GLOBALG.A.P. standards have now been adopted by over 200 000 certified producers in more than 135 countries farming over 4 million hectares. Standards cover 700 different plant and aquatic species. GLOBALG.A.P. works with over 400 partners, including many leading global food manufacturers and retailers, across Europe, Africa, Asia, North and South America.

GLOBALG.A.P.'s flagship standard is the Integrated Farm Assurance (IFA) standard (latest version IFA V6) (https://www.globalgap.org/what-we-offer/solutions/). There are different versions of the IFA Standard covering fruits and vegetables, flowers and ornamentals and aquaculture products, as well as other products such as combinable crops, hops, tea, compound feed and plant propagation material. IFA is a holistic standard with core topics including food safety; traceability; worker health, safety and welfare; responsible use of fertilisers and biostimulants; water management; responsible use of plant protection products and integrated pest management; energy efficiency; waste management and protection of biodiversity and habitat. Within the IFA, there are over 200 Principles and Criteria (P&C):

- Principles are documented key requirements and outcomes.
- Criteria are specific practices and supporting evidence to demonstrate compliance that key requirements and outcomes are being met.

Demonstration of producer compliance with these P&C is essential to achieving IFA certification. The producers are audited annually by accredited and independent, third-party certification bodies.

GLOBALG.A.P. has also developed a number of add-on or 'bolt-on' standards that are focused particularly on environmental sustainability and social responsibility, including the following:

- GLOBALG.A.P. Farm Sustainability Assessment (GGFSA): allows producers to combine achieving IFA certification with compliance with the Farm Sustainability Assessment (FSA) programme of the Sustainable Agriculture Initiative (SAI) (https://www.globalgap.org/what-we-offer/solutions/ggfsa/);
- Impact-Driven Approach to Sustainability (IDA): facilitates digital recording of input consumption data and sustainability metrics (https://www.globalgap.org/what-we-offer/solutions/ida/);
- BioDiversity: covering topics such as soil management, land restoration measures and integrated pest management (https://www.globalgap.org/what-we-offer/solutions/biodiversity/).;
- Sustainable Programme for Irrigation and Groundwater Use (SPRING): covering topics such as water use and extraction rates, legal compliance and protection of water sources as well as watershed management (https://www.globalgap.org/what-we-offer/solutions/spring/); and
- GLOBALG.A.P.'s Risk Assessment on Social Practice (GRASP): covers topics such as labour and human rights, worker representation, protection of children and young workers (https://www.globalgap.org/what-we-offer/solutions/grasp/).

Whilst GLOBALG.A.P.'s flagship IFA embeds elements of economic, environmental and social sustainability, there is recognition of the need for a dedicated environmental sustainability standard. An Environmental Sustainability Working Group (ESWG) was established in 2022 and tasked with developing a new standard, the Environmental Sustainability Solution (ESS), for launch in 2025 (https://www.globalgap.org/about/standard-setting/environmental-sustainability-solution/). The EWSG was asked to develop a comprehensive approach to achieving environmental sustainability for the farming industry in way that avoided duplication with existing GLOBALG.A.P. standards, harmonised as far as possible with existing sustainability programmes, and which supported the cost-efficient adoption of responsible farming practices. The current timetable for development is as follows:

- 2023 Quarter 2: concept development; drafting of principles and criteria;
- 2023 Quarter 3: first public consultation for draft normative documents;
- 2023 Quarter 4: review and integration of feedback in collaboration with focus groups and technical committees;
- 2024 Quarter 2: second round of public consultation and field trials to test performance and suitability;
- 2024 Quarter 3: review and finalisation of the normative documents; final round of public consultation;
- 2024 Quarter 4: recommendation of the final normative documents by the technical committees and adoption by the GLOBALG.A.P. Advisory Board;
- 2024 Quarter 1: prelaunch preparations including proofreading, training development and integration into the GLOBALG.A.P. IT systems; and
- 2025 Quarter 2: launch of the Environmental Sustainability Solution (ESS).

In its current iteration the draft modules for the ESS are the following:

- Module 1 Enabling the transition;
- Module 2 Transparency;
- Module 3 Horizon scanning;
- Module 4 Water;
- Module 5 Biodiversity;
- Module 6 Greenhouse gases;
- Module 7 Footprinting;
- Module 8 Agricultural regeneration; and
- Module 9 Food loss, food surplus and food waste.

Table 8 draws together illustrative examples of proposed Principles and Criteria (P&Cs) for modules 4–8 in the draft ESS released for initial public consultation

Table 8 Environmental Sustainability Solution (ESS): examples of draft modules and proposed principles and criteria (P&Cs) within modules

ESS module	Examples of proposed P&Cs
4 Water	Crop irrigation plan matching irrigation programmes to crop/soil requirements. Watershed governance Establishment of buffer zones. Water risk assessment Erosion risk management Water management programme Measurement of water consumption Wastewater management
5 Biodiversity	Biodiversity governance Establishing a biodiversity baseline Identifying protected/endangered species Supporting regional biodiversity initiatives Biodiversity action plan (e.g. landscape elements/corridors such as trees, hedges, ponds; providing nesting, feed and water sources; adapting management to meet wildlife needs (e.g. minimising disruption during breeding); Soil nutrient management plan Optimising fertiliser use (e.g. establishing soil nutrient content; recording applications; proper storage; measures to avoid run-off) Control fertiliser/pesticide use Integrated pest management (IPM) plan Management of invasive species Monitoring (e.g. mapping landscape elements, aerial photos) Peat management Use of crop rotations Maintaining soil structure and avoiding erosion
6 Greenhouse gases	Identifying different GHG emission sources and baseline levels Action plan to reduce emissions (incl. energy use, fertiliser use, avoiding crop losses, optimising soil health) Promoting soil carbon sequestration/carbon farming (e.g. crop residue management, cover crops, more diverse rotations, min. or no-till, ecosystem restoration) Reducing nutrient losses from fertiliser application Monitoring e.g. regular measurement of total GHGs emitted or captured; net balance
7 Footprinting	Establishing footprint type (e.g. carbon or water) Baseline data Data and methodology for calculating footprint

ESS module	Examples of proposed P&Cs
8 Agricultural regeneration	Evidence no forests have been cleared since 2020 (e.g. maps, aerial photos) Forest restoration programmes Managing fire risks Soil characterisation study Soil health risk assessment (e.g. risks of compaction, erosion, nutrient loss) Soil testing (e.g. penetrometer to test physical structure, chemical analysis, measuring health of soil microbiome; sampling regime) Soil health management plan (incl. measuring and optimising soil carbon) Cropping action plan (e.g. objectives, methods (e.g. rotations, cover crops), choice of crops/varieties, planting and management) (e.g. to avoid biotic/abiotic threats, including documenting planting, harvesting etc.) Crop nutrition action plan (e.g. use of synthetic/organic fertilisers, soil amendments (e.g. lime), organic amendments (e.g. manure, compost, use of biostimulants; timing, application method and timing) Minimising nitrogen losses Tillage action plan (e.g. type and timing of tillage operation (e.g. depth)) Risk assessment for organic fertiliser use Manure, slurry, compost management Grazing action plan Use of precision agriculture techniques (e.g. GIS, proximal and remote sensing, soil mapping, variable rate application, yield and other performance monitoring Measuring performance (e.g. crop yields)
9 Food losses	Identify unharvested or harvested but unutilised crop Identify waste, co-products Re-direct, repurpose or recycle Plan to reduce surplus/waste

in autumn 2023. Although subject to further review, revision and further rounds of public consultation, this draft (with a focus on process-based measures of performance) provides a particularly detailed and rigorous framework from a globally respected body in defining what sustainable farming will need to look like in the future, as well as the specific measures needed to achieve it.

8 Conclusion

The carbon cycle has always been a crucial element of agri-food production. However, the rise in atmospheric CO_2 from around 280 parts per million (ppm) in 1780 pre-industrial revolution to 417 ppm in 2022 has led to a focus on whether carbon farming to increase nature-based CO_2 sequestration can provide opportunities to halt and then reverse this trend. The aim of this chapter has been to consider carbon farming in terms of standards, practices, reporting and verification within the context of wider agricultural practices. Carbon codes, such as the UK WCC and PCC, are emerging as integral components of carbon trading in the UK and play a critical role in quantifying and certifying the carbon sequestered through land use change. They allow carbon farmers to generate tradable carbon credits, contributing to efforts in mitigating climate change while also ensuring high environmental integrity and engaging with local communities in a transparent and collaborative manner. An approach to embedding carbon farming within farming systems is the development and adoption of standards for regenerative agriculture and environmental sustainability. The chapter ends with a review of the development of these standards, with both third-party and company-specific examples. As stated earlier, agricultural practices need to transition to address climate action and biodiversity action whilst still supplying food when there is predicted to be an additional 2 billion people on the planet by 2050. The flexibility of the framing of carbon farming and regenerative agricultural practices may be a strength, allowing farming businesses to choose the particular set of regenerative practices that suit their needs. However, in the long term, if there is weak implementation, poor transparency or limited impact, this will lead to concerns over effectiveness and potential greenwashing. Initiatives by organisations such as the SAI Platform and GLOBALG.A.P. will be key in providing the clarity and rigour needed for a real impact. The development of new farming practices by academics and farmers working together is also key to establishing a clear foundation of authoritative knowledge on how particular regenerative agricultural practices work, how to measure their effects and optimise their implementation.

9 Where to look for further information

- *A Soil Owner's Manual* by Jon Stika.
- *Dirt to Soil* by Gabe Brown.
- *Growing a Revolution: Bringing Back to Life* by David Montgomery.
- *For the Love of Soil* by Nicole Masters.
- *Holistic Management* by Allan Savory.
- *The Carbon Farming Solution: A Global Toolkit of Perennial Crops and Regenerative Agriculture Practices for Climate Change Mitigation and Food Security* by Eric Toensmeier
- https://chelseagreen.co.uk/book/the-carbon-farming-solution/.
- *Understanding and Fostering Soil Carbon Sequestration* by Cornelia Rumpel (Ed.)
- https://bdspublishing.com/_webedit/uploaded-files/All%20Files/Open%20Access/9781801463256.pdf.

10 References

Ajani, J. I., Keith, H., Blakers, M., Mackey, B. G. and King, H. P. (2013). Comprehensive carbon stock and flow accounting: a national framework to support climate change mitigation policy. *Ecological Economics* 89, 61–72.

Basso, B. (2022). Techno-diversity for carbon farming and climate resilience. *Italian Journal of Agronomy* 17(4), 1–3 .

Brunyee, J. and Semple, D. (2022). Regenerative agriculture: context, definitions and drivers of change. Report 1: the GREAT project. Available at: https://www.greatglos.co.uk/post/great-report-1-regenerative-agriculture-context-definitions-and-drivers-of-change.

BS 8001:2017. *Framework for Implementing the Principles of the Circular Economy in Organisations. Guide*. The British Standards Institution, London.

BS EN ISO (International Standards Organisation) 14008:2020. *Monetary Valuation of Environmental Impacts and Related Environmental Aspects*. The British Standards Institution, London.

BS EN ISO (International Standards Organisation) 14044:2006+A1:2020. *Environmental Management. Life Cycle Assessment. Requirements and Guidelines*. The British Standards Institution, London.

BS EN ISO (International Standards Organisation) 14064-1:2019 (2019). *Greenhouse Gases. Part 1: Specification with Guidance at the Organisational Level for Quantification and Reporting of Greenhouse Gas Emissions and Removals*. The British Standards Institution, London.

BS EN ISO 14067:2018. *Greenhouse Gases: Carbon Footprint of Products: Requirements and Guidelines for Quantification*. The British Standards Institution, London.

Carbon Trust (2023). Product carbon footprint label. Available at: https://www.carbontrust.com/what-we-do/assurance-and-labelling/product-carbon-footprint-label (Accessed 3 June 2023).

CCI (Carbon Cycle Institute) (2023). Carbon farming. Available at: https://www.carboncycle.org/what-is-carbon-farming/#:~:text=Carbon%20Farming%20is%20a%20whole,and%2For%20soil%20organic%20matter (Accessed 29 April 2023).

Dales, J. (1968). *Pollution, Property and Prices*. University of Toronto Press, Toronto.

DEFRA (2023). *Harmonisation of Carbon Accounting Tools for Agriculture*. Department for Environment, Food and Rural Affairs (DEFRA). Project no.: SCF0129. Available at: https://sciencesearch.defra.gov.uk/ProjectDetails?ProjectId=20967.

Nod efinitive (2017). Farm models/tools to help farmers reducing emissions. EIP-AGRI Focus Group Reducing livestock emissions from Cattle farming. Available at: https://ec.europa.eu/eip/agriculture/sites/default/files/eip-agri_fg_livestock_emissions_final_report_2017_en.pdf.

Dumbrell, N. P., Kragt, M. E. and Gibson, F. L. (2016). What carbon farming activities are farmers likely to adopt? A best–worst scaling survey. *Land Use Policy* 54, 29–37.

EC (European Commission) (2023). Carbon farming. Available at: https://climate.ec.europa.eu/eu-action/sustainable-carbon-cycles/carbon-farming_en (Accessed 29 April 2023).

Environment and Forestry Directorate (2018). Engaging communities in decisions relating to land: guidance. Available at: https://www.gov.scot/publications/guidance-engaging-communities-decisions-relating-land/ (Accessed 4 August 2023).

European Academies Science Advisory Council (EASAC) (2022). *Regenerative Agriculture in Europe: A Critical Analysis of Contributions to European Union Farm to Fork and Biodiversity Practices. EASAC Policy Report 44*. Available at: www.easac.eu.

Eve, M., Pape, D., Flugge, M., Steele, R., Man, D., Riley-Gilbert, M. and Biggar, S. (2014). *Quantifying Greenhouse Gas Fluxes in Agriculture and Forestry: Methods for Entity-Scale Inventory*. USDA Technical Bulletin 1939. United States Department of Agriculture, Washington, DC.

Francis, C. A., Harwood, R. R. and Parr, J. F. (1986). The potential for regenerative agriculture in the developing world. *American Journal of Alternative Agriculture* 1(2), 65–74.

Giller, K. E., Hijbeek, R., Andersson, J. A. and Sumberg, J. (2021). Regenerative agriculture: an agronomic perspective. *Outlook on Agriculture* 50(1), 13–25.

Gliessman, S. (2007). *Agroecology: The Ecology of Sustainable Food Systems*. CRC Press, New York.

Green, A. (2022). *Carbon Farming*. Available at: https://www.cla.org.uk/news/carbon-farming/ (Accessed 27 May 2023).

Green Finance Institute (2023). Financing a farming transition. Available at: https://www.greenfinanceinstitute.co.uk/gfihive/farming-finance/ (Accessed 5 August 2023).

The Greenhouse Gas Protocol (2011). *Product Live Cycle Accounting and Reporting Standard*. Available at: https://ghgprotocol.org/product-standard. (Accessed 27 May 2023). ISBN 9781569737736.

The Greenhouse Gas Protocol: A Corporate Accounting and Reporting Standard (2015). (Revised Edition). WRI/WBCSD (World Resources Institute/World Business Council for Sustainable Development). Available at: https://ghgprotocol.org/sites/default/files/standards/ghg-protocol-revised.pdf (Accessed 29 May 2023).

Hepburn, C. (2007). Carbon trading: a review of the Kyoto mechanisms. *Annual Review of Environment and Resources* 32(1), 375–393.

Higashi, S. Y., de Queiroz Caleman, S. M., Manning, L., De Aguiar, L. K. and Monteiro, G. F. A. (2023). Factors influencing Brazilian sugar and ethanol refineries' failure. *RAUSP Management Journal* 58, 55–68 .

Hurley, P. D., Rose, D. C., Burgess, P. J. and Staley, J. T. (2023). *Barriers and Enablers to Uptake of Agroecological and Regenerative Practices. Project Report SCF0321 for Defra*. Cranfield University and Centre for Ecology and Hydrology, UK.

International Labour Organization (2015). *Guidelines for a Just Transition Towards Environmentally Sustainable Economies and Societies for All*. Switzerland. Available at: https://www.ilo.org/wcmsp5/groups/public/@ed_emp/@emp_ent/documents/publication/wcms_432859.pdf (Accessed 5 August 2023).

IPCC (Intergovernmental Panel on Climate Change) Fourth Assessment Report: Climate Change (2007a). *Section 2.10.2 Direct Global Warming Potentials*. Available at: https://archive.ipcc.ch/publications_and_data/ar4/wg1/en/ch2s2-10-2.html (Accessed 27 May 2023).

IPCC (Intergovernmental Panel on Climate Change) Fourth Assessment Report: Climate Change (2007b). *Working Group I: the Scientific Basis. Working Group I: the Scientific Basis*. Available at: https://archive.ipcc.ch/ipccreports/tar/wg1/016.htm (Accessed 27 May 2023).

ISO 14065:2020 (2020). General principles and requirements for bodies validating and verifying environmental information. Available at: https://www.iso.org/standard/74257.html (Accessed 4 August 2023).

Jain, P., Raina, G., Sinha, S., Malik, P. and Mathur, S. (2021). Agrovoltaics: step towards sustainable energy-food combination. *Bioresource Technology Reports* 15, 100766.

John, R. S. and Mahto, R. V. (2021, June). Agrovoltaics farming design and simulation. In: 2021 IEEE 48th Photovoltaic Specialists Conference (PVSC) (pp. 2625–2629). IEEE.

Jose, S., Walter, D. and Mohan Kumar, B. (2019). Ecological considerations in sustainable silvopasture design and management. *Agroforestry Systems* 93(1), 317–331.

Kuhnert, M., Vetter, S. H. and Smith, P. (2022). Measuring and monitoring soil carbon sequestration. In: Rumpel, C. (Ed.). *Understanding and Fostering Soil Carbon Sequestration*. Burleigh Dodds Science Publishing, Cambridge.

Lal, R. (2015). Sequestering carbon and increasing productivity by conservation agriculture. *Journal of Soil and Water Conservation* 70(3), 55A-62A.

Manning, L. (2023). Being resilient in challenging times in food supply chains. *Engineering Proceedings* 40(1), 1.

McDonald, H., Frelih-Larsen, A., Lóránt, A., Duin, L., Andersen, S. P., Costa, G. and Bradley, H. (2021). Carbon farming. Making agriculture fit for 2030. European Parliament. Policy Department for Economic, Scientific and Quality of Life Policies Directorate-General for Internal Policies. PE 695.482. November 2021. Available at: https://www.europarl.europa.eu/RegData/etudes/STUD/2021/695482/IPOL_STU(2021)695482_EN.pdf (Accessed 27 May 2023).

Nabuurs, G.-J., Mrabet, R., Abu Hatab, A., Bustamante, M., Clark, H., Havlík, P., House, J., Mbow, C., Ninan, K. N., Popp, A., Roe, S., Sohngen, B. and Towprayoon, S. (2022). Agriculture, Forestry and Other Land Uses (AFOLU).. In: Shukla, P. R., Skea, J., Slade, R., Al Khourdajie, A., van Diemen, R., McCollum, D., Pathak, M., Some, S., Vyas, P., Fradera, R., Belkacemi, M., Hasija, A., Lisboa, G., Luz, S., Malley, J. (Eds).*IPCC, 2022: Climate Change 2022: Mitigation of Climate Change. Contribution of Working Group III to the Sixth Assessment Report of the Intergovernmental Panel on Climate Change.*. Cambridge University Press, Cambridge and New York.

Nair, P. R., Kumar, B. M. and Nair, V. D. (2021). *An Introduction to Agroforestry: Four Decades of Scientific Developments*. Springer, Cham.

Newton, P., Civita, N., Frankel-Goldwater, L., Bartel, K. and Johns, C. (2020). What is regenerative agriculture? A review of scholar and practitioner definitions based on processes and outcomes. *Frontiers in Sustainable Food Systems* 4, 577723.

Nicholls, C., Altieri, M. A. and Vazquez, L. (2016). Agroecology: principles for the conversion and redesign of farming systems. *Journal of Ecosystem and Ecography* 5, 010.

NOAA (National Oceanic and Atmospheric Administration) (2023). *Greenhouse Gases Continue to Increase Rapidly in 2022*. Available at: https://www.noaa.gov/news-release/greenhouse-gases-continued-to-increase-rapidly-in-2022 (Accessed 3 June 2023).

NOS (National Ocean Service) (n.d.). *What Is Coastal Blue Carbon?* Available at: https://oceanservice.noaa.gov/ecosystems/coastal-blue-carbon/ (Accessed 27 May 2023).

PAS 2050:2011. *Specification for the Assessment of the Life Cycle Greenhouse Gas Emissions of Goods and Services*. The British Standards Institution, London.

PAS 2060:2014. *Specification for the Demonstration of Carbon Neutrality*. The British Standards Institution, London.

Paudel, S., Baral, H., Rojario, A., Bhatta, K. P. and Artati, Y. (2022). Agroforestry: opportunities and challenges in Timor-Leste. *Forests* 13(1), 41.

Paustian, K., et al. (2017). Field and farm-scale assessment of soil greenhouse gas emission mitigation using COMET-Farm. In: Delgado, J. A., et al. (Eds). *Precision Conservation: Geospatial Techniques for Agricultural and Natural Resources Conservation, Agronomy Monographs* (Vol. 59). ASA/CSSA/SSSA, Madison, WI.

Paustian, K., Collier, S., Baldock, J., Burgess, R., Creque, J., DeLonge, M., Dungait, J., Ellert, B., Frank, S., Goddard, T., Govaerts, B., Grundy, M., Henning, M., Izaurralde, R. C., Madaras, M., McConkey, B., Porzig, E., Rice, C., Searle, R., Seavy, N., Skalsky, R., Mulhern, W. and Jahn, M. (2019). Quantifying carbon for agricultural soil management: from the current status toward a global soil information system. *Carbon Management* 10(6), 567–587.

PD ISO/TS 14074:2022. Environmental management. Life cycle assessment. Principles, requirements and guidelines for normalization, weighting and interpretation.

Pearson, S., Brewer, S., Manning, L., Bidaut, L., Onoufriou, G., Durrant, A., Leontidis, G., Jabbour, C., Zisman, A., Parr, G., Frey, J. and Maull, R. (2023). Decarbonising our food systems: contextualising digitalisation for net zero. *Frontiers in Sustainable Food Systems* 7, 1094299.

Pozza, L. E. and Field, D. J. (2020). The science of soil security and food security. *Soil Security* 1, 100002.

Rackley, S. A. (2010). *Carbon Capture and Storage*. Butterworth-Heinemann. ISBN 978-1-85617-636-1. doi: 10.1016/C2009-0-19306-6.

Rocha, A. (2022). *GWP*: A Better Way of Measuring Methane and Hw It Impacts Global Temperatures*. Available at: https://clear.ucdavis.edu/explainers/gwp-star-better-way-measuring-methane-and-how-it-impacts-global-temperatures (Accessed 4 August 2023).

Rogelj, J., Shindell, D., Jiang, K., Fifita, S., Forster, P., Ginzburg, V., Handa, C., Kheshgi, H., Kobayashi, S., Kriegler, E. and Mundaca, L. (2018). Mitigation pathways compatible with 1.5 C in the context of sustainable development. In *Global Warming of 1.5 C* (pp. 93–174). Intergovernmental Panel on Climate Change. Available at: https://www.ipcc.ch/site/assets/uploads/sites/2/2019/05/SR15_Chapter2_Low_Res.pdf (Accessed 4 August 2023).

Schreefel, L., Schulte, R. P. O., de Boer, I. J. M., Schrijver, A. P. and van Zanten, H. H. E. (2020). Regenerative agriculture: soil is the base. *Global Food Security* 26, 100404.

Sechi, V., van Belle, J., Fritz, C., Tilak, A., Geurts, J., Roehrig, N., Nailon, P., Cartmell-Done, K., Liu, W., Smits, T., De Boever Niall ÓBrolcháin, M., Morley, T., Field, C., Kennedy, J., Johnson, S., Caporn, S., Halevy,C., Ryan, J., Eakin, M., Fernandez, F., Bain, C., Domegan, C., McGuinness, S., McCorry, M. and Crushell, P. (2021). *Towards a Carbon Credit & Blue Credit Scheme for Peatlands: Whitepaper*. Available at: https://repository.ubn.ru.nl/bitstream/handle/2066/245729/245729.pdf?sequence=1 (Accessed 4 August 2023)

Scottish Government (2022). Interim principles for responsible investment in natural capital. Scottish Government. Available at: https://www.gov.scot/publications/interim-principles-for-responsible-investment-in-natural-capital/.

Sharma, M., Kaushal, R., Kaushik, P. and Ramakrishna, S. (2021). Carbon farming: prospects and challenges. *Sustainability* 13(19), 11122.

Smith, P., Soussana, J., Angers, D., Schipper, L., Chenu, C., Rasse, D. P., Batjes, N. H., van Egmond, F., McNeill, S., Kuhnert, M., Arias-Navarro, C., Olesen, J. E., Chirinda, N., Fornara, D., Wollenberg, E., Álvaro-Fuentes, J., Sanz-Cobena, A. and Klumpp, K. (2020). How to measure, report and verify soil carbon change to realise the potential of soil carbon sequestration for greenhouse gas removal. *Global Change Biology* 26(1), 219–241.

Talebian, S. H., Jahanbakhsh, A. and Maroto-Valer, M. M. (2023). Carbon resilience calibration as a carbon management technology. *Frontiers in Energy Research* 11, 1089778.

The National Trust for Scotland and the UK National Committee of the IUCN Peatland Programme Peatland Code Version 2.0. Available at: https://www.iucn-uk-peatlandprogramme.org/sites/default/files/2023-03/Peatland%20Code%20V2%20-%20FINAL%20-%20WEB_2.pdf (Accessed 4 August 2023).

Tiseo, I. (2023). *Coverage of Largest Emissions Trading Systems (ETS) Worldwide as of 2022*. Available at: https://www.statista.com/statistics/1315109/largest-ets-markets-by-coverage/ (Accessed 5 August 2023).

United Kingdom Accreditation Service (UKAS) (2023). Accreditation: supporting net zero policies. Available at: https://www.ukas.com/resources/latest-news/accreditation-supporting-net-zero-policies/ (Accessed 4 August 2023).

United Kingdom (UK) Carbon Land Registry. Available at: https://www.woodlandcarboncode.org.uk/uk-land-carbon-registry (Accessed 4 August 2023).

United Kindgom (UK) Peatland Code Registry (2023). Available at: https://www.iucn-uk-peatlandprogramme.org/peatland-code/introduction-peatland-code/peatland-code-registry (Accessed 5 August 2023).

United Nations Climate Change (2016). *The Paris Agreement*. United Nations. Available at: https://unfccc.int/process-and-meetings/the-paris-agreement (Accessed 5 August 2023).

Vallero, D. A. (2019). *Air Pollution Calculations: Quantifying Pollutant Formation, Transport, Transformation, Fate and Risks*. Elsevier. https://doi.org/10.1016/C2017-0-02742-8 [Accessed 27 May 2023].

Wallace, K. (2021). How do you explain woodland carbon credits to your client? *Solicitors Journal* 164(10), 64–65.

Wang, X. and Lo, K. (2021). Just transition: a conceptual review. *Energy Research and Social Science* 82, 102291.

Wezel, A., Herren, B. G., Kerr, R. B., Barrios, E., Gonçalves, A. L. R. and Sinclair, F. (2020). Agroecological principles and elements and their implications for transitioning to sustainable food systems: a review. *Agronomy for Sustainable Development* 40(6), 40.

Whittaker, C., McManus, M. C. and Smith, P. (2013). A comparison of carbon accounting tools for arable crops in the United Kingdom. *Environmental Modelling and Software* 46, 228–239.

Woodland Carbon Code (2022). *Woodland Carbon Code: Requirements for Voluntary Carbon Sequestration Projects*. London. Available at: www.woodlandcarboncode.org.uk.

Woodland Carbon Code (2023). *WCC Statistics* (30 June 2023). Available at: https://www.woodlandcarboncode.org.uk/uk-land-carbon-registry/wcc-statistics#number (Accessed 4 August 2023).

Chapter 7

Biodiversity and nature credit markets: opportunities and challenges for landowners and land managers

Jo Treweek, eCountability Ltd, UK

1 Introduction

Biodiversity is in a crisis of decline and extinction at all geographic scales of consideration and efforts to tackle this have had demonstrably poor success, with successive national and global targets being missed. Despite this poor track record, international biodiversity agreements have responded with elevated levels of ambition. The UN Convention on Biological Diversity's (CBD) 'new deal for nature' claims to be based on 'ambitious but plausible' global goals of no net loss by 2030, and 10–15% biodiversity net gain (BNG) by 2050 (CBD, 2020). Global calls for action are reflected in new national policies and non-market mechanisms for safeguarding and enhancing biodiversity. In 2022, the UK was one of many other countries that committed to bring 30% of their territories (land and sea) under management for nature by 2030. The European Union (EU) Biodiversity Strategy for 2030[1] and its associated Action Plan include a new Nature Restoration law and ambitious targets to reverse the degradation of ecosystems. The Strategy is seen as a core part of the European Green Deal and 'green recovery' following the COVID-19

1 https://www.eea.europa.eu/policy-documents/eu-biodiversity-strategy-for-2030-1

http://dx.doi.org/10.19103/AS.2024.0133.12

pandemic and emphasises the need for better recognition of the values of biodiversity and natural capital both to society and to business, arguing that it constitutes enlightened self-interest to operate in a biodiversity-compatible manner.

The World Economic Forum's latest annual global risk report shows loss of biodiversity, ecosystem collapse and shortage of natural resources ranking in the top three longer-term risks (10 years plus) for private sector respondents, suggesting that businesses are generally well aware of the need to tackle biodiversity decline. Nevertheless, it is estimated that $78–91 billion per year is spent on actions to benefit biodiversity globally from public and private sources, compared with $500 billion on activities that are potentially harmful to biodiversity (OECD, 2020). There is a significant shortfall in the level of investment needed to reverse historic losses of biodiversity and restore it to viable and sustainable levels. While the public sector has a major part to play in filling the gap, it is generally acknowledged that finance from the private sector will be needed to reach the scale of investment needed for nature recovery (WCL, 2023). Hence the Kunming-Montreal Global Biodiversity Framework, the 'Paris Agreement for nature' (signed by almost 200 countries in 2022), called on signatories to come up with 'innovative' financing initiatives (such as biodiversity credits) to secure the levels of investment needed to recover nature.

What happens in the agriculture sector is particularly important because farming is a major cause of biodiversity loss worldwide and also depends fundamentally on biodiversity for its future viability. The global agrifood system emits one-third of all carbon emissions[2] is the leading source of methane emissions and uses around 70% of fresh water. This has prompted interest in 'climate-smart' agriculture, regenerative farming and a range of other more sustainable approaches to land use.

Agri-environment Schemes are an important and long-established mechanism for tackling biodiversity decline in the farmed landscape, which have been in use for decades and are essentially based on 'Payment for Ecosystem Services' (PES). In Europe and the UK, for example, farmers receive public money in return for undertaking food production using more sustainable methods or implementing management with specific environmental benefits. Despite some successes, however, it is evident that these schemes have failed to achieve sufficient reductions in rates of biodiversity decline in farmed landscapes (Box 1).

2 https://www.carbonbrief.org/food-systems-responsible-for-one-third-of-human-caused-emissions/

Box 1

In the UK, in common with many other countries, biodiversity declines have continued in the farmed landscape despite decades of agri-environment support for more sustainable and wildlife-friendly practices:

- The number of farmland birds declined by more than half since 1970.
- Only 7% of the UK's woods are in good ecological condition and woodland bird species have declined by 34% since 1970.
- Over a third of nutrient pollution in rivers is caused by agricultural run-off.
- Farming produces more than 11% of UK greenhouse gas (GHG) emissions and if the current trajectory is maintained, agriculture will be the third largest source of emissions in the UK by 2030.

After Wildlife and Countryside Link. Nature 2030. Five urgent reforms to meet natural environment targets in the next Parliament[3].

The Wildlife and Countryside Link in the UK (WCL, 2023) suggested that investment in nature-friendly farming and land management would need to be doubled to around GB£6 billion a year to turn this situation around by 2030. The same organisation makes a strong case for increased national investment in reviving nature, simply because it is a precondition of food security in future. They argue that escalating costs of fuel, synthetic fertilisers and chemical pesticides, combined with declines in essential ecosystem services such as pollination, combine to make 'intensive farmland management a poor long-term business choice'.

Agri-environment schemes (AESs) can be expected to continue to play an important part in nature recovery efforts because they offer farmers and landowners access to public funds to continue producing food whilst also working to boost biodiversity or improve ecosystem health and are relatively flexible and short-term. At the same time, they are unlikely to achieve the scale of transformative changes in land use that are needed to tackle the biodiversity crisis. The 'innovative' financing initiatives called for under the Paris Agreement are intended to meet this need.

An increasingly wide range of environmental credits are being traded and in a variety of ways. Carbon markets dominate, but there is growing interest in biodiversity and other nature-credit markets, as well as carbon markets that incorporate biodiversity and vice versa. Some merit serious consideration

3 https://www.wcl.org.uk/assets/uploads/img/files/Nature_2030_Report_18.07.2023.pdf

by landowners and land managers interested in diversifying into farming for nature. The taxonomy of biodiversity credit markets shown in Box 2 was developed by Nature Finance (The Future of Biodiversity Credit Markets, 2023) and provides a useful overview of the spectrum of markets and finance mechanisms that has developed. Demand for biodiversity or nature credits may come from a philanthropic desire of individuals or organisations to invest in nature or re-wilding (such as 'Cartier for Nature'[4]) or it may be driven by the need to comply with statutory requirements (e.g. obligations under England's new Environment Act requiring BNG); access finance (requirement to offset residual impacts in Performance Standard 6 of the International Finance Corporation, IFC 2012); or to demonstrate to shareholders that businesses have an ethical approach to the management of their operations and impacts and/or to ensure a sustainable future for business. One UK-based habitat bank (the Environment Bank) is even offering 'Nature Shares' to investors, with accompanying monitoring and verification of outcomes and the Landbanking Group (2024) has issued a consultation paper on the concept of nature equity, a new asset class 'that brings nature on the balance sheet'. All approaches potentially offer farmers, landowners or land managers opportunities to generate credits for sale.

Box 2 Biodiversity or nature credit markets (after The Future of Biodiversity Credit Markets, 2023)[5]

- **Philanthropic claims/certificate markets**: an entity makes a financial contribution/investment to protect or regenerate a defined biodiversity landscape or seascape and this contribution/investment is certificated. Biodiversity credits can increase confidence in a donation's effectiveness. They underpin the 'right to brag' but don't have tradable financial value and their value can't be asserted as part of legal compliance requirements.
- **In-setting credit markets**: investments in actions for sustainable management of nature, often along food value chains. Actions to enhance ecosystem services within nature-dependent supply chains confer resilience and may also generate biodiversity (and potentially climate) co-benefits. Such investments are largely profit-seeking. 'Inset credits' can be placed on a balance sheet as a financial asset and can potentially be traded if there are direct financial returns and/or the value chain-linked financial returns are of transferable value to a third party.

4 https://websitefactory.cartier.com/akamai-storage/videos/Cartierfornature/Cartier-for-nature-annual-report-2022-2023-digital.pdf
5 https://www.naturefinance.net/wp-content/uploads/2023/02/TheFutureOfBiodiversityCreditMarkets.pdf

- **Biodiversity-linked carbon offset markets:** One of the many shortfalls in carbon offset markets is the lack of system-wide consideration of their broader biodiversity impacts. Recognising that generating carbon credits can either enhance or unintentionally depreciate biodiversity, a second generation of carbon credits is emerging which incorporate biodiversity considerations. An open question is whether this will significantly alter the carbon credit valuation or/and lead to changes in the management of the underlying nature asset? The stacking of different kinds of ecosystem services, and its role in achieving multiple goals (biodiversity, climate, sustainable development) will shape future valuation and financing considerations in this kind of market (and also the likely attractiveness to landowners).
- **Biodiversity offset markets**: these may be voluntary or regulated. After the mitigation hierarchy has been applied to management of impacts, residual losses of biodiversity may remain. Biodiversity offsets compensate for these residual losses so that ecological equivalence or 'no net loss' of biodiversity is achieved. An increasing number of countries has legislation requiring developers in certain sectors to offset the residual damage associated with their operations. Biodiversity credits are the certified proof that the business has complied with regulatory requirements. In this case, legal and financial liabilities are met, but credits aren't monetised through trade. This could change in future, if auctions emerge and create a more competitive market space for delivery and sale of offset credits.
- **Biodiversity financial assets:** there is a growing demand by the global asset management sector for financial assets that adequately value nature within portfolios and help diversify and mitigate climate and nature risks. This is as part of their efforts to try and meet environmental and social governance or ESG and impact investment criteria. Biodiversity credits, if properly designed, issued, valued, and traded (i.e. 'securitised'), could answer some of the needs of these fast-growing asset management trends and therefore become a new, significant financial asset class. Landowners would potentially have the option to generate such credits.

This chapter provides an overview of some of the 'nature markets' and finance mechanisms, which may provide farmers and land managers with new revenue streams, as well as offering opportunities to draw private finance into conservation. We start by considering the main drivers behind these new markets and go on to review some of the opportunities and challenges they may bring.

2 Drivers behind biodiversity offsets and nature credit schemes

The values of nature and biodiversity have not been well internalised economically in the past, but this is changing rapidly, driven by wider recognition of the risks of failure (World Economic Forum, 2024), strengthening global agreements, new national laws and regulations and new public and private sector sustainability policies (Box 3).

Box 3 Key drivers behind development of nature credit schemes and markets

Drivers behind the development of nature credit schemes and markets include:

- global agreements such as the Paris Agreement;
- national sustainability policies/statutory requirements;
- private sector demands for environmental credits to offset their impacts; and
- sustainable business/finance and associated disclosure and reporting requirements (Fig. 1).
- investment opportunity (trading green bonds and nature credits); and
- philanthropic motivation.

Figure 1 shows how new disclosure and reporting requirements for organisations to be accountable for their impacts on biodiversity are catalysing new approaches, standards, and arrangements for transparent disclosure.

This is not to suggest that businesses have only just woken up to their responsibilities when it comes to biodiversity. The critical dependence of sustainable business on a healthy environment has been recognised for a long time, and there are well-established practices in the private sector around mainstreaming biodiversity in environmental assessment and management approaches, such as Ecological Impact Assessment (Treweek, 1990; CIEEM, 2018). These have gradually transitioned from an emphasis on damage limitation towards outcome-oriented approaches. Biodiversity offsets are an example of such an approach, now used routinely as an additional step in the mitigation hierarchy to compensate for biodiversity losses that would otherwise occur post-development, despite actions to avoid and limit damage to ecosystems. Businesses have been encouraged to develop biodiversity offsets to comply with legal requirements or to access finance. They may also use

Figure 1 Overview of global biodiversity standards on reporting (Global Reporting Initiative)

them voluntarily to manage biodiversity-related risks, to address shareholder concerns or as part of efforts to secure licenses to operate in sensitive or biodiversity-rich environments. Biodiversity offsets typically come into play in quite subscribed circumstances in which obvious, direct and irreversible losses of biodiversity occur, for example as a result of a mining or other infrastructure development project. They require achievement of tightly defined outcomes, with a strict requirement to demonstrate ecological equivalence between what is lost and what is gained (Maron *et al.,* 2024), the break-even point generally referred to as 'no net loss'. No net loss must be achieved as a minimum, though some of the international performance standards that underpin financial loans to infrastructure projects (such as IFC Performance Standard 6) require a net gain outcome to give greater assurance for certain threatened ecosystems and species. To deliver no net loss, biodiversity must be enhanced elsewhere to 'offset' the impact caused by the activity, for example an infrastructural development project.

Biodiversity offsets have proved challenging to integrate into business activities that are spatially and temporally dispersed or have complex, international supply chains, such as those that typify agri-business or the fashion industry and so activities haven't gained so much traction in these sectors. This is one reason behind the emergence of alternative approaches under the umbrella of 'Nature Positive', which has been gathering pace as a unifying movement for businesses wanting to internalise biodiversity values into their corporate strategies and operations. The Leaders' Pledge for Nature calls for a nature-positive future to be achieved by 2030 and several global 'Fortune 100' companies have revealed aspirations to contribute (Maron *et al.*, 2024). The push for Nature Positive has strengthened interest in biodiversity or nature credits as a way to channel private sector investments. They allow biodiversity losses and gains to be balanced in a more devolved manner: gains may even be delivered in different countries from where losses have occurred though this is controversial and there is increasing recognition of the need for 'guard rails' to enhance credibility.

On the demand side for biodiversity or nature credits, there are many inter-related drivers and rapid proliferation of approaches, standards and rules which can be complex and challenging for landowners and land managers to navigate.

On the supply side of nature markets, farmers and landowners will have varied motivations for embarking on a journey of land use transformation from conventional farming, with a primary emphasis on food production, to farming of/for nature. Before embarking on significant land use change, it is important to weigh up the pros and cons of the various options available and to have clear goals and objectives in mind.

Possible motivations for farmers to embark on farming nature for new nature markets include the following:

1 **Making marginal land productive**
Biodiversity offsets and markets for nature credits provide opportunities to use and rehabilitate land that may be marginal for productive or profitable farming. Land that is less fertile has difficult topography or is unproductive without drainage can be allocated to newly created or enhanced wildlife habitat, generating biodiversity credits that can be sold onto the private market or through habitat banks.

2 **Alternative income streams or diversifying sources of income**
Costs of production may be increasing and profit margins declining, requiring new economic strategies and business models. Conventional food commodities and products may have volatile markets, especially with climate change, and new markets for nature may provide ways to continue practicing farming or forestry with a mixed and perhaps more resilient portfolio. The Woodland Carbon Code in the UK, for example, provides for production of both timber and carbon.

3 **Changes in level or type of subsidy**
It may be necessary to seek new income streams if public subsidies are reducing or being withdrawn. In some countries (including England and countries within the EU), public subsidies are transitioning away from support for food production towards greater support for environmental goods and services.

4 **Retirement and succession planning**
Options available through nature markets may allow farmers to leave farming and take their entire land holdings out of agricultural production.

New nature markets can also provide opportunities to transfer land to younger generations even if they are not interested in farming it, allowing farmers to retire without having to close their business or sell their land.

5 **Personal motivation**

Deep personal interest and satisfaction can be gained from restoring nature. New markets for nature may make this a possibility by providing more farmers and landowners with opportunities to sustain an adequate level of income despite removing land from intensive food production.

Finally, governments and non-governmental organisations (NGOs) have an interest in promoting nature markets as part of their efforts to catalyse greater levels of private investment in conservation, as needed to reverse biodiversity decline and meet ambitious global and national restoration targets. National biodiversity offset and credit schemes offer opportunities for land managers specialising in conservation management, for example, England's BNG policy has catalysed new interest in habitat banking among wildlife organisations such as the Wildlife Trusts and the Eden Project in Cornwall, which has created a new company (the Eden Project Wildflower Bank) to focus on creation of diverse habitats prescribed under the BNG Scheme. As well as providing potential new revenue streams, biodiversity gains delivered through offset and credit schemes offer a means of drawing private investment into conservation and contributing to nature recovery.

3 Options for landowners and land managers

For landowners/ land managers interested in allocating land to nature recovery, nature positive or other environmental enhancements, options include the following:

- Subsidies under agri-environment schemes (AESs) (receiving payments for sustainable management or for delivery of ecological services);
- Entering into agreements to provide biodiversity offsets, involving explicit efforts to balance losses of biodiversity due to a development with commensurate, demonstratable gains. There may be opportunities to generate and sell biodiversity units (BUs) through statutory/national BNG or offset schemes (such as BNG in England) or to engage directly with businesses seeking to compensate for their own impacts. While farmers are at liberty to enter into private agreements with developers, many are likely to opt to work through habitat banks which take responsibility for designing and implementing the specialist management that might be needed and administer any regulatory provisions for tracking outcomes;

- Sale of biodiversity, nature or other environmental credits directly to private sector developers or in voluntary global markets; and
- Producing more than one type of environmental credit, for example entering land into schemes intended to generate carbon credits which may also benefit biodiversity (or vice versa).

These options are reviewed here, with an emphasis on implications for the landowner and land manager.

3.1 Agri-environment schemes

AES provide funding to farmers and land managers to farm in a way that benefits various aspects of the environment within the farmed landscape. AES are a major source of nature conservation funding within the European Union (EU) and play a part in conserving the genetic diversity of livestock, protecting diversity of agro-ecosystems, and producing food with a lower environmental and ecological footprint than intensive farming. The initial purpose of AES in Europe was to protect threatened habitats or landscapes (Batáry *et al.*, 2015) and farmland areas designated as having 'high nature value' (Lomba *et al.*, 2014). Over time, the emphasis changed to prevention of species' loss, especially farmland birds, across agricultural land. Farmers might be paid to plant crops to produce winter food for wild birds or to restore hay meadows, for example. More recently, AES have also been used to improve and maintain ecosystem services, such as pollination and biocontrol (Ekroos *et al.*, 2014), for example retaining buffer strips on the margins of arable fields to provide habitat for pollinators and natural predators of insect pests. Scotland's Agri-Environment Climate Scheme (AECS) has an explicit objective of supporting land management that will contribute to climate change adaptation and mitigation and many current schemes have objectives reduce water pollution, enhance natural flood attenuation, improve soil health, enhance access to the countryside and protect cultural landscapes and heritage as well as creating habitat for wildlife.

Most public AES support positive management without any strict requirement to demonstrate achievement of specific outcomes. Payments are usually predicated on evidence that the farmer/land manager has either refrained from undertaking certain activities or has implemented management intended to benefit a specific aspect of the environment. In the language of 'pressure-state-response' indicators, they focus on providing support for management responses intended to tackle loss of biodiversity from the farmed landscape. Hitherto, payments have not been contingent on monitoring of 'state' indicators though this may change.

Options available to farmers in England under the new Environmental Land Management Scheme (ELMS) are summarised in Box 4.

Box 4 Agri-environment schemes and initiatives under England's Environmental Land Management scheme (ELMS)

The current Basic Payment Scheme (BPS) in England (provided as a basic subsidy to all farmers) will be replaced with 'delinked' payments in 2024. The new Environmental Land Management scheme (ELMS) has been rolled out in England (post Brexit), under the concept of 'public money for public goods'.

Sustainable Farming Initiative

Payments under the SFI (launched in summer 2022) are intended to ensure that farming activities are performed in a more environmentally sustainable way, allowing food to be produced alongside environmental goods and services. They do not require any actions that explicitly target biodiversity gain or nature recovery, but they may achieve a degree of this incidentally. SFI agreements last for 3 years. They include the following standards:

- Arable and horticulture soils;
- Improved grassland soils;
- Moorland;
- Nutrient management;
- Integrated pest management;
- Hedgerows;
- Arable and horticultural land;
- Improved grassland; and
- Low/no input grassland.

Landscape Recovery

Landscape Recovery funds have been allocated to a small number of longer-term, larger-scale, bespoke projects (22 projects were chosen for the first round of the scheme in summer of 2023). Most involved groups of farmers' land managers collaborating to deliver a range of environmental benefits at scale across farmed landscapes. These benefits include river restoration, farmland biodiversity, water quality, flood attenuation and carbon sequestration. Some of them include targeted measures to benefit threatened species such as the water vole, otter, pine marten, lapwing, great crested newt, European eel, and marsh fritillary.

Countryside Stewardship

Countryside Stewardship (CS) provides financial incentives for farmers, foresters and land managers to look after and improve the environment by carrying out management that targets enhancement of biodiversity, improved habitat or natural flood management. Grants are available for capital works, to develop certain management plans or to undertake specific management actions. 'Higher Tier' grants support multi-year management and purchase of capital items needed to safeguard or enhance 'the most environmentally important sites'. They are awarded on a competitive basis and are usually allocated to support targeted conservation management, such as creating or restoring habitats to enhance populations of pollinators or deliver improved natural flood attenuation.

3.2 Biodiversity offsets

Box 5 Definition of biodiversity offsets

Biodiversity offsets are 'measurable conservation outcomes of actions designed to compensate for significant residual adverse biodiversity impacts arising from project development after appropriate prevention and mitigation measures have been taken' (Business and Biodiversity Offset Programme, 2009). The overall intent is for affected biodiversity to be no worse off at a given scale of evaluation than it was prior to the development taking place.

Biodiversity offsets are routinely used as part of the mitigation hierarchy in environmental impact assessments carried out for infrastructure development. Farmers may be able to generate credits through targeted land management and sell them to developers through private agreements or through statutory schemes in some countries.

The Mitigation Hierarchy has guided biodiversity input to environmental impact and risk assessments for proposed new infrastructure developments for a long time. Offsets evolved as a final step in the mitigation hierarchy, coming into play only when developers have made other appropriate efforts to avoid or minimise impacts or to restore damaged ecosystems. When used as an explicit step in the mitigation hierarchy, offsets should deliver at least 'no net loss' of biodiversity and preferably go further, achieving a net gain (see Fig. 2).

Schemes requiring residual impacts on biodiversity to be offset are now established in many countries and bring large areas of land under a form of conservation management or permanent legal protection each year to compensate for residual losses of biodiversity to infrastructural development. In

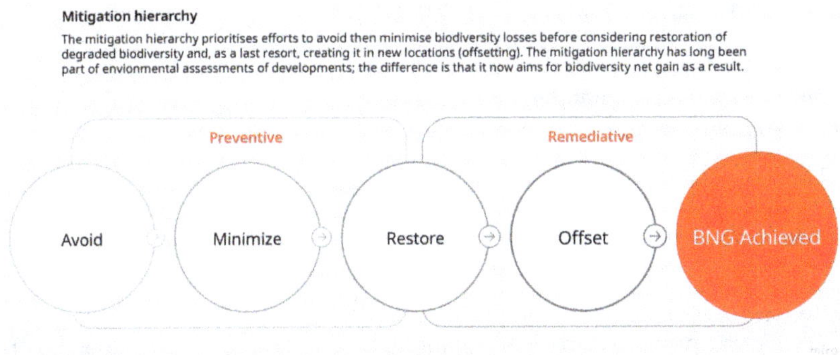

Figure 2 Mitigation hierarchy diagram (Little book of biodiversity net gain, BSI, 2023)

2016, The Biodiversity Consultancy reported that more than 100 countries had national government policies requiring, guiding or enabling the use of offsets or were developing them. Some countries also had sub-national offset policies, notably provincial or state policies in Australia, Canada and South Africa. On the other hand, the European Union has a cross-national policy on offsets.

Offsets are used voluntarily by developers to compensate for their own impacts, or they may be a statutory requirement under state or national law. They are also identified as a mechanism for safeguarding and enhancing important habitats by international finance institutions. For conservation organisations, biodiversity offsets offer one way to secure additional funding for conservation efforts and invest it in the highest priority and most cost-effective conservation activities at a landscape or regional scale. Biodiversity offset systems open up new markets for farmers and land managers to become offset providers. Many offset systems also allow conservation gains to be delivered by specialist providers with the necessary expertise and capacity (e.g. habitat banks). Finally, opportunities may arise for brokers to match supply and demand and negotiate prices for credits.

A range of implementation mechanisms are in use for biodiversity offsets worldwide, including:

1 purchase of credits from conservation banks (e.g. USA, Canada, Australia);
2 purchase of credits directly from offset provider (e.g. England);
3 establishment of conservation easements (e.g. USA, Australia);
4 custom-built offsets implemented with or by third parties such as NGOs or private companies (e.g. Madagascar);
5 payment into government conservation funds (e.g. Brazil's funds supporting protected areas conservation); or
6 an organisation develops its own offsets 'in house'.

3.2.1 England's statutory biodiversity net gain policy/scheme

Following its departure from the EU, and as part of the Government's 25 Year Environment Plan, England established new policies and rules for biodiversity. The Environment Act (2021) mandates a requirement for BNG within the planning process and marks a transition from a previous policy – emphasis on no net loss to a net gain of biodiversity.

BNG is an approach to terrestrial and intertidal development that mandates all developments requiring consent under the Town & Country Planning Act (1990) (TCPA) and all 'Nationally Significant Infrastructure' (NSIP) developments to deliver a minimum 10% net gain in biodiversity. The Environment Act (2021) sets out the framework for BNG requirements, with secondary legislation, policy and guidance supporting it.

Under the Environment Act, infrastructure developers must provide a plan to deliver BNG with their applications for planning consent, based on the mitigation hierarchy and a prescribed habitat-based metric, described further. If residual impacts will occur despite efforts to deliver BNG within the boundaries of their development site, infrastructure developers must purchase off-site BUs from landowners directly or from habitat banks. The number of 'biodiversity units' required to compensate for losses is determined using a metric first proposed by Treweek and Butcher (2010) then adopted by Defra and Natural England and now published as Biodiversity Metric 4.1 (Panks *et al.*, 2023).

There isn't a universally agreed single measure of biodiversity. Reasons include the fact that biodiversity is multi-dimensional, dynamic in space and time and highly context specific. In common with the majority of offset schemes in use worldwide, England's metric uses habitat as a proxy for overall biodiversity. Habitats are relatively immobile compared with species populations, they develop relatively predictably in response to environmental conditions and are normally used as the primary focus for planning biodiversity conservation. Furthermore, although expansion and enhancement of habitats will not necessarily guarantee enhanced outcomes for their associated species populations, many species are threatened by loss and decline of habitat and can be expected to benefit from increased extents of suitable land under positive management for conservation.

To derive BUs for impact and offset areas, hectares of habitat of a particular type are scored for their intrinsic distinctiveness, their condition (which in UK habitats is often closely linked to suitability of management) and the strategic importance of the location concerned (whether it is where biodiversity will be lost or where it will be gained). The resulting BUs are then adjusted further using multipliers that account for the time taken for different habitats to reach target levels of condition ('time to target' multiplier) and the uncertainty and risk of failure (uncertainty multiplier). The metric framework was developed with pre-assigned distinctiveness scores for England's characteristic habitats, classified using a new classification for the UK (the UK Habitat Classification or UKHab[6]), with some adaptations.

A metric calculator tool developed by Natural England calculates the change in BUs resulting from a project or development by subtracting the number of pre-intervention or baseline BUs from the number of post-intervention BUs (i.e. those projected to exist with the development in place). The metric (area × distinctiveness × condition) can be applied to all UK sites/habitats and is fungible. However, as emphasised by Treweek *et al.* (2010), it represents a 'lowest common denominator'. Delivering gains using this metric will not address the needs of high-priority biodiversity features for which habitat is not a good surrogate, nor will it guarantee positive outcomes for the full range of species of conservation concern in England, not all of which are well catered for through existing protection mechanisms. BNG will contribute to nature recovery in England, but additional measures will be needed to achieve it fully. Also, as used in the English system, only direct impacts of development are accounted for using the metric. There is no requirement to compensate for cumulative and indirect effects, so the BNG Scheme doesn't address development impacts in their entirety.

Landowners wishing to produce and sell BUs under the BNG Scheme may benefit from undertaking several preparatory steps, some of which are listed in Box 6. Some of these may require inputs from ecologists, farm business advisors and lawyers which can be costly and time consuming.

Box 6 Typical preparatory steps that landowners might need to take to produce and sell BUs under England's BNG scheme.

	Step	Description
1	Local demand assessment	Evaluate whether there is demand locally for BUs or via national schemes (e.g. England's statutory credit scheme), and what types of credits are most likely to be needed, reflecting anticipated habitat losses due to development.
2	Land assessment	Assess suitability of land for BU-generation. Which habitat types is the land suitable for? This may require soil sampling/assessment and collection of data on the water table, hydrology. Assessments of historical maps and information can also be useful to identify habitats that may be characteristic of the landscape.

6 www.ukhab.org

	Step	Description
3.	BU baseline	To demonstrate gains in BUs a verifiable baseline is needed. Survey existing habitats and land uses and calculate the BUs currently represented on the proposed offset areas using the Biodiversity Metric.
3	Land use change scenarios	Explore potential gain in BUs from different BNG scenarios. Map out alternative ways to transform the land and assess the viability of each scenario (e.g. in terms of alignment with local demand, revenue potential, costs, implementation feasibility/risks, biodiversity potential and natural capital). Maximum credit-uplift may not translate into the best outcomes for nature and farmers may not want to put their entire land holding into BNG. An integrated assessment is needed to understand implications of BNG for farm operations overall.
4	Land management plan	Outline necessary land interventions for desired scenario in more detail, including initial habitat creation/ restoration and long-term management. A formal habitat management plan will be required prior to finalising any agreement to sell BUs.
5	Detailed management costings, financial modelling, and BU pricing	Calculate costs of the planned intervention, incl. capital expenditures, long-term maintenance, as well as opportunity, contingency and insurance costs. Create financial model to contrast potential costs with potential revenues and to set a feasible BU price.
6	Legal entity	Create a suitable legal entity via which to sell BUs, establish responsibilities/ liabilities and safeguard the land over intended time period (30x years).
7	Buyer engagement & sales	Identify, engage, and negotiate with potential buyers and refine land management plan and financial model/pricing according to buyers' specific needs.
8	Create/restore habitat	Implement the planned land-use changes and create/restore habitat in line with the (refined) land management plan (e.g. via a phased approach). This may require specialist knowledge and new or different equipment.
9	Long-term management and monitoring	Engage in long term habitat monitoring and management to assure that the agreed upon habitat attains and maintains its appropriate condition (incl. respective human resource needs). Defra (in England) has developed a template that can be used.

3.3 Biodiversity credit markets

A biodiversity credit is a tradable security representing a unit of biodiversity improvement, which is brought about by some form of land-management change. Credits can be traded in biodiversity or nature credit markets, designed 'to deliver equitable, nature positive outcomes' (Nature Finance, 2023). Oversight and verification are important to ensure that biodiversity or nature do benefit in tangible ways and commensurate with the traded value. Australia has put a national regulatory framework in place for biodiversity credit markets to support this. There has been recent interest in nature as a form of equity, with the intention of rewarding land stewards for protecting or restoring ecosystems through outcome-based payments, while enabling investors to gain assured proof of effective nature stewardship. Such arrangements might be attractive to landowners interested in managing their land to enhance natural capital (including nature) and who want flexibility about what kinds of credit to sell.

A consultation paper issued by The Landbanking Group (2024) envisages that nature equity assets could be exchanged through an online platform and that transactions would be recognised on the balance sheets of both land stewards and investors. This would allow nature remediation to be regarded as a capital investment rather than a Corporate Social Responsibility (CSR) expense on Profit and Loss statements, making it more straightforward for businesses to secure funds needed to carry it out during their strategic financial planning activities.

The Landbanking Group's (2024) consultation paper's proposals are predicated on three important tests: the nature test, the equity test and the nature equity test.

3.3.1 Capital maintenance requirement ('Nature Test')

This test is met if investment into nature capital is secured, allowing it to grow and deliver future goods and services (whereas continued exploitation without investment will result in its depreciation). The consultation paper quotes Gunhild Stordalen, (Executive Chair of 'Eat Foundation') as saying, 'Even now, we see the disruption of major food supply chains and the failure of major crops. The reason is clear: our agricultural system degrades soils, water bodies and ecosystems at large. Like any other infrastructure, the biological systems that support our food sector require investments. These investments are not only needed, they are feasible and attractive.'

3.3.2 Contribution reward requirement ('Equity Test')

This test is met if markets reward those who, through their land use choices and work maintain and create natural capital, and those who, through their payments, catalyse these improved economic choices. This would be achieved through binding contracts between Nature Equity sellers ('land stewards') and Nature Equity investors ('buyers').

3.3.3 Value accrual requirement ('Nature Equity Test')

This test is met if a financial transaction over ecosystem outcomes allows the accrual of balance sheet value over time both for the seller (land steward) and the buyer (natural capital investor). Safeguards are needed for assurance that an equitable arrangement is made, for example, unique and immutable identification of underlying measurements, transparent valuation methodologies and clear contractual rights with defined durations. For investors, the contractual rights to measurable outcomes for nature would be recorded as intangible assets on the left side of the balance sheet.

Box 7 summarises the concept, which creates a potential role for farmers and land managers as 'land stewards' responsible for delivering gains in natural capital and receiving payments in return. It is not entirely clear how natural capital gains would be valued.

Box 7 Consultation on nature equity assets (Landbanking Group, 2024)

A contract would be signed between an investor and a land steward to link payments with a unit of biophysical nature preservation or uplift. The contract would also establish specific monitoring reporting and verification rules. A land steward might be a person (a farmer or land manager) legally entitled to take land use decisions, whether as an owner or tenant, who undertakes to provide natural capital. Buyers might be businesses with an interest in protecting their value chain, demonstrating positive contributions, and accruing value as a 'hedge' or store of value (just like gold or real estate). This contract does not represent a title deed. It solely contains the measured capacity of nature to produce outcomes such as harvests, carbon sequestration, water holding, water purification, and biological regeneration in the future. In return, investors receive a right of significant value, for example, exposure to safe physical assets in their supply chain, solid and generally recognised proof of climate or nature remediation, reliable delivery on government requirements, and finally a holding right on an intrinsic, scarce value.

There are several initiatives emerging to secure private finance for conservation projects through investment platforms. Projects for Nature, for example, connects businesses and other donors with nature recovery projects that have been screened by experts and are delivered by organisations such as the Wildlife Trusts in the UK. The UK's Environment Bank (www.environmentbank.com) is offering 'Nature Shares' for sale to businesses as a mechanism to achieve their nature positive targets and undertakes to deliver 'impactful uplifts for biodiversity and natural capital in a way that can tangibly be quantified for stakeholders', underpinned by robust ecological methodologies and science-based reporting tools. The Environment Bank aims to display current data, live visuals and audio feeds to 'bring projects to life' for internal and external audiences, an approach which is also intended to provide evidence to feed into corporate reporting on environmental, social and governance (ESG) goals.

3.4 Comparing options

Markets for nature offer farmers attractive opportunities to diversify their businesses into farming for nature and natural capital in a variety of ways. They are proliferating at such a bewildering rate that farmers may struggle to make decisions about the best options to choose, though there is guidance available, and a new generation of advisors is emerging. For example, the GFI has developed a 'Farmers' toolkit for assessing nature market opportunities' which provides free advice to UK farmers to brief them on options to consider and guide them in the process of engaging in nature markets, should they decide it is right for them.[7] The toolkit identifies some common and valid concerns that farmers may have summarised in Box 8. These are all valid concerns, gradually being resolved as markets develop and mature, but nevertheless challenging to address due to variability between markets and schemes. Lack of clarity around liability for failure to deliver gains is a particular concern, as is uncertainty about tax implications. In some cases, managing cash flow to support long-term management may also be challenging and is one argument for using habitat banks or nature credit brokers to assist with legal and financial aspects of market engagement. In future, standards could have a part to play in achieving a more level playing field.

Box 8: Farmers' common concerns about participating in nature markets (Green Finance Initiative)

1. Size of land
Do I have enough land to participate in nature markets?

2. Length of time commitment
How long am I committing my land for? What are my options for an early exit?

7 https://www.greenfinanceinstitute.com/gfihive/farming-toolkit/

3. Conflict with food production
Will selling nature credits displace or decrease my food production?

4. Conflict with agri-environment schemes
Can I enter into nature market agreements alongside public grant schemes?

5. Unclear liabilities
What am I responsible for, and what will I be liable for if the environmental improvements or interventions fail?

6. Development costs
What will I have to pay for at the start, and how can I be sure this will be repaid by the trade?

7. Land tenure
I'm a tenant farmer. Am I able to participate?

8. Timing of sales
Should I wait until markets become more developed or prices increase?

9. Future requirements of farmers
Am I putting myself at risk by selling my ecosystem services now, given there may be future environmental requirements from agricultural buyers or the government?

10. State of natural capital
What if my natural capital is already in good condition? Will I miss out on payments?

11. Loss of tax benefits
How will my nature market project affect my wider tax position?

Table 1 below identifies some of the options available to farmers in England and the UK, comparing them in terms of their duration and the types of metrics used to define credits or units of exchange. This is not intended to be comprehensive, but emphasises the wide range of schemes now available to landowners. As nature markets begin to crystallise, a few pathfinder projects are 'testing the water' to understand how solutions are delivered in practice. This includes developing collaborative corporate structures, new legal mechanisms for binding land-use (such as conservation covenants) business plans and long-term finance models for landscape transition, best methods of habitat establishment and management, and strategies for monitoring and evaluation, including novel techniques which deploy new technologies such as environmental DNA or acoustics. An overview of one such project is provided in the next section.

3.5 Case study: Wendling Beck Project, Norfolk

The Wendling Beck Environment Project (www.wendlingbeck.org) is a landscape scale (~2000 acres) project in Norfolk, England, looking to transform land use from arable farming to generation of biodiversity and a variety of

Table 1 Comparison of options available to farmers in England/UK.

Scheme	Type of scheme	Summary	Timeframe	Metrics and prices	Requirements
Agri-environment schemes	Public subsidy	Farmers paid to farm in a more nature-friendly or environmentally sustainable manner	3 to 5 years	Cost of sympathetic management and/or cost of profit foregone	Short-term so doesn't always benefit nature in the long-term
Biodiversity Net Gain	Compliance-driven in England under the Environment Act.	Property developers pay farmers (directly or via a habitat banking company) to create or restore habitats that deliver biodiversity uplift, contributing to their requirement to deliver 10% Biodiversity Net Gain	30 years	Measured in Biodiversity Units determined using statutory metric.	Projects or management to deliver biodiversity uplift need a detailed habitat management plan and periodic monitoring and validation of Biodiversity Units delivered.
Water quality payments	Compliance-driven	Water companies pay farmers to reduce phosphate and nitrate run-off, which can deliver operational cost savings to them.	1 to 10 years	Units of reduction in levels of phosphate/ nitrate.	Contingent on need and context (not all areas will have this option).
Nutrient Neutrality (example of water quality payments)	Compliance-driven agreement applicable in England (can be combined with BNG payments).	Property developers pay landholders to reduce their nitrogen or phosphate run-off into local waterways, to mitigate the impact of new property developments. Landowners opt in to make land available for nature-based solutions.	80 to 125 years	Nutrient credits (1kg of total nitrogen or total phosphate).	Entering into negotiations, identifying solutions, and agreeing credits and management costs can be a protracted and time-consuming process.
Nature Credits	Variety of agreements	Tradeable units of biodiversity enhancement brought about by land management change are generated and sold	Varies	Various metrics and credit types (biodiversity, natural capital)	Vary

(Continued)

Table 1 (*Continued*)

Scheme	Type of scheme	Summary	Timeframe	Metrics and prices	Requirements
Carbon (including the Woodland Carbon Code, the Peatland Carbon Code and the possibility of a soil carbon code).	Voluntary agreement through global markets	Various metrics/ credits. Landowners create woodlands or restore peatlands to contribute to carbon sequestration. They receive a guaranteed income on delivery of verified Carbon Units. Under the Woodland Carbon Code payments are every 5 or 10 years up to 2055/56 and prices are agreed through an auction.	30 to 100 years	Carbon credits (e.g. Woodland Carbon Units)	Vary. Under the Woodland Carbon Code, projects must be certified and registered and verified according to Woodland Carbon Code requirements.
Natural Flood Management	Bespoke voluntary agreements	Businesses (e.g. water companies) or public bodies – pay farmers/landowners to install interventions that help slow the flow of water across a catchment and reduce property damage/disruption costs from flooding.	10 years plus	Measures financed might include reinstating natural hydrological function or geomorphological processes, or minimising run-off by improving soil infiltration	Nature-based solutions to flooding can take a long time to design and need specialist input. Contingent on need and context (not all areas will have this option).
Philanthropic or voluntary contributions	Voluntary agreements and donations	Individuals or businesses choose to donate funds to 'good' or approved projects designed to benefit nature/ biodiversity	Varies	Not necessarily measured. Using credits can help verify any claims.	Sometimes accused of greenwash if donors make exaggerated and unverifiable claims, about outcomes.

natural capital benefits (freshwater, flood attenuation, carbon and amenity), whilst still providing a sustainable way of living for the four landowners and creating employment for others. The project is a collaboration between private landowners, local government planning authorities, environmental NGOs and Anglian Water (Fig. 3). It has identified the sale of BUs as a key revenue stream, alongside payments for other ecosystem services and revenues from regenerative farming of livestock and blackcurrants. Government grants and support from The Nature Conservancy helped to support initial transition and essential preparatory work, including a comprehensive baseline habitat survey and visioning and scenario-building exercises to help the landowners develop a masterplan. This involved comparing and contrasting likely ecological outcomes and the costs of delivery over time, as well as considering implications for current agricultural activities and revenue streams. Spade-in-ground work started in Autumn 2021, with credit sales commencing in early 2024. Assuming the BNG market performs as anticipated, the project will be fully invested and all habitat creation will have started by 2030. Goals articulated in the masterplan include the following:

- Allowing nature to thrive by creating the right conditions and reintroducing species that have declined and disappeared through time, then trusting nature to 'do what nature does';
- Creating and restoring habitats to support native species that occurred in the past;
- Enhancing wellbeing: helping to reconnect nature with society and allowing people to experience a wilder landscape and become closer to nature;
- Building resilience for the future: supporting nature-driven processes, at scale to regenerate biodiversity, water and soil, help reverse biodiversity loss and limit climate change;

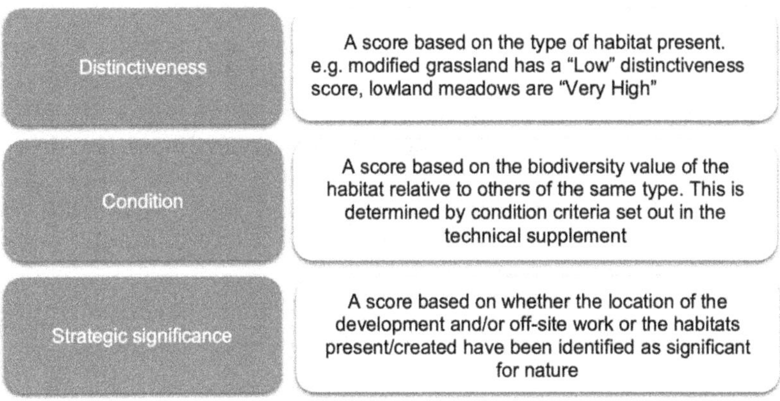

Figure 3 Dillington Carr SSSI (Wendling Beck Project). Source: Photo Daniel Casson.

- Maintaining food production through regenerative practices and grass-fed cattle and sheep, which will also be used to manage high distinctiveness of habitat; and
- Creating a model which is both financially and environmentally resilient and encourages others to follow a similar journey, creating connectivity within landscapes and delivering these benefits at a scalethat really makes a difference.

The project will deliver around 4000 BUs under England's BNG Scheme, and 250 000 + tonnes of carbon sequestration. More than 5 km of rivers will be restored and it is also exploring opportunities associated with policies on nutrient neutrality and natural flood managementand, improving public access for people and communities and engaging volunteers to take part in the monitoring of nature recovery.

4 Governance and regulation

Experience gained from voluntary carbon markets has highlighted the risk of 'phantom' credits being traded without evidence of investment or any assurance of a tangible outcome, and there are issues with transparency, integrity and accountability on both demand and supply sides of nature credit markets which are not yet fully resolved. A high-level Taskforce on Nature Markets was established in March 2022 to work on solutions to some of the problems encountered in early stages of nature credit market operation, such as traceability and low prices being paid to the landowners and managers who deliver credits, compared with the revenues gained when they are traded.[8]

Markets for biodiversity/nature credits vary in the extent to which they link losses and gains, whether spatially, temporally or in terms of responsible parties. For any trade-off between impacts and gains, there is a risk that certain biodiversity losses will be compensated through uncertain and insufficient gains, sometimes with significant consequences for people depending on biodiversity for their livelihoods and human rights. When biodiversity offset schemes are used, it is important to guard against inappropriate use of offsets and the risk that they might be used as a 'license to trash', without due emphasis on avoidance of impacts (Maron *et al.*, 2024). Offsets involve certain losses of biodiversity being traded for uncertain gains, with gains being implemented in a different location ('off-site') from the affected biodiversity and often involving uncertain methods and time delays. There are concerns that offsets might be used (inappropriately) to legitimise losses of biodiversity that cannot be replaced, allowing a gradual shift to more restorable habitats. To guard against this, offsets should not be

8 [https://www.naturefinance.net/wp-content/uploads/2023/02/TheFutureOfBiodiversityCreditMarkets.pdf]

used for impacts on high distinctiveness habitats which need to be conserved *in situ* for them to persist, or for any habitat which takes so long to establish and mature that it is effectively irreplaceable in any reasonable human timeframe (generally referred to as irreplaceable habitats). Offsets should also not be used for impacts on populations of species at risk of extinction or ecosystems at risk of collapse. In such circumstances, avoidance or prevention of impacts is the only acceptable mitigation option. Going further than the break-even point of no net loss can help to buffer the resilience of the affected features. Nevertheless, offsets are often seen as a controversial mechanism not helped by the lack of systematic planning for biodiversity in most countries which limits scope for genuine consideration of alternative development locations and also strategic, upstream planning of development so that biodiversity outcomes can be optimised. An adapted version of the mitigation hierarchy, the 'conservation hierarchy' (Arlidge *et al*., 2018) has been proposed to help identify and plan locations for delivery of offsets in advance so that they make the best possible contribution to nature recovery at a wider geographic scale.[9]

Another deepening controversy surrounds the widespread adoption of 'Nature Positive' approaches. Figure 4 presents an illustration of how 'Nature Positive' can be applied to make meaningful contributions to the global biodiversity crisis, versus one in which it is misapplied as 'greenwash', by-passing strict requirements to achieve ecological equivalence (see Fig. 4).

Various efforts have been made to establish principles and standards to promote or ensure good practice in both offset schemes and nature credit markets. The Business and Biodiversity Offset Programme (BBOP) developed principles on biodiversity offsets with wide collaboration and 'buy in' from the full spectrum of players in the offset process, which are considered to embody good international practice (Box 8). Although widely accepted, these principles can be challenging to apply in practice. The ability to conform with principles 3 and 5 depends crucially on approaches to spatial planning for biodiversity in the relevant implementation context, for example, and it can take a long time and a lot of dedicated effort to secure effective stakeholder participation in offset design. Principles intended to ensure that nature markets deliver genuine benefits to nature, climate and society have were introduced in 2023 by the UK's Wildlife Trusts, the Royal Society for the Protection of Birds (RSPB), The Woodland Trust and the National Trust in collaboration with Finance Earth and Federated Hermes. The 'Nature Markets Principles' encompass seven key project development principles from the supply-side perspective including science-based nature recovery, environmental and social safeguarding, additionality, permanence and financial prudence, seeking co-benefits, verifiability, and transparency. There is therefore some overlap with BBOP's principles, and it is

9 https://conservationhierarchy.org/what-is-conservation-hierarchy/

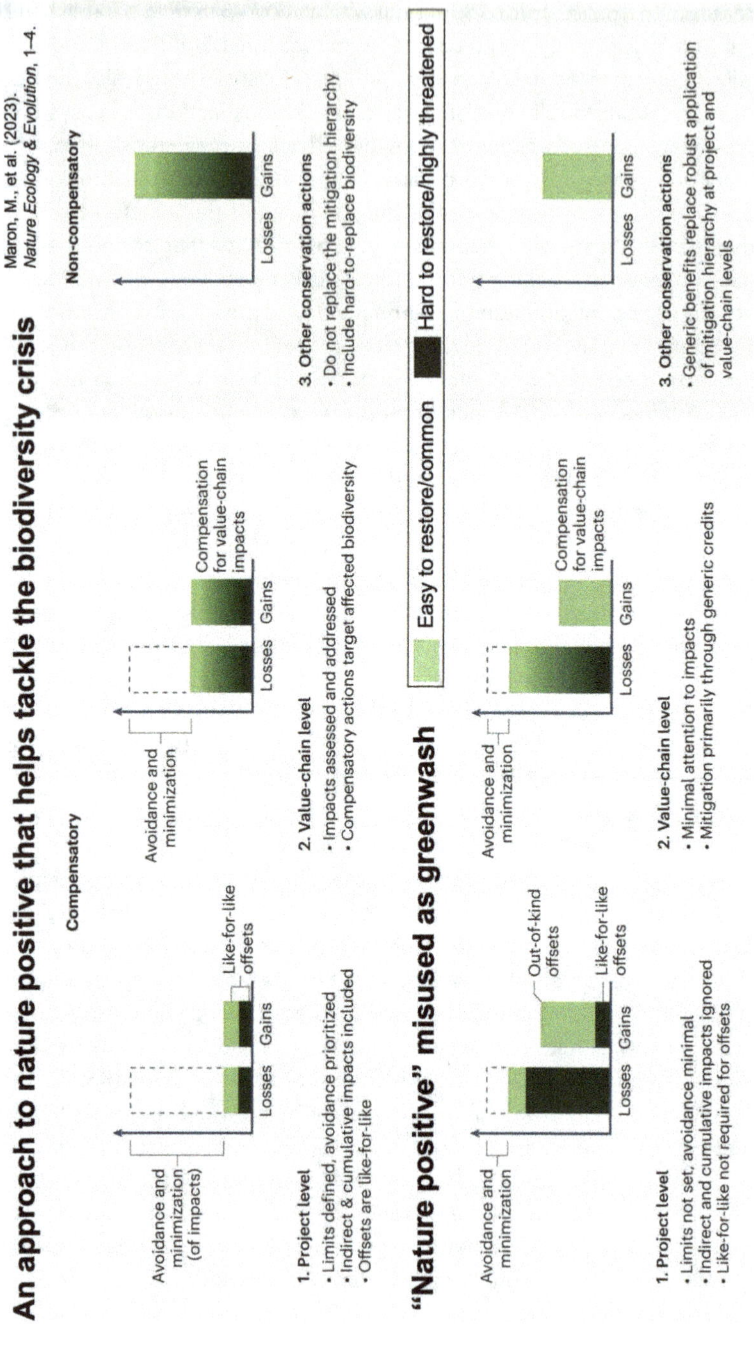

Figure 4 Scores included in England's BNG Metric Source: Adapted from Panks et al. (2023).

possible that similar challenges will be encountered with respect to achieving and demonstrating additionality and co-benefits. Whichever principles apply, those delivering offsets or credits will have to demonstrate conformance by providing evidence of gain from monitoring.

Box 8 BBOP principles on biodiversity offsets

Biodiversity offsets are measurable conservation outcomes resulting from actions designed to compensate for significant residual adverse biodiversity impacts arising from project development after appropriate prevention and mitigation measures have been taken. The goal of biodiversity offsets is to achieve no net loss and preferably a net gain of biodiversity on the ground with respect to species composition, habitat structure, ecosystem function and people's use and cultural values associated with biodiversity.

The BBOP principles establish a framework for designing and implementing biodiversity offsets and verifying their success.

1. ***Adherence to the mitigation hierarchy:*** A biodiversity offset is a commitment to compensate for significant residual adverse impacts on biodiversity identified after appropriate avoidance, minimisation and on-site rehabilitation measures have been taken according to the mitigation hierarchy.
2. ***Limits to what can be offset:*** There are situations where residual impacts cannot be fully compensated for by a biodiversity offset because of the irreplaceability or vulnerability of the biodiversity affected.
3. ***Landscape context:*** A biodiversity offset should be designed and implemented in a landscape context to achieve the expected measurable conservation outcomes taking into account available information on the full range of biological, social and cultural values of biodiversity and supporting an ecosystem approach.
4. ***No net loss:*** A biodiversity offset should be designed and implemented to achieve *in situ*, measurable conservation outcomes that can reasonably be expected to result in no net loss and preferably a net gain of biodiversity.
5. ***Additional conservation outcomes:*** A biodiversity offset should achieve conservation outcomes above and beyond results that would have occurred if the offset had not taken place. Offset design and implementation should avoid displacing activities harmful to biodiversity to other locations.
6. ***Stakeholder participation:*** In areas affected by the project and by the biodiversity offset, the effective participation of stakeholders should be ensured in decision-making about biodiversity offsets, including their evaluation, selection, design, implementation, and monitoring.
7. ***Equity:*** A biodiversity offset should be designed and implemented in an equitable manner, which means the sharing among stakeholders of the rights and responsibilities, risks and rewards associated with a project and offset in a fair and

balanced way, respecting legal and customary arrangements. Special consideration should be given to respecting both internationally and nationally recognised rights of indigenous peoples and local communities.

8. ***Long-term outcomes:*** The design and implementation of a biodiversity offset should be based on an adaptive management approach, incorporating monitoring and evaluation, with the objective of securing outcomes that last at least as long as the project's impacts and preferably in perpetuity.
9. ***Transparency:*** The design and implementation of a biodiversity offset, and communication of its results to the public, should be undertaken in a transparent and timely manner.
10. ***Science and traditional knowledge:*** The design and implementation of a biodiversity offset should be a documented process informed by sound science, including an appropriate consideration of traditional knowledge.

5 Monitoring and reporting

To counter some of the issues identified in the previous section, all offsetting and nature credit schemes require some form of monitoring and reporting to give assurance to public or private purchasers of credits (and company shareholders) that the gains they have paid for will be, or have been, delivered. England's BNG scheme, for example, includes a requirement for offsite BNG initiatives to be entered into a national net gain register and for monitoring reports to be submitted regularly (Rampling *et al.*, 2023). Beyond such registration requirements, assurance is needed that agreed outcomes are achieved 'on the ground' in accordance with contractual agreements. Accessing guaranteed prices for woodland carbon units under the UK's Woodland Carbon Code, for example, requires the woodland to be validated and verified to the Woodland Carbon Code. Box 9 explains the difference between monitoring, verification and validation in this context.

Box 9 Definitions of monitoring, verification and validation

Monitoring: the process of generating the evidence needed to identify and measure changes over time, relative to a defined starting point for generation of credits (or a dynamic baseline that represents future outcomes in the absence of actions to enhance biodiversity).

Verification: the process of establishing the accuracy and validity of gains relative to a baseline that can be shown to represent the situation at a particular time and in a particular location.

Validation: the process of supporting the intended claim, by demonstrating that credits were delivered in line with agreements.

For monitoring to be meaningful, it must generate observations of change relative to a starting point and needs to align with a verified, or verifiable, baseline that reflects the starting point for generation of credits. This is a complex area because some responsibilities lie with credit purchasers to carry out their own due diligence and others lie with those responsible for delivering the credits. Government may also have an interest in monitoring and evaluation to establish the effectiveness of statutory credit schemes at delivering policy objectives. In BNG or other biodiversity offset schemes, habitat banks take on at least some responsibility for monitoring and reporting.

The Science Based Targets initiative champions science-based target setting as a means of strengthening investor confidence. On the supply side of nature markets targets can also structure monitoring so that different actors in the market can track outcomes for carbon or nature. Many certified schemes such as the UK's Woodland Carbon Code require use of standardised methods for measuring gains and demonstrating that they are additional to what would have been the case without investment through the Scheme. To conform with statutory requirements, or conditions in voluntary or legal agreements, landowners may need to commission or carry out monitoring for various reasons, including:

- demonstrating that regulatory or credit-sale requirements are being met in line with schedule and are additional to what would have occurred without the relevant payment scheme;
- reviewing the effectiveness of habitat creation and restoration management in achieving target outcomes to support adaptive management;
- detecting adverse trends (domination of habitats by invasive and competitive 'non-target' species) in a timely manner to allow a management response; and
- establishing whether management to deliver credits is also delivering other environmental benefits or contributing to nature recovery, for example the following:
    - land managed for biodiversity also delivers demonstrable/quantifiable gains in soil carbon and/or above-ground carbon; and
    - habitats restored to generate nature credits also provide flood attenuation, for example, by averting flood risk to infrastructure.

The need for systematic monitoring and reporting can place quite onerous demands on farmers and land managers, requiring new skills or expenditure on specialist technical services. Those choosing to engage in varied markets may have to deal with multiple, distinct but overlapping requirements, placing considerable demands on their time and demanding specialist inputs that may not be readily available.

6 Conclusion and future trends

Agriculture faces increasing challenges from climate change, competition for land and increasing costs of production. In many parts of the World, there is declining productivity associated with extensive soil compaction, erosion or salinisation and a host of ecological challenges such as colonisation by invasive and persistent weeds, pollination failure and water shortage. Integrating nature with agriculture is seen as essential to make food production more resilient in the face of climate change and economically more sustainable for farmers. Efforts to promote nature recovery, nature markets and initiatives aiming to promote more sustainable practices in the agriculture sector are likely to gather pace and coalesce.

Some of the goals of Climate Smart Agriculture, for example, are readily translated into mechanisms and units of gain that could potentially be valued and traded: a reduced level of pesticide use could be used as a proxy for averted risk to biodiversity or potentially calibrated against gains in populations of indicator species. There is likely to be growing interest in hydroponics and other production methods that intensify production on a smaller footprint and may release more land for farming of nature. Many forms of regenerative farming also rebuild soil organic matter, restore degraded soils, and restore vegetated margins in ways which can be measured and translated into units of gain which can in theory be tracked and/or traded along supply chains or certified. This requires traceability, which is likely to become increasingly important. Both embody efforts to enhance the longer-term productivity of the land in the face of environmental challenge, whether climate change or decline in fertility due to loss of the soil biota.

The Broadway Initiative, Finance Earth and the Green Finance Institute (Young *et al.*, 2022) envisioned high-integrity markets for nature-based environmental services becoming a major driver of nature recovery across the UK. They envisioned that, by 2030, farmers would routinely participate in markets to supply nature-based projects, that businesses would become active buyers in markets for environmental services to meet both regulated and voluntary obligations and that governments and philanthropic organisations would be strategic buyers of environmental services to help meet local environmental improvement targets where there is insufficient private sector demand. In England, BNG offers a substantial new opportunity to some farmers, especially those in locations where demand for BUs is high, as a result of development pressure. To meet this demand, opting to enter into agreements with habitat banks can remove some of the complexity involved in finalising agreements. Where there is less development pressure, BNG may offer less scope for generation and sale of BUs and other options may be preferable.

Figure 5 Options for habitat resporation on farmland.

Farmers are presented with a growing number of options to consider that allow them to benefit economically from farming for nature on some or all of their land. Some of these are illustrated in Fig. 5. They are not necessarily straightforward to adopt, but the guidance and support needed to engage in nature markets efficiently and profitably is becoming more accessible. Many of these options can be integrated with ongoing production and may benefit it, as well as many of the ecosystem services which have declined so dramatically in most countries.

7 Where to look for further information

- Green Finance Initiative: farming toolkit: https://www.greenfinanceinstitute.com/gfihive/farming-toolkit/
- Taskforce on nature-related financial disclosure: Taskforce on Nature-related Financial Disclosures (TNFD)
- Science-based targets initiative: Science Based Targets initiative #ScienceBasedTargets #ForNature
- UK Habitat Classification: http://ukhab.org
- Wildlife and Countryside Link

8 References

Batáry, P., Dicks, L. V., David Kleijn, D. and Sutherland, W. J. (2015). The role of agri-environment schemes in conservation and environmental management. *Conservation Biology* 29(4): 88–98.

Business and Biodiversity Offsets Programme (BBOP). (2009). *Biodiversity Offset Design Handbook*. Washington, D.C., BBOP. www.forest-trends.org/biodiversityoffsetprogram/guidelines/odh.pdf

British Standards Institute. Little Book of Biodiversity Net Gain. (2023). https://www.bsigroup.com/en-GB/standards/bsi-knowledge/sustainability/the-little-book-of-biodiversity-net-gain/

CIEEM. (2018). Guidelines for Ecological Impact Assessment (EcIA). https://cieem.net/resource/guidelines-for-ecological-impact-assessment-ecia/

Ekroos, J., Olsson, O., Rundlöf, M., Wätzold, F. and Smith, H. G. (2014). Optimizing agri-environment schemes for biodiversity, ecosystem services or both? *Biological Conservation* 172: 65–71.

Landbanking Group. (2024). *Nature equity consultation paper: a new asset class that brings nature on the balance sheet*. The Landbanking Group. https://25771685.fs1.hubspotusercontent-eu1.net/hubfs/25771685/Resources/NatureEquityConsultationPaper_WebVersion.pdf

Lomba, A., Guerra, C., Alonso, J., Honrado, J. P., Jongman, R. and McCracken, D. (2014). Mapping and monitoring High Nature Value farmlands: challenges in European landscapes. *Journal of Environmental Management* 143: 140–150.

Maron, M., Quétier, F., Sarmiento, M., ten Kate, K., Evans, M. C., Bull, J. W., Jones, J. P. G., zu Ermgassen, S. O. S. E., Milner-Gulland, E. J., Brownlie, S., Treweek., J. and von Hase, A., (2024). 'Nature positive' must incorporate, not undermine, the mitigation hierarchy. *Nature Ecology and Evolution* 8: 14–17.

OECD. (2020). A Comprehensive Overview of Global Biodiversity Finance.

Panks *et al.* (2021). Natural England Biodiversity Metric 3.1 User Guide.

Panks *et al.* (2023). Natural England Biodiversity Metric 4.1 User Guide.

Rampling, E. E., zu Ermgassen, S. O. S. E., Hawkins, I. and Bull, J. W. (2023). Achieving biodiversity net gain by addressing governance gaps underpinning ecological compensation policies. *Conservation Biology* 00: e14198. https://doi.org/10.1111/cobi.14198

The Future of Biodiversity Credit Markets. (2023) Governing High Performance Biodiversity Credit Markets. Consultation Paper, March 2023. https://www.naturefinance.net/wp-content/uploads/2023/02/TheFutureOfBiodiversityCreditMarkets.pdf

Treweek, J., Temple, H. and Butcher, B. (2010). "Biodiversity offsets: possible methods for measuring biodiversity losses and gains for use in the UK". *In Practice* 69: 29–32. Romsey, Chartered Institute of Ecology and Environmental Management.

Wildlife and Countryside Link. (2023). Cop 15 one year on – where are we now? https://www.wcl.org.uk/assets/uploads/img/files/2023_COP15_one_year_on_Report.pdf

William, N., Arlidge, S., Bull, J. W., Addison, P. F. E., Burgass, M. J., Gianuca, D., Gorham, T. M., Jacob, C., Shumway, N., Sinclair, S. P. Watson, J. E. M. Chris Wilcox, Milner-Gulland, E J. (2018). A global mitigation hierarchy for nature conservation. *BioScience* 68(5): 336–347. https://doi.org/10.1093/biosci/biy029

World Economic Forum. (2024) Global Risks Report. WEF, Davos. https://www.weforum.org/publications/global-risks-report-2024/digest/

Young, D., Aboobakar, A., Curtis, T., Draisey, Z., Fitton, R., Grundmann, L., Higgs, R., Howard, B., Macedo, C., McAleese, L., Pinkerton, V., Shah, R., Tremolet, S. and Twining, S. (2022). *Financing Nature Recovery UK: Scaling Up High-Integrity Environmental Markets Across the UK*. London, United Kingdom.

Chapter 8

The role of standards in improving the sustainability of livestock production: current challenges and future perspectives

E.M. De Olde, Wageningen University & Research, The Netherlands; and K. McGarr-O'Brien, Wageningen University & Research, The Netherlands and Teagasc, Ireland

1 Introduction

Sustainability standards are well known for commodities such as coffee, cotton and aquaculture. Recent literature reviews on sustainability standards show the emphasis in academic literature on forestry, crop production (e.g. coffee, cotton, palm) and fisheries/aquaculture (Dietz et al., 2022, Traldi, 2021). Sustainability standards have been developed for the livestock sector in recent decades, but research in this area remains limited (De Olde and Busch, 2022).

http://dx.doi.org/10.19103/AS.2024.0133.34

This can be contrasted with the substantial environmental and social impact of the livestock sector, which contributes about 56–58% of the food sector's global agricultural greenhouse gas (GHG) emissions and, given the amount of land used for livestock production, is responsible for up to 78% of terrestrial biodiversity loss in Europe (Leip et al., 2015; Poore and Nemecek, 2018). Moreover, 40% of global arable land is currently used for feed production, and one-third of the global cereal production is used for animals (Mottet et al., 2017).

Farm labour conditions, rural livelihood, public health impact and animal welfare are examples of social issues that are increasingly being raised in food supply chains. Of these issues, animal welfare is the most frequently discussed in the literature. Several recent studies have compared standards underpinning animal welfare labels and revealed a high degree of heterogeneity in requirements between different schemes (Sandøe et al., 2020, 2022; Heinola et al., 2021).

It is surprising that, despite the urgency of addressing the environmental impact and social issues related to livestock production, research and practice relating to standards in this sector is still in its infancy. Sustainability standards and programmes in livestock production are, however, now developing rapidly and continuously to address the multitude of challenges in the sector, partly in response to the high level of societal debate about livestock production and pressure from non-governmental organisations (NGOs). This trend, however, has resulted in a multiplicity of standards, with no consensus on a generic framework, making it difficult to compare standards (More et al., 2017).

As the urgency for more sustainable livestock farming practices increases, actors within the sector are increasingly looking into ways to communicate and improve the sustainability performance of their products. In this chapter, we describe current developments in standards both from the scientific literature and interviews with dairy industry actors in the Netherlands and Ireland, to shed light on where these developments might lead. We use a transition theory framework to show the current position of sustainable development in the context of livestock production standards and envision future directions for (livestock) sustainability standards.

2 Current research on livestock sustainability standards

A comprehensive insight into the role and effectiveness of certification and standards in livestock production is lacking (De Olde and Busch, 2022). For instance, livestock standards are not considered in the reporting of the United Nations Forum on Sustainability Standards and the International Trade Centre's report on the State of Sustainable Markets (UNFSS, 2022; Willer et al., 2022). Moreover, the International Trade Centre Standards Map, a comprehensive database of 322 standards on a wide range of sectors, does not specify standards focused on animal products such as milk, dairy or eggs (ITC, 2021).

Despite this lack of attention, there are scientific publications published on the issue. To develop an overview, a search on Scopus for articles of any date published in English related to certification or labels in relation to animal production (i.e. dairy, beef, pigs, poultry) at the farm level was carried out. Through reading the title and abstract, relevant papers were selected (i.e. those literature sources with a focus on the role and/or potential of certification, labels or standards in livestock production) and this resulted in a total of 98 publications. The publications mainly focused on specific livestock sectors including dairy (n = 39), pigs (n = 14), broiler (n = 9), laying hens (n = 7) and other (9). Twenty further publications focused on animal production in general.

Animal welfare is the most commonly discussed sustainability theme relating to certification in livestock production (n = 34). This includes studies that describe consumers' perspectives, attitudes, and willingness to pay (n = 14), studies that compare animal welfare programmes (n = 10), and other themes (e.g. focused on farmer or stakeholder perspectives, market developments, or studies describing the development of standards). While food safety, food quality and quality assurance programmes were key themes in earlier publications, e.g. Payne et al. (1999), in recent years the attention to other topics such as animal welfare, sustainability and biodiversity is becoming more prominent, e.g. Luhmann et al. (2016); Olsen et al. (2023). Organic production, as a well-known international standard, is the focus of 13 studies which describe the development of certification in different contexts, but also cover economic consequences and impact on other sustainability themes.

Overall, one-third of the publications focus on the preferences of consumers, whereas the perspective of farmers, those that must meet the criteria to be certified, is only addressed in 6 of the 98 studies e.g. van Dijk et al. (2018). While consumers ultimately must be willing to pay for a product produced under a certain standard, often associated with a price premium, it is the farmers that need to ensure they comply with the criteria of a standard. Farmers may need to invest to meet these criteria (i.e. compliance costs) or pay a fee for participation without a guarantee of these investments resulting in a viable business model (More et al., 2017). It is important to note that these extra costs, even if offset by a price premium, can limit access to those markets which have lower regulatory and/or consumer expectations (Vogeler, 2019). Moreover, the steady increase in the number of standards, as well as competition and lack of uniformity between different standards, can create an additional administrative and cost burden for farmers who are not always able to make rapid changes to meet standard requirements such as investment in improving housing systems. Luhmann et al. (2016) also found that despite the overall willingness of future-oriented dairy farmers to participate in a sustainability standard, challenges in implementation are a topic of concern which highlights the need to involve farmers in the development and implementation of standards.

A total of 13 publications made comparisons between standards, either based on comparing the criteria put forward by each standard (n = 9), by comparing certified farms against different standards or by comparing certified farms with those that are not (n = 4). The number of studies in this last category is relatively small although these kinds of studies can potentially provide insight into the effectiveness of standards in improving on-farm conditions.

In comparing on-farm implementation of standards, it can be difficult to attribute improvements to specific initiatives (Mullan et al., 2016). Moreover, when comparing compliance scores of different welfare programmes, Stull et al. (2005) found that these differences are likely due to differences in the design of the standard, rather than differences in animal welfare levels per se. Heterogeneity in standards makes it difficult to compare standards both on paper and in terms of measurable outcomes on farms (More et al., 2017, Heinola et al., 2021). However, ways of assessing the scientific validity and effectiveness of standards are key to understanding and improving the role of standards in moving towards more sustainable modes of production.

Overall, we find that studies looking into standards in livestock production focus mainly on animal welfare and consumer perspectives. Animal welfare is seen as a key topic on which market players are trying to differentiate themselves, resulting in competition and lack of overall coherence and consistency (Vogeler, 2019; Ingenbleek et al., 2013). This competition relies on consumers being aware of and able to distinguish different standards, whilst perceptions of what constitutes good welfare may vary widely.

3 Understanding developments in dairy sustainability standards using transition theory: the sustainable market transformation model (SMTM)

The review of the literature in this area highlighted multiple challenges in the development, implementation and credibility of livestock standards, including competition and associated variability in standards and the involvement of farmers in their development. In this section, we will look in more detail at how these challenges are addressed in current developments in dairy standards and position these developments using transition theory. We go on to discuss how two sustainability standards originating from the Netherlands and Ireland are addressing these challenges.

Transition theory is a theoretical framework that describes a pathway with a number of phases, along which sectors/industries may make transitions towards increased sustainability (see, e.g. Geels and Schot, 2007). There are numerous versions of transition theory, in which multiple actors, each with

Published by Burleigh Dodds Science Publishing Limited, 2024.

different priorities, may influence and work towards sustainable transitions (van der Minne et al., 2021). One transition theory model which is applicable to farming systems and food supply chain transitions is the 'sustainable market transformation model' (SMTM) proposed by Simons and Nijhof (2020). In the SMTM, after a 'problem' (e.g. the need for a sustainability transition) has been identified, there are four phases of transition, linked by cause and effect chains, each with its own unique characteristics and set of actors influencing transition possibilities (Fig. 1; Simons and Nijhof, 2020).

In the first phase of transition, called 'inception', actors start to recognise sustainability issues and, partly due to increased public pressure, initiate experiments, pilot projects or innovations to explore actions they can take to address and improve sustainability (Nijhof et al., 2022). In the second phase, referred to as 'competitive advantage', first movers start to recognise that their approaches from the first phase can be used to differentiate themselves and function as a competitive advantage (Tetrault Sirsly and Lamertz, 2007, Nijhof et al., 2022). Actors start to compete in the market which can result in efficiency gains and innovative solutions. Other companies may follow in response, especially when customers start to look for more sustainable products. The third phase, 'synergy', is based on the idea that actors move away from competition and start to collaborate. Collaborations can take place through platforms and in multi-stakeholder collaboration, for instance, to develop a sector strategy. Support of NGOs and governments for frontrunners is important in this phase (Nijhof et al., 2022). In the fourth phase of institutionalisation, as sustainability solutions and measures start to become the new normal, they are formalised either through top-down legislation or bottom-up industry alignment, or a combination of these. Laggards are faced with the consequences of still applying unsustainable practices and risk losing market access. Finally, while a new normal is being accepted, new sustainability issues are raised, and the first phase begins again in order to address these new concerns (Nijhof et al., 2022). In this way, the iterative SMTM framework illustrates continuous action toward identifying and improving sustainability issues.

In the following section, we describe how different initiatives in the dairy sector can be positioned within this framework.

4 Understanding developments in dairy sustainability standards using the SMTM framework: practical examples

The starting point (e.g. identified 'problem') for this transition is the widespread recognition of sustainability challenges within the dairy sector, and a need to increase dairy sustainability. In the following paragraphs, we provide examples of developments across the phases of the SMTM framework. Figure 1 illustrates

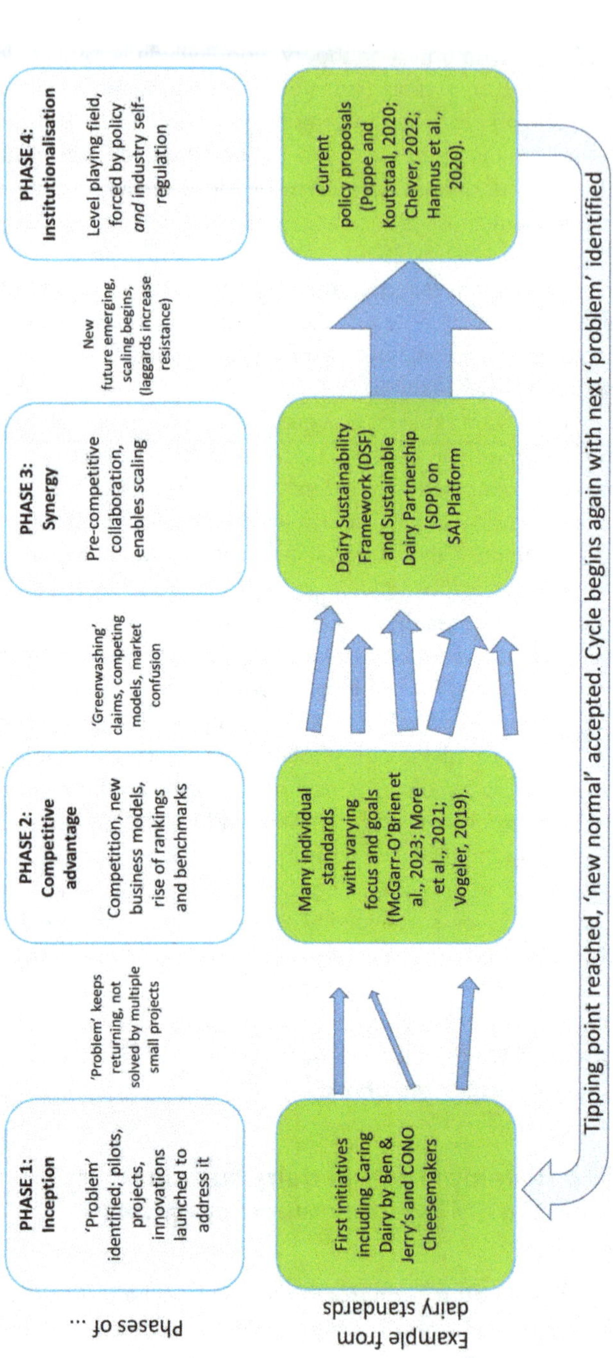

Figure 1 Phases of the sustainable market transformation model (adjusted from Simons and Nijhof (2020)), and their application to dairy sustainability standards.

the current position of the dairy sector's sustainability standards on their journey to a sustainability transition.

Caring Dairy can be seen as one of the first dairy programmes on sustainability and presents an example of an initiative from the inception phase. In 2004, ice cream maker Ben & Jerry's initiated the project for their European production operation in the Netherlands (Van Calker et al., 2005). Their aim was to ensure more sustainable dairy production and align this with Unilever's sustainable agriculture policy (Unilever bought the Ben & Jerry's brand in 2000). The project has turned into the Caring Dairy programme in which 400 dairy farmers in the Netherlands participate to improve their on-farm sustainability (Ben & Jerry's, 2023). Over time, they also recognised the value of sustainability as a marketing tool: *'Responding to stakeholder preferences not only matters for the continuity of a company but also presents market opportunities'* (Hoekstra et al., 2013, p. 197).

The second phase, characterised by competition, can be recognised by the wide variety of current standards and labels existing in the dairy sector. For example, in the study of McGarr-O'Brien et al. (2023), 19 dairy standards in Europe and North America were identified and analysed. Differences between standards can result from differences in geographical scope, sustainability focus or ambition levels. Individual retailers are also increasingly launching programmes to differentiate themselves and show consumers they are responding to sustainability concerns. In the Netherlands, for example, competition between sustainability programmes of retailers has been observed: *'... Albert Heijn made an agreement in 2017 with a private dairy processor to be able to offer 'distinguishable' milk. This resulted in the 'Better for Cow, Nature and Farmer'-milk which by now has already been available for two years. But this milk has lost its distinguishable character through the introduction of PlanetProof-milk from FrieslandCampina which has helped other retailers to make a step forward.' 'That's why Albert Heijn has to go beyond again'* (Janssen, 2020). This proliferation of standards may create confusion for consumers and potential problems of lock-in for farmers once they make on-farm changes to participate in a particular standard (McGarr-O'Brien et al, 2023).

Signs of developments towards the third phase can be recognised in different initiatives such as the development of the Dairy Sustainability Framework (DSF), linked to the Sustainable Agriculture Initiative (SAI) platform (DSF, 2023a). The DSF was created and is led by members of the dairy sector, and its members account for over 30% of global milk production. The aim is to harmonise sustainability approaches throughout the dairy sector. To that end, 11 sustainability criteria and associated indicators are included in the framework. The framework could contribute to reducing the number of individual standards and audits farmers need to comply with. The Sustainable Dairy Partnership (SDP) is a business-to-business programme for milk processors and dairy customers that builds upon the DSF (DSF, 2023a). Members of the DSF and SDP monitor and report their progress on the 11 sustainability criteria (DSF, 2023a). Using the same criteria facilitates international comparison and reporting on the progress made along the dairy supply chain. With regard to pre-competitiveness, a representative of SDP member Danone stated: *'Through SAI platform we actually have a platform where we can have that discussion in a safe space and with a lot of honesty and transparency so we can find solutions that address everybody's needs I would say'* (SAI Platform, 2020).

Meanwhile, the first proposals to move towards the fourth phase of institutionalization are being made, especially within the European context, for instance by the Dutch Council for the Environment and Infrastructure (2019). They have advised linking public eco-scheme payments with private sustainability initiatives using performance indicators linked to a point system. They argue: *'As the management and oversight of these schemes will be in the hands of private organizations (accredited and supervised by government), implementation will be lean, bureaucracy kept to the minimum and the possibilities for synergy increased.'* (p. 2). A recent study requested by the European Parliament's Committee on Agriculture and Rural Development has recommended *'to encourage the use of relevant certifications' schemes within the CAP national strategic plans to achieve the EU sustainability objectives'* (Chever, 2022, p. 8). They also argue that each scheme should be transparently assessed in terms of their requirements and the method of implementation to prevent any risk of greenwashing. Besides greenwashing, another challenge to linking public and private sustainability schemes is whether or how comprehensively multiple sustainability objectives/criteria of, e.g. EU strategies are addressed by current standards (Chever, 2022). Although they show the challenges involved, these developments can be seen as examples of how actors are considering how institutionalization of current developments could be achieved.

Based on our insights into the development of standards in the dairy sector, we argue that current efforts are mainly focused on the second and third phase of sustainability transformation. A relatively large number of standards are developing and trying to differentiate themselves, for example by emphasising different sustainability themes or ambition levels. This competitive focus is unlikely to continue, due to pressure from the market to prove sustainability credentials, as well as proposals to align or link private livestock standards with public sustainability goals. Some livestock standards have already submitted themselves to scrutiny from national accreditation boards advising on audit and certification processes (e.g. Irish National Accreditation Board (INAB)). However, most standards remain self-governing without official regulation or any sector-wide alignment. We discuss the challenges involved in the sustainability transformation later in the chapter.

5 How dairy sustainability standards work: two case studies

A framework such as the SMTM allows identification of key actors and the dairy sector's current trajectory towards a sustainability transition. However, to better understand the standards and how they can increase dairy sustainability, it is necessary to investigate individual standards in more detail. For example, who governs the standards, what are their specific aims, and what challenges do they face? What are the sustainability topics they address, and how is success measured? Additionally, to understand the sustainability of the sector as a whole, is there an interaction between the various stakeholders who are steering dairy sustainability and, if so, by what mechanisms and with what goals? All these considerations influence the ability of standards to improve dairy sector sustainability.

One way to explore these questions is through in-depth interviews with those involved with particular standards. Interviews were therefore undertaken with representatives of two sustainability standards:

- Origin Green in Ireland
- Foqus Planet in the Netherlands, Belgium and Germany

Further details of these standards are shown in Box 1. Results from these interviews reveal emerging developments in current dairy sustainability standards and potential future pathways. Similarities between challenges and opportunities across sectors mean it may also be possible to apply these insights to other sectors, which we discuss further in the chapter.

Box 1: Two dairy sustainability standards

Foqus planet (FrieslandCampina dairy cooperative) and Origin Green (Bord Bia, Ireland's food and drinks board) represent two large sustainability standards in which the majority of Dutch and Irish dairy farmers, respectively, participate. Utilising market-based sustainability standards is occurring in both the Netherlands and Ireland, but in different ways. In Ireland a whole-sector, top-down sustainability standard was launched for the dairy sector by the Irish Food Board, Bord Bia, in 2013. There are also additional Irish organic and private sustainability programmes for dairy production. In the Netherlands, the sustainability standards that have emerged in the dairy sector are mainly driven by individual dairy cooperatives, processors, retailers, or NGOs, without a collective approach such as that in Ireland.

Foqus planet

Foqus planet is a quality assurance and sustainable development standard of the largest Dutch dairy cooperative FrieslandCampina, which is the eighth largest global dairy cooperative. FrieslandCampina processes its own milk under a variety of brands, and also sells milk and dairy products to business customers including large brands such as Danone (FrieslandCampina, 2023b). The 9927 member dairy farms in the Netherlands, Belgium and Germany together deliver 9.5 billion kg of milk per annum (FrieslandCampina, 2023a).

Foqus planet was first introduced as a quality assurance programme for the milk being produced by FrieslandCampina's dairy farms. Within the basic Foqus planet requirements, which are mandatory for all FrieslandCampina cooperative members, there are over 200 criteria. In 2012, Foqus planet also began a sustainable development standard which runs in parallel to, but is distinct from, the quality assurance programme. In 2015, the sustainable development section was further developed from reward on effort to reward on results. The sustainable development section of Foqus planet allows farmers to develop themselves on a minimum of two out of three sustainability themes: animal health and welfare, biodiversity and climate. For these themes, eight indicators with thresholds have been developed, and farmers can earn a premium based on their performance on these indicators. In addition, there is a fourth theme on outdoor grazing for which a premium can be obtained. Through different administrative and on-farm audits, the performance of the members is checked. Moreover, there is an independent auditor checking this compliance control of FrieslandCampina. Foqus planet is reviewed and updated annually with input and permission from farmers of the FrieslandCampina cooperative, among other stakeholders (Focus Planet Representative, 2022).

Origin Green

Nearly 100% of milk sold to Irish processors is enrolled in the Origin Green sustainability programme developed and managed by the Irish Food Board, Bord Bia (Origin Green, 2022). Approximately 16 500 dairy farms participate in the standard, and membership is a precondition for farmers to supply their milk to processors in Ireland (Bord Bia, 2022). Origin Green is operationalised for dairy production by the Sustainable Dairy Assurance Scheme (SDAS) audit process, which is itself audited annually by the Irish National Accreditation Board (INAB) (Bord Bia, 2022). The SDAS audits are conducted by auditors who are self-employed but trained by Bord Bia. There are 170 criteria required by the audit.

Origin Green's dairy audit SDAS was introduced in 2013 in anticipation of the lifting of EU dairy quotas in 2015, which resulted in a 50% increase in production (NDC, 2022). This meant the need to establish a competitive advantage for Irish dairy products in the global market, based on their quality and sustainability, to support export sales.

Today, Ireland exports approximately 90% of its dairy products (NDC, 2022). Origin Green aligns itself with 15 of 17 of the United Nations sustainable development goals (SDGs) and has been a member of the UN Global Compact since 2018 (Origin Green, 2022). SDAS has until now been operating without major updates to the first version of the standard. This is due to change, however, in the next few years with the planned introduction of a four-year cycle of reviewing and updating the standard after input from multiple stakeholders, including farmer representatives, and a new process of review by a panel of experts. Farmer input is sought prior to any changes to the SDAS audit, this includes on-farm testing of new criteria to evaluate how their assessment would be carried out in a real-world environment (Origin Green Representative, 2023).

Individual research interviews in late 2022/early 2023 with one Foqus planet and one Origin Green representative revealed common features in their collaborative, multi-stakeholder approach, and the importance of involving farmers in creating and improving standards. Each representative also addressed how the standards balance inherent trade-offs, as well as challenges the standards face. Notable challenges faced by both standards are the increasing external pressure being applied by consumers, business customers and governments for sustainability standards to continue improving, as well as increasing demands to supply proof of sustainability claims (although the specific sustainability claims are distinct to each individual standard). In the following section, we will describe these trade-offs and challenges in more detail.

5.1 Addressing trade-offs

There are inherent trade-offs in developing and implementing standards such as Origin Green and Foqus planet, where the aim is to increase sustainability across multiple sustainability topics and dimensions (e.g. social, environmental, and economic sustainability). Trade-offs include the need to make continuous improvements to standards while minimising the burden on farmers which frequent changes to standard criteria may create (De Olde et al., 2018; Bush et al., 2013). Interviewees for both Origin Green and Foqus planet provided their perspectives on the balancing act of keeping standards scientifically valid at the same time as being practical for farmers to implement.

In terms of accessibility for farmers, the Foqus planet representative described the need to keep sustainability criteria achievable for farmers, stating '*...if you want to keep dairy farmers motivated...then you should not do something that cannot be accessible to (farmers), because then of course no* *one will move.'* FrieslandCampina have received feedback that farmers can feel 'bombarded' by constant and frequently changing demands. Involving farmer-owners of the cooperative in the annual review of Foqus planet was emphasized as being important. Similarly, the Origin Green representative stated that *'It needs to be a strong, robust standard, but a standard everyone can live with.'* Testing of criteria and how they would be operationalised on the farms, with farmers, was an important focus for Origin Green to ensure accessibility. Continuous improvement of standards was recognised as important by both standards, as the Origin Green representative stated, *'It's a response to regulation, legislation, it's a response to market need…we mightn't like change but change is inevitable.'*

In terms of ensuring scientific validity, both standards rely on legal requirements for some of the sustainability topics they address (e.g. food safety, animal health, animal welfare, commitments to climate targets under the Paris Agreement). The Origin Green representative mentioned it is easier to prove scientific validity on criteria that are readily quantifiable. Each indicator must be (scientifically) addressed separately, as different experts must be consulted for each (e.g. biodiversity requires ecologists, animal welfare requires veterinarians, GHG reduction requires experts in calculating carbon footprints). The Foqus planet representative noted that, even amongst experts, there is not always a consensus on the most appropriate measure, e.g. in the case of biodiversity and animal welfare indicators.

Different tools are used to verify the performance of farmers against the eight indicators in the Foqus planet scheme, including the Annual Nutrient Cycling Assessment (Van Dijk et al., 2022) and KalfOK for young stock-rearing quality (Santman-Berends et al., 2018). Although individual types of measurement can be scientifically validated, the selection of the criteria used by different schemes cannot. Criteria are included because they fit the company's strategy and react to factors such as national policy and feedback from a company's business customers and consumers.

5.2 The demand for continuous change and impact

A common feature identified by both the Foqus planet and Origin Green representatives is an acknowledgement of escalating scrutiny of their standards over the last few years. This sense of increasing pressure is coming from multiple sources, including increasing consumer awareness on livestock sustainability issues, government policy changes aimed at increasing agricultural sustainability (e.g. the Climate Action Plan in Ireland (Government of Ireland, 2023), the vision on Circular Agriculture in the Netherlands (Netherlands, 2019), and the European Farm to Fork Strategy (European Union, 2020)), and pressure from NGOs focused on sustainability issues. The Foqus planet representative stated

'We know that NGOs also have goals. Often the same goals we pursue, but differ in the way we look at the path to take to achieve the goal. Yes, we hope that there will also be appreciation for what you already do. And they do keep us on our toes. We think it is important to keep an eye on what is going on in society.'

This increased awareness of the importance of addressing sustainability issues in livestock production also translates in an increase in sustainability demands from customers: *'everyone has demands and we see more and more that these demands also increase' (Foqus Planet Representative, 2023).* Meeting the demands of different stakeholders and customers can be challenging and can result in duplication of effort. Sustainability ambitions articulated, e.g. by consumers or NGOs can sometimes get ahead of what is feasible for farmers or supported by research. Ensuring farmers are on board when introducing changes in a standard is key, particularly in a cooperative like FrieslandCampina: *'Sometimes our members feel overwhelmed by the speed of changes and number of demands. It was not always clear to them what challenge we were facing and what goal we were pursuing and why. They wanted to be involved in finding a solution for our challenges.'* For that reason, changes in the governance were made to better involve farmers in the development and integration of new standard requirements.

One important pressure being felt by both standards is the need to 'prove' they are making a sustainable impact. This was a clear piece of feedback from Origin Green's most recent INAB audit, as they explained in reference to upcoming changes: *'…there was quite a bit of pushback in terms of the assessors saying…you're gathering a lot of info, doing a lot of assessments, doing a lot of measurements but you're not necessarily following through and saying 'do x, y, or z' on the back of that… we're quite keen that within the standard this time we're actually going to have things in there that promote more sustainable practices at farm level. And not just promote but require farmers to implement them…'* (Origin Green Representative, 2023).

Both standards are confronted with administrative challenges given the large number of farmers involved, involving both physical and administrative audits. For instance, for Origin Green, operating the SDAS audit includes over 16 500 dairy farms across Ireland.

5.3 Opportunities

An opportunity recognised by representatives of both standards is that they are making a contribution to improving sustainability, and that there is a strong social contract to use standards to continue to improve sustainability in dairy farming. The Origin Green representative stated that even 3 or 4 years ago, the momentum towards increasing sustainability was not so well-established. Now, due in part to legislation (e.g. Ireland's Climate Action Plan), as well as consumer sentiment, the feeling towards sustainability has changed. It now feels like there is more alignment with the entire agriculture industry in Ireland traveling in the same direction. It is assumed there will therefore be less resistance from farmers to introducing changes. This sentiment was also echoed by the Foqus planet representative, where industry-wide discussions on the next best steps for standards are taking place, e.g. through participating in the SAI Platform's Sustainable Dairy Partnership (of which Origin Green is also a member) (SAI Platform, 2023).

Both interviewees emphasized the collaborative nature of their sustainability standards. The Foqus planet representative explained *'...that should be the strength of a cooperative...that you also include dairy farmers very much in the entire decision-making process and explain why it is necessary... And so, when there are valid arguments from dairy farmers, things are really being adjusted based on what dairy farmers themselves indicate.'* In the research and development stage of the standards, the Origin Green representative stated *'We look at other schemes and standards in operation [in] other countries around the world, we look at private standards, public standards. We look at what's coming back from the marketplace...'.* In doing so, developments in the perceptions of consumers and business customers are considered as well. The importance of addressing customer views was also expressed by the Foqus planet representative. For instance, they work together with their customers to benchmark standards with the Foqus planet programme. Moreover, they collaborate with customers to study the impact of certain practices (e.g. a pilot in collaboration with Barry Callebaut chocolate on the use of feed additives to reduce GHG emissions (FrieslandCampina, 2023c). Harmonising the use of indicators was also identified as key to aligning sustainability initiatives across organisations (Foqus Planet Representative, 2022).

5.4 Lessons learned

Through discussions with representatives of two large dairy sustainability standards, it was apparent that a balancing act is undertaken by both organisations (FrieslandCampina and Bord Bia). Differing demands coming from a variety of stakeholders (e.g. ease of use and benefit for farmers, validity and continuous improvement for researchers and consumers, a unique selling point for customers) provide pressures that shape the development and updating of standards. Additional insights gained from the interviews include the common experience of increasing pressure from multiple stakeholders to continually improve sustainability and to provide proof of effectiveness. Opportunities for collaboration across the sector and a desire for alignment of sustainability goals and methods were also identified and pointed to possible dairy sustainability standard development in the future.

Published by Burleigh Dodds Science Publishing Limited, 2024.

6 Future developments: competition versus convergence

In our analysis of developments in dairy standards, we identified several future directions for sustainability standards involving both opportunities as well as risks in moving towards a more sustainable agricultural sector. This section looks at the degree to which individual standards seek to compete with each other and the degree to which they are converging towards a common industry-wide standard

Using the SMT framework, we have argued in this chapter that current efforts in dairy standards are mainly focused on the second and third phases of transformation (i.e. competitive advantage and synergy). There may be conflicting pressures on how far individual standards focus on competition or convergence. Standards that want to be recognised for their impact, to prove their claims are not 'greenwashing' (e.g. through strong internal governance and measurement), would have an interest in progressing to phase three and establishing convergence. As an example, the CEO of FrieslandCampina argued: *'You should not want to compete on sustainability'* (Van Rossum, 2023). However, not all standards may desire to collaborate across the entire sector. Standards which are primarily driven by market demand and attempts to differentiate products may wish to continue operating within the competitive phase (phase two).

Those responsible for individual standards may need incentivisation from external sources to work in a more collaborative way with other organisations or the public sector. Sources of incentivisation could be legislation or pressure from policymakers to regulate sustainability initiatives, or a perceived benefit to participating in a sector-wide consensus defining dairy sustainability. Sustainability ranking and benchmarking are considered as supporting actions towards the third phase (Nijhof et al., 2022). An example is the Business Benchmark on Farm Animal Welfare (BBFAW) that analyses reported management practices, processes and reporting on farm animal welfare by the 150 largest food companies globally (BBFAW, 2020). In response, food companies, like Unilever, report their sustainability ratings and ranking on their website (Unilever, 2023).

While the interest to move from phase 2 to phase 3 may vary between standards, the pressure to move to a more aligned approach with some form of regulatory oversight is becoming stronger, in particular to maintain consumer confidence (More et al., 2021; Boström et al., 2015; EC, 2010). As FrieslandCampina's CEO stated in an interview: *'Because of all the labels, the consumer is not able to see the forest through the trees'* (Van Rossum, 2023). Proposals to introduce a minimum requirement for standards have therefore been made (e.g. Remkes, 2022).

In terms of convergence, both standards (Foqus planet and Origin Green) showed some degree of public–private alignment through the inclusion of both public and private goals in their programmes. Further alignment of public goals in private initiatives is seen as a way to limit the administrative burden on farmers to comply with requirements from different actors. Alignment of private standards and public policy is, however, no guarantee that farmers will be well supported. Indeed, sustainability 'front-runner' dairy farmers in the Netherlands, when interviewed, were concerned about the investments required to conform with sustainability legislation, about the risk that requirements could change from year to year, and the financial and business continuity risks involved (Engelberts et al., 2021). This instability has caused some farmers to hold negative views on how policy supports their businesses (Engelberts et al., 2021). This is also reflected in the interview with Foqus planet, where the need for a common 'dot on the horizon' (long-term vision for dairy sustainability) was emphasised. Any arrangement between public policy and private sustainability standards will need to consider the importance of long-term planning and stability for farmers, as well as the necessity of earning trust and buy-in from farmers in the process of creating and participating in the standard. In the end, the change being demanded by society and governments has to be achievable in practice by farmers and willingly implemented by them.

Moving towards convergence and ultimately institutionalisation puts pressure on the entire sector to change, as institutionalisation establishes as a new normal in which unsustainable behaviour is no longer accepted (Simons and Nijhof, 2020). Currently, we observe that some standards that already include all dairy farmers within the organisation's remit, such as every cooperative member, or all farmers within a country, as seen in the interviews with the two standard representatives in this chapter. We also recognise that there are more exclusive standards focused instead on establishing niche markets with price premiums for a segment of products, often concentrating on a specific sustainability issue (e.g. through product labels highlighting animal welfare). The starting point for both approaches is very different. For the more inclusive standards, farmer acceptance of the value of the standard and the feasibility of compliance with it key while, for the more niche exclusive approach, consumer trust and demand is critical. In the latter group, farmers are often self-selective and already align with the ambitions and values of the standard. To establish synergies between public ambitions and private standards, an inclusive approach would be needed that offers everyone the opportunity to transition towards increased sustainability. *'We can create real impact if we can elevate the entire sector and improve it.'* according to the CEO of FrieslandCampina (Van Rossum, 2023).

7 Future developments: stability versus continuous improvement and harmonisation versus differentiation

Moving toward phase 4 would mean that sustainable practices have become embedded in institutions and have established a new normal (Simons and Nijhof, 2020). While sustainable dairy standards have not yet reached phase 4, it's also important to recognise that, even when commonly accepted sustainable practices become embedded, this might, in turn, result in the identification of new sustainability issues (Nijhof et al., 2022). As such, sustainability transformation can be regarded as a continuous process (as also indicated by the concept of sustainable development) for two main reasons. First, as Nijhof et al. (2022) argue: *'sustainable market transformation ends when the initial sustainability problem is solved or, at least, the level playing field is increased when it comes to sustainability.'* Indeed, in many cases there will not necessarily be an absolute state in which a sustainability issue is solved, but rather a relative improvement. And second, sustainability as a concept is open to many interpretations and methods of implementation, depending on the perceptions of those involved, their context and time (Bell and Morse, 2008). This suggests that which environmental, economic and social priorities are considered important, and which should be addressed in standards, varies between stakeholders and may evolve over time. For instance, as the concept of regenerative agriculture is currently gaining momentum, the importance of soil health and biodiversity are increasingly recognised as important sustainability themes to be included in standards (Schreefel et al., 2020), while other organisations have even developed specific regenerative standards, e.g. Regenerative Organic Certified (Regenerative Organic Alliance, 2023).

The development of standards will be an iterative and ongoing process. This presents challenges for everyone involved. Developers of standards will be required to include new sustainability issues, and farmers are likely to have to comply with an increasing list of criteria. This may be at odds with the stability that farmers need for long-term planning. Moreover, an increase in the comprehensiveness of standards might come at the expense of practicality of implementation by farmers (De Olde et al., 2018). A continuous proliferation of and increase in the complexity of sustainability standards might also undermine consumers' understanding of and trust in standards.

Harmonisation is seen as an important step in the alignment of standards towards becoming institutionalised. In particular, harmonisation of terminology, criteria and monitoring approaches is key, for instance, to allow comparison between standards. As early as 2010, the European Commission published guidelines for best practices in voluntary certification schemes (EC, 2010). However, these recommendations do not seem to have progressed

(More et al., 2021). While harmonisation is important, there are also some risks involved in harmonising approaches for standards. First, harmonisation does not guarantee comprehensiveness of standards. Instead, it can focus on only agreeing upon a selected set of 'usual' criteria, rather than striving for a comprehensive approach to sustainability. Second, harmonised or generic approaches may fail to address the context-specific nature of sustainability. For instance, while animal welfare is a key issue in North-West Europe, it is not the main sustainability concern in many other parts of the world (Garcia and McGlone, 2022). It's therefore important that harmonisation efforts still allow for a diversity of perspectives and context-specific values and practices.

As an example, the SAFA guidelines of the FAO, aimed at harmonising sustainability assessments in food and agriculture (FAO, 2014), require a detailed contextualisation process in order to define what would qualify as sustainable in a specific context. Sustainability is therefore defined through how it is operationalised and translated into practice. In the context of aquaculture, (Osmundsen et al., 2020) even stated: *'Sustainability indicators determine what sustainable aquaculture has come to be.'* Standards play an important role in defining what is regarded as sustainable and, in that sense have an important responsibility in recognising the values of those affected. In doing so, they will have to balance 'individual', context-specific approaches with harmonised, generic approaches (Gasso et al., 2015).

8 Future developments: measuring impact

Given the significant contribution of the livestock sector to environmental problems such as GHG emissions, those working in the industry are exploring ways to reduce their environmental impact. Large food companies such as Nestlé and Danone have formulated commitments to reduce their climate impact by 50% by 2030 and achieve net zero emissions by 2050 (Nestlé, 2023; Danone, 2023). Such commitments are then translated to their producers, putting the focus on measuring the sustainability performance of farms. This results in increased pressure on standards to provide proof of actual sustainability improvements. However, recent studies evaluating the efficacy of standards in improving sustainability showed rather mixed results (Dietz et al., 2022; Traldi, 2021; Oya et al., 2018).

To improve the reliability of standards to contribute to sustainable development, a different approach may be needed. While historically target- and practice-based indicators have been commonly used in standards (Konefal et al., 2023), a more recent shift towards including performance-based indicators can be observed (Hatanaka et al., 2022). Performance-based indicators measure the performance or impact of farming practices through direct observations, measurements or calculations. Target- and practice-based indicators focus more on management. Target-based indicators refer to having plans or policies in place (e.g. having water use records), while practice-based indicators focus on whether certain practices or processes have been undertaken (e.g. planting flower strips or implementing seasonal grazing) (FAO, 2014). Target- and practice-based indicators are easier to check in an audit since they only require checking whether a farmer has a plan in place or can provide evidence of applying certain practices. However, the link between having a target or applying a practice and actual sustainability improvement in practice is not always clear or verifiable.

Performance-based indicators, however, often require on-farm observations and measurements over a period of time and are therefore associated with higher investment in time and costs. Additionally, variability in e.g. weather events can add complexity to the verification of a farm's performance against identified indicators. Nevertheless, performance-based indicators are perceived as more relevant as they give more insight into the performance and impacts of a farm (FAO, 2014), and can also drive continuous improvement (Hatanaka et al., 2022). Such measurements give farmers insight into what works, enabling them to make changes in their management to improve their performance (Konefal et al., 2023).

However, a shift to performance-based indicators comes with its own risks. First of all, it may require organisations to be more selective in the sustainability issues they want to address in their standards, given the associated time and cost investment involved in measurement (GlobeScan, 2022). Reports from the Dairy Sustainability Framework, e.g. show that members prioritise GHG emissions and animal care whereas water quality, on-farm waste management and working conditions receive the least priority (DSF, 2023b). Whilst focussing on measuring performance in a few priority areas is understandable, the concept of sustainable development requires an integrative approach that encompasses environmental, economic and social dimensions, as promoted by the UN's Sustainable Development Goals (SDGs). A focus on a limited set of sustainability themes could result in trade-offs, for instance, actions to improve animal welfare potentially increasing GHG emissions. As highlighted by Osmundsen et al. (2020) '... *The mutual dependence between issues and impacts in the environment, economic, culture, and governance domains needs to be highlighted in order to create solutions that are truly sustainable.*'

Second, performance-based approaches may contribute to the idea that standards can be considered as neutral or objective. Standards can help to shape what is considered as sustainable, from the selection of sustainability themes, the experts involved, what is measured, the compliance approach and the audit procedure. As formulated by Amundsen (2022): '*This idea of standards being neutral and objective derives from a technical understanding of certification, which is centred on the perception that sustainability is a technical outcome to be achieved.'* In particular, when using performance-based indicators, there is a risk that those issues that are not easily quantified, assessed or audited are disregarded which may undermine sustainable development more broadly defined. While a shift towards performance-based approaches can help to validate and 'prove' the impact of standards, it's important to recognise that *'The use of numbers gives an illusion of certainty and precision that is often at odds with the uncertainty that characterizes many complex and multidimensional policy issues, such as sustainability ...'* (Kovacic, 2018). This also holds for specific issues such as GHG emissions, where assumptions made in Life Cycle Assessments (LCAs), such as the allocation method and estimating emission factors, have an important impact on the results (Kyttä et al., 2022). Recognising the uncertainty, ambiguity and complexity involved is therefore of critical importance to ensure the validity of and trust in standards as tools for sustainable development into the future.

Ultimately, there is a difference in creating and showing impact. It is important to recognise these differences, especially as not all impacts, or everything that is valuable in terms of sustainable agriculture, can be measured. Alrøe et al. (2017) argue that, in addition to performance-based approaches such as the use of indicators and benchmarks, there are value-based approaches which *'aim at communicating and mediating sustainability values to enable coordinated and cooperative action to transform the food system'.* The development of Participatory Guarantee Systems may be regarded as an example of this approach, in which producers are certified through the active participation of stakeholders. Loconto and Hatanaka (2018) consider it as an alternative to third-party certification that is more *'democratic and hence, produces forms of sustainability that incorporate the lived experiences of people around the world'.* Such approaches enable a more local, accessible and adaptive approach in which producers and consumers are linked (Jacobi et al., 2023). Value-based approaches are regarded as better facilitating knowledge exchange and trust-building (IFOAM, 2022). Nonetheless, the benefits of value-based approaches are difficult to communicate in the market and may lack credibility (Alrøe et al., 2017). While value-based approaches may have great potential to create impact, their capacity to show and report their impact in terms of contributions to industry commitments will be challenging.

9 Conclusion

Through this chapter we have explored the role of sustainability standards in livestock production. We have applied the lens of transition theory to recognise the progress and challenges involved in this sustainability transformation. We recognise that standards are confronted with, in some situations, conflicting pathways for development, all with their own opportunities and risks. This means that there is no ultimate solution to address these challenges, and attention to uncertainty, context specificity and inclusiveness of stakeholder perspectives is key. Standards, through their selection of issues and criteria, define what is regarded as sustainable, and which practices are implemented. As a result, standards can play an important role, in particular when aligned with public goals. This, however, also comes with great responsibility. Responsibility in being transparent about their approach, assumptions and limitations. Responsibility to keep an eye on undesirable trade-offs that may occur. And, responsibility in involving stakeholders, and especially farmers, in the development of standards for a sustainable future.

10 Where to look for further information

The following articles provide a good overview of the subject:

- McGarr-O'Brien, K., Herron, J., Shalloo, L., De Boer, I. J. M. and De Olde, E. M. 2023. Characterising sustainability standards in dairy production. *Animal* 17(7) doi: https://doi.org/10.1016/j.animal.2023.100863
- De Olde, E. M. and Busch, G. 2022. Opinion paper: Role of standards in the transformation of the livestock sector. *Animal* 16(8) doi: https://doi.org/10.1016/j.animal.2022.100588

More information on the two case study standards can be found here:

- Foqus planet by FrieslandCampina (https://www.frieslandcampina.com/owned-by-farmers/foqus-planet)
- Origin Green by Bord Bia (https://www.origingreen.ie) and the Sustainable Dairy Assurance Scheme (SDAS) (https://www.bordbia.ie/farmers-growers/get-involved/become-quality-assured/sustainable-dairy-assurance-scheme-sdas/).

Key research in this area can be found at the following organisation:

- Wageningen University & Research (www.wur.eu).

11 References

Alrøe, H., Sautier, M., Legun, K., Whitehead, J., Noe, E., Moller, H. and Manhire, J. 2017. Performance versus values in sustainability transformation of food systems. *Sustainability*, 9, 332.

Amundsen, V. S. 2022. From checking boxes to actual improvement: a new take on sustainability certification. *Aquaculture* [Online], 548. https://dx.doi.org/10.1016/j.aquaculture.2021.737672

BBFAW. 2020. *Business Benchmark on Farm Animal Welfare* [Online]. Available: https://www.bbfaw.com/ [Accessed 1 February 2023].

Bell, S. and Morse, S. 2008. *Sustainability Indicators: Measuring the Immeasurable?*, London, UK, Earthscan.

Ben & Jerry's. 2023. *Caring Dairy* [Online]. Available: https://www.benjerry.co.uk/values/how-we-do-business/caring-dairy [Accessed 15 March 2023].

Bord Bia. 2022. *Sustainable Dairy Assurance Scheme (SDAS)* [Online]. Available: https://www.bordbia.ie/farmers-growers/get-involved/become-quality-assured/sustainable-dairy-assurance-scheme-sdas/ [Accessed 21 November 2022].

Boström, M., Jönsson, A. M., Lockie, S., Mol, A. P. J. and Oosterveer, P. 2015. Sustainable and responsible supply chain governance: challenges and opportunities. *Journal of Cleaner Production* [Online], 107. https://dx.doi.org/10.1016/j.jclepro.2014.11.050.

Bush, S. R., Toonen, H., Oosterveer, P. and Mol, A. P. J. 2013. The 'devils triangle' of MSC certification: balancing credibility, accessibility and continuous improvement. *Marine Policy*, 37, 288–293.

Chever, G. L. 2022. *Research for AGRI Committee-Farm Certification Schemes for Sustainable Agriculture, State of Play and Overview in the EU and in Key Global Producing Countries* [Online]. Brussels. Available: https://www.europarl.europa.eu/RegData/etudes/STUD/2022/699633/IPOL_STU(2022)699633_EN.pdf [Accessed 15 March 2023].

Danone. 2023. *Danone Announces an Ambitious Plan to Reduce Its Methane Emissions* [Online]. Available: https://www.danone.com/media/press-releases-list/danone-announces-an-ambitious-plan-to-reduce-its-methane-emissions.html [Accessed 1 April 2023].

De Olde, E. M. and Busch, G. 2022. Opinion paper: role of standards in the transformation of the livestock sector. *Animal*, 16, 100588.

De Olde, E. M., Sautier, M. and Whitehead, J. 2018. Comprehensiveness or implementation: challenges in translating farm-level sustainability assessments into action for sustainable development. *Ecological Indicators*, 85, 1107–1112.

Dietz, T., Biber-Freudenberger, L., Deal, L. and Börner, J. 2022. Is private sustainability governance a myth? Evaluating major sustainability certifications in primary production: a mixed methods meta-study. *Ecological Economics* [Online], 201. Available: https://www.sciencedirect.com/science/article/pii/S0921800922002087 [Accessed 1 November 2022].

DSF. 2023a. *Dairy Sustainability Framework* [Online]. Available: https://dairysustainabilityframework.org/ [Accessed 20 February 2023].

DSF. 2023b. *DSF Annual Reporting for the 2021 Calendar Year* [Online]. Available: https://dairysustainabilityframework.org/wp-content/uploads/2023/01/DSF-2021-Reporting.pdf [Accessed 15 February 2023].

Dutch Council for the Environment and Infrastructure. 2019. *European Agricultural Policy: Working Towards Circular Agriculture* [Online]. Available: https://en.rli.nl/sites/default/files/190703_rli_advice_european_agricultural_policy._working_towards_circular_agriculture.pdf [Accessed 1 April 2023].

EC. 2010. *Commission Communication – EU Best Practice Guidelines for Voluntary Certification Schemes for Agricultural Products and Foodstuffs* [Online]. European Commission. Available: https://eur-lex.europa.eu/LexUriServ/LexUriServ.do?uri=OJ:C:2010:341:0005:0011:en:PDF [Accessed 20 April 2023].

Engelberts, L., Van Rheede, A., Kievit, H. and Nijhof, A. 2021. Appreciating multiple realities in the transformation towards a sustainable dairy sector: an explorative study from the inside-out perspective. *Agronomy* [Online], 11.

European Union. 2020. *Farm to Fork Strategy: For a Fair, Healthy and Environmentally-Friendly Food System* [Online]. Available: https://food.ec.europa.eu/system/files/2020-05/f2f_action-plan_2020_strategy-info_en.pdf [Accessed 10 January 2023].

FAO. 2014. *Sustainability Assessment of Food and Agriculture Systems (SAFA) Guidelines*. Version 3. Food and Agricultural Organization (FAO), Rome, Italy.

Foqus Planet Representative. 2022. *RE: Foqus Planet Dairy Sustainability Standard Interview*. Wageningen: Type to Animal Production Systems Group Wageningen University & Research.

Frieslandcampina. 2023a. *The Cooperative: For Farmers by Farmers* [Online]. Available: https://www.frieslandcampina.com/owned-by-farmers/the-cooperative/ [Accessed 18 January 2023].

Frieslandcampina. 2023b. *Foqus Planet* [Online]. Available: https://www.frieslandcampina.com/owned-by-farmers/foqus-planet/ [Accessed 15 January 2023].

Frieslandcampina. 2023c. *Lower Emissions from Cow Feed to Chocolate* [Online]. Available: https://www.frieslandcampina.com/stories/barry-callebaut/ [Accessed 29 April 2023].

Garcia, A. and Mcglone, J. J. 2022. Animal welfare and the acknowledgment of cultural differences. *Animals* [Online], 12. [Accessed 15 February 2022].

Gasso, V., Oudshoorn, F. W., De Olde, E. and Sørensen, C. A. G. 2015. Generic sustainability assessment themes and the role of context: the case of Danish maize for German biogas. *Ecological Indicators* [Online], 49. Available: https://www.sciencedirect.com/science/article/pii/S1470160X14004968 [Accessed 1 February 2015].

Geels, F. W. and Schot, J. 2007. Typology of sociotechnical transition pathways. *Research Policy* [Online], 36. Available: https://www.sciencedirect.com/science/article/pii/S0048733307000248 [Accessed 1 April 2007].

Globescan. 2022. *GlobeScan SustainaAbility Survey- Sustainability Leaders* [Online]. Available: https://globescan.wpenginepowered.com/wp-content/uploads/2022/06/GlobeScan-SustainAbility-Leaders-Survey-2022-Report.pdf [Accessed 28 April 2023].

Government of Ireland. 2023. *Climate Action Plan 2023* [Online]. Available: https://www.gov.ie/en/publication/7bd8c-climate-action-plan-2023/ [Accessed 20 January 2023].

Hannus, V., Venus, T. J. and Sauer, J. 2020. Acceptance of sustainability standards by farmers – empirical evidence from Germany. *Journal of Environmental Management* [Online], 267. https://dx.doi.org/10.1016/j.jenvman.2020.110617.

Hatanaka, M., Konefal, J., Strube, J., Glenna, L. and Conner, D. 2022. Data-Driven Sustainability: Metrics, Digital Technologies, And Governance In Food And Agriculture*. *Rural Sociology*, 87, 206–230.

Heinola, K., Kauppinen, T., Niemi, J. K., Wallenius, E. and Raussi, S. 2021. Comparison of 12 different animal welfare labeling schemes in the pig sector. *Animals* [Online], 11.

Hoekstra, G., Kroft, C. and Van Calker, K. J. 2013. A case study of marketing sustainability. In P. De Jong (ed.) *Sustainable Dairy Production*. First ed. John Wiley & Sons, Ltd. http://dx.doi.org/10.1002/9781118489451.ch1.

IFOAM. 2022. *Participatory Guarantee Systems (PGS)* [Online]. Available: https://www.ifoam.bio/our-work/how/standards-certification/participatory-guarantee-systems [Accessed 20 November 2022].

Ingenbleek, P. T. M., Harvey, D., Ilieski, V., Immink, V. M., De Roest, K. and Schmid, O. 2013. The European market for animal-friendly products in a societal context. *Animals* [Online], 3.

ITC. 2021. *International Trade Centre standards map* [Online]. Available: https://www.standardsmap.org/en/identify [Accessed 3 March 2023].

Jacobi, J., Toledo Vásquez, D. G., Solar Alvarez, J. M. and Bürgi Bonanomi, E. 2023. "First we eat and then we sell": participatory guarantee systems for alternative sustainability certification of Bolivian agri-food products. *Agroecology and Sustainable Food Systems* [Online], 47. https://dx.doi.org/10.1080/21683565.2022.2131692.

Janssen, C. 2020. *Albert Heijn milk will become even more sustainable and more expensive* [Online]. Available: https://www.foodlog.nl/artikel/ah-melk-wordt-nog-duurzamer-en-wat-duurder/ [Accessed 6 March 2023].

Konefal, J., De Olde, E. M., Hatanaka, M. and Oosterveer, P. J. M. 2023. Signs of agricultural sustainability: a global assessment of sustainability governance initiatives and their indicators in crop farming. *Agricultural Systems,* 208, 103658.

Kovacic, Z. 2018. Conceptualizing numbers at the science–policy interface. *Science, Technology, & Human Values* [Online], 43. https://journals.sagepub.com/doi/abs/10.1177/0162243918770734.

Kyttä, V., Roitto, M., Astaptsev, A., Saarinen, M. and Tuomisto, H. L. 2022. Review and expert survey of allocation methods used in life cycle assessment of milk and beef. *The International Journal of Life Cycle Assessment* [Online], 27. https://dx.doi.org/10.1007/s11367-021-02019-4.

Leip, A., Billen, G., Garnier, J., Grizzetti, B., Lassaletta, L., Reis, S., Simpson, D., Sutton, M. A., De Vries, W., Weiss, F. and Westhoek, H. 2015. Impacts of European livestock production: nitrogen, sulphur, phosphorus and greenhousegas emissions, land-use, water eutrophication and biodiversity. *Environmental Research Letters* [Online]. https://dx.doi.org/10.1088/1748-9326/10.

Loconto, A. and Hatanaka, M. 2018. Participatory guarantee systems: alternative ways of defining, measuring, and assessing 'sustainability'. *Sociologia Ruralis* [Online], 58. https://dx.doi.org/10.1111/soru.12187.

Luhmann, H., Schaper, C. and Theuvsen, L. 2016. Future-oriented dairy farmers' willingness to participate in a sustainability standard: evidence from an empirical study in Germany. *International Journal on Food System Dynamics,* 7, 243–257.

Mcgarr-O'Brien, K., Herron, J., Shalloo, L., De Boer, I. J. M. and De Olde, E. M. 2023. Characterizing sustainability standards in dairy production. *Animal*, 17(7). https://doi.org/10.1016/j.animal.2023.100863.

More, S. J., Hanlon, A., Marchewka, J. and Boyle, L. 2017. Private animal health and welfare standards in quality assurance programmes: a review and proposed framework for critical evaluation. *Veterinary Record* [Online], 180. https://dx.doi.org/10.1136/vr.104107.

More, S. J., Marchewka, J., Hanlon, A., Balzani, A. and Boyle, L. 2021. An evaluation of four private animal health and welfare standards and associated quality assurance programmes for dairy cow production. *Food Policy* [Online], 105. Available: https://www.sciencedirect.com/science/article/pii/S0306919221001494 [Accessed 1 December 2021].

Mottet, A., De Haan, C., Falcucci, A., Tempio, G., Opio, C. and Gerber, P. 2017. Livestock: on our plates or eating at our table? A new analysis of the feed/food debate. *Global Food Security* [Online], 14. Available: https://www.sciencedirect.com/science/article/pii/S2211912416300013 [Accessed 1 September 2017].

Mullan, S., Szmaragd, C., Cooper, M., Wrathall, J., Jamieson, J., Bond, A., Atkinson, C. and Main, D. 2016. Animal welfare initiatives improve feather cover of cage-free laying hens in the UK. *Animal Welfare* [Online], 25. https://dx.doi.org/10.7120/09627286.25.2.243.

NDC. 2022. *National Dairy Council* [Online]. Available: https://ndc.ie/from-the-ground-up/a-natural-source-of-pride/ [Accessed 29 April 2023].

Nestlé. 2023. *Climate action: tackling climate change can't wait and neither can we* [Online]. Available: https://www.nestle.com/sustainability/climate-change [Accessed 15 March 2023].

Netherlands, G. O. T. 2019. Plan of action-supporting the transition to circular agriculture. Ministry of Agriculture, Nature and Food Quality.

Nijhof, A., Wins, A., Argyrou, A. and Chevrollier, N. 2022. Sustainable market transformation: a refined framework for analyzing causal loops in transitions to sustainability. *Environmental Innovation and Societal Transitions* [Online], 42. Available: https://www.sciencedirect.com/science/article/pii/S2210422422000107 [Accessed 1 March 2022].

Olsen, J. V., Andersen, H. M.-L., Kristensen, T., Schlægelberger, S. V., Udesen, F., Christensen, T. and Sandøe, P. 2023. Multidimensional sustainability assessment of pig production systems at herd level – The case of Denmark. *Livestock Science,* 270, 105208.

Origin Green. 2022. *Producers: Famers and Growers* [Online]. Available: https://www.origingreen.ie/who-is-involved/producers/ [Accessed 20 November 2022].

Origin Green Representative. 2023. *RE: Origin Green Dairy Sustainability Standard Interview.* Wageningen: Type to Animal Production Systems Group Wageningen University & Research.

Osmundsen, T. C., Amundsen, V. S., Alexander, K. A., Asche, F., Bailey, J., Finstad, B., Olsen, M. S., Hernández, K. and Salgado, H. 2020. The operationalisation of sustainability: sustainable aquaculture production as defined by certification schemes. *Global Environmental Change,* 60, 102025.

Oya, C., Schaefer, F. and Skalidou, D. 2018. The effectiveness of agricultural certification in developing countries: a systematic review. *World Development* [Online], 112. Available: https://www.sciencedirect.com/science/article/pii/S0305750X18303012 [Accessed 1 December 2018].

Payne, M., Bruhn, C. M., Reed, B., Scearce, A. and O'Donnell, J. 1999. On-farm quality assurance programs: a survey of producer and industry leader opinions. *Journal of Dairy Science* [Online], 82. Available: https://www.sciencedirect.com/science/article/pii/S002203029975469X [Accessed 1 October 1999].

Poore, J. and Nemecek, T. 2018. Reducing food's environmental impacts through producers and consumers. *Science* [Online], 360. https://dx.doi.org/10.1126/science.aaq0216.

Poppe, K. and Koutstaal, H. 2020. Eco-schemes and private sustainability initiatives: creating synergies. In *EuroChoices,* 19, 36–40. https://doi.org/10.1111/1746-692X.12250.

Regenerative Organic Alliance. 2023. *Regenerative Organic Alliance* [Online]. Available: https://regenorganic.org/ [Accessed 18 March 2023].

Remkes, J. 2022. *What Can Be Done: Out of the Impasse and a Start for Perspective* [Online]. Available: https://www.rijksoverheid.nl/documenten/rapporten/2022/10/05/wat-wel-kan [Accessed 20 April 2023].

Sai Platform. 2020. *Sustainable Dairy Partnership-Why Are Companies Coming Together to Be a Part of the SDP?* [Online]. Available: https://vimeo.com/438125260 [Accessed 20 March 2023].

Sai Platform. 2023. *SAI Platform Sustainable Dairy Partnership (SDP)* [Online]. Available: https://saiplatform.org/sdp/ [Accessed 6 January 2023].

Sandøe, P., Hansen, H. O., Forkman, B., Van Horne, P., Houe, H., De Jong, I. C., Kjær, J. B., Nielsen, S. S., Palmer, C., Rhode, H. L. H. and Christensen, T. 2022. Market driven initiatives can improve broiler welfare – a comparison across five European countries based on the Benchmark method. *Poultry Science* [Online], 101. Available: https://www.sciencedirect.com/science/article/pii/S0032579122001146 [Accessed 1 May 2022].

Sandøe, P., Hansen, H. O., Rhode, H. L. H., Houe, H., Palmer, C., Forkman, B. and Christensen, T. 2020. Benchmarking farm animal welfare–a novel tool for cross-country comparison applied to pig production and pork consumption. *Animals* [Online], 10. https://dx.doi.org/10.3390/ani10060955.

Santman-Berends, I. M. G. A., Brouwer, H., Ten Wolthuis-Bronsvoort, A., De Bont-Smolenaars, A. J. G., Haarman-Zantinge, S. and Van Schaik, G. 2018. Development of an objective and uniform scoring method to evaluate the quality of rearing in Dutch dairy herds. *Journal of Dairy Science* [Online], 101. https://dx.doi.org/10.3168/jds.2018-14460.

Schreefel, L., Schulte, R. P. O., De Boer, I. J. M., Schrijver, A. P. and Van Zanten, H. H. E. 2020. Regenerative agriculture – the soil is the base. *Global Food Security* [Online], 26. Available: https://www.sciencedirect.com/science/article/pii/S2211912420300584 [Accessed 21 Sepetmber 2020].

Simons, L. and Nijhof, A. 2020. *Changing the Game: Sustainable Market Transformation Strategies to Understand and Tackle the Big and Complex Sustainability Challenges of Our Generation*. http://dx.doi.org/10.4324/9780429052613.

Stull, C. L., Reed, B. A. and Berry, S. L. 2005. A comparison of three animal welfare assessment programs on California dairies. *Journal of Dairy Science* [Online], 88. https://dx.doi.org/10.3168/jds.s0022-0302(05)72828-9.

Tetrault Sirsly, C.-A. and Lamertz, K. 2007. When does a corporate social responsibility initiative provide a first-mover advantage? *Business & Society* [Online], 47. https://doi.org/10.1177/0007650307299221.

Traldi, R. 2021. Progress and pitfalls: a systematic review of the evidence for agricultural sustainability standards. *Ecological Indicators* [Online], 125. Available: https://www.sciencedirect.com/science/article/pii/S1470160X21001552 [Accessed 1 June 2021].

UNFSS. 2022. Voluntary sustainability standards: sustainability agenda and developing countries, opportunities and challenges. *Fifth Flagship Report of the United Nations Forum on Sustainability Standards* [Online]. Available: https://unfss.org/wp-content/uploads/2022/10/UNFSS-5th-Report_14Oct2022_rev.pdf [Accessed 15 March 2023].

Unilever. 2023. *Sustainability Ratings and Rankings* [Online]. Available: https://www.unilever.com/planet-and-society/sustainability-reporting-centre/sustainability-ratings-and-rankings/ [Accessed 10 March 2023].

Van Calker, K. J., Hooch Antink, R. H. J., Beldman, A. C. G. and Mauser, A. 2005. *Caring Dairy: a Sustainable Dairy Farming Intiative in Europe* [Online]. Available: https://ageconsearch.umn.edu/record/24234/files/cp05va03.pdf [Accessed 15 December 2022].

Van Der Minne, C., Hekkert, M., Nijhof, A., Simons, L., Loorbach, D. and Termeer, K. 2021. Houvast voor duurzame vernieuwers. Vier perspectieven op transitiedenken en doen, Whitepaper. Het Groene Brein, The Hague, the Netherlands. 44 pages, Available at: https://drift.eur.nl/app/uploads/2024/02/Houvast-voor-duurzame-vernieuwers_Vier-perspectieven-op-transitiedenken-en-doen.pdf.

Van Dijk, L., Elwes, S., Main, D. C. J., Mullan, S. M. and Jamieson, J. 2018. Farmer perspectives on welfare outcome assessment: learnings from four farm assurance scheme consultation exercises. *Animal Welfare* [Online], 27. Available: https://www.cambridge.org/core/article/farmer-perspectives-on-welfare-outcome-assessment-learnings-from-four-farm-assurance-scheme-consultation-exercises/5413BFE10E54E63FF76FBA9270F0A61A.

Van Dijk, W., De Boer, J. A., De Haan, M. H. A., Mostert, P., Oenema, J. and Verloop, J. 2022. *Calculation Rules of the Annual Nutrient Cycling Assessment 2020; Background Information about Farm-Specific Excretion Parameters (update of ANCA report 2019)*. Wageningen: Wageningen Livestock Research.

Van Rossum, M. 2023. *FrieslandCampina CEO: 'Don't Just Embrace the Cuddly Toy Farm'* [Online]. Available: https://www.nieuweoogst.nl/nieuws/2023/01/24/topman-frieslandcampina-niet-alleen-knuffelboerderij-omarmen?utm_source=nieuwsbrief&utm_medium=email&utm_campaign=nieuwsbrief-24-01-2023&utm_term=1D9F2E10-2F11-4421-9046-1E026F2C456F [Accessed 15 April 2023].

Vogeler, C. S. 2019. Market-based governance in farm animal welfare–A comparative analysis of public and private policies in Germany and France. *Animals* [Online], 9.

Willer, H., Sampson, G., Larrea, C., Schlatter, B., Bermudez, S., Duc Dang, T., Rüger, M. and Meier, C. 2022. *The State of Sustainable Markets 2022: Statistics and Emerging Trends* [Online]. Available: https://intracen.org/media/file/13366 [Accessed 15 April 2023].

Chapter 9

Developing good agricultural practices (GAP) across global agri-food supply chains: challenges and opportunities

Louise Manning, University of Lincoln, UK; and John G. Keogh, McGill University, Canada

1 Introduction

Good agricultural practices (GAP) are a 'collection of principles to apply for on-farm production and post-production [post-harvest or post-slaughter] processes, resulting in safe and healthy food and non-food agriculture products, while taking into account economic, social and environmental sustainability' (UN FAO, 2016, p. 1). GAP's four 'pillars' are as follows (Bamunuarachchi et al., 2019):

- Food safety and quality;
- Economic viability;
- Environmental sustainability; and
- Social acceptability.

GAP standards can be designed to be agricultural enterprise specific, e.g. individual standards for horticulture, specific crops (e.g. cereals) or for particular species in livestock production; and compliance with these standards can be verified and certified by third-party organisations (Pandit et al. 2017). Some GAP standards are universal, e.g. those related to minimising food contamination,

http://dx.doi.org/10.19103/AS.2024.0133.13

more generally, whilst others relate to the food safety or environmental risks associated with particular crops, farming systems and/or post-production activities.

GAP codes and standards can be developed by regulators, non-governmental organisations (NGOs) or market actors, by individual stakeholders, or through multi-stakeholder collaboration. GAP standards are, therefore, policy instruments that are based on available knowledge and understanding of farming practices and their potential impact(s) (Speedy, 2004). GAP standards set normative requirements which farmers must comply with in order to demonstrate acceptable practice. However, the social acceptance of on-farm standards, practices and the associated technologies, whether by farmers, processors, retailers or consumers, is mediated by a number of factors, including social acceptance of the practices and technologies themselves and social factors such as age or gender, emotional, attitudinal and cognitive elements (El Bilali et al. 2021). This implies that the definition of acceptable farming practices, such as standards for animal welfare or environmental protection, is shaped by social and cultural contexts. Additionally, it is influenced economically by what the market is willing to pay for these practices.

Regulators can compel farmers to comply with GAP through either penalties (risk of prosecution), or through incentives for compliance, e.g. public payments often described as subsidy payments (Hobbs, 2003). Voluntary market-orientated GAP standards define specific criteria with a requirement for organisations to either comply with or exceed legal minimum standards, and these are often linked to a certification process (Burrell, 2011). Private GAP standards can be business-to-business (B2B) focused where there is no identification on the finished product, e.g. GlobalGAP[1], or they can be business-to-consumer (B2C) where the product packaging includes a logo or symbol identifying consumers that products have been produced to a given standard (Burrell, 2011), e.g. in the United Kingdom (UK), the Red Tractor Standard. These logos, or visual packaging cues, communicate compliance with particular safety/quality, ethical, social or environmental requirements directly to consumers (Grunert et al., 2014; Rees et al., 2019). There can be broader market standards, e.g. organic farming standards that are set by individual national organisations (such as the Soil Association in the UK) and overseen by an international body such as IFOAM – Organics International, or private standards that are animal welfare or worker welfare focused, e.g. social standards related to 'fair' trade.

B2B GAP standards underpin business risk management activities, and frame market access requirements that show a supplier is 'fit to supply'. Whilst some organisations have developed their own GAP standards that their

1 https://www.globalgap.org/uk_en/

suppliers need to comply with, an alternative procurement approach is to require potential and existing suppliers to demonstrate that they hold third-party certification (TPC) to private farm assurance standards. Pre-procurement and post-procurement protocols are developed by organisations to ensure that suppliers have a lower food safety or pollution risk associated with their activities or have a higher likelihood of complying with prescribed standards (Burrell, 2011). These private standards also promote trade liberalisation through private governance systems that provide opportunities for the trading of the global food supply that is assured to be safe and to be produced to a range of ethical standards (Van der Grijp et al., 2005).

Good farming can be perceived from this perspective as being linked to *correctness,* planting crops at the correct depth, in the correct way, in terms of *efficiency,* i.e. good farming is productive and efficient, minimising waste (Burton et al., 2008); or in terms of *virtue*, i.e. good versus bad decisions or behaviour (Haggerty et al. 2009). Some suggest that social and cultural aspects frame what good farming and being a good farmer is perceived to be in practice (Thomas et al., 2019), which may or may not include demonstrating compliance with prescriptive standards that can be audited. However, others argue that GAP involves the 'establishment of guidelines or standards for agricultural producers and post-farm handlers, the monitoring of these standards, and the communication of these standards through credible quality signals to downstream firms, consumers and the public in general' (Hobbs, 2003, p. 1).

What is perceived as being a 'good farming practice' can vary between stakeholders, e.g. between farmers and veterinarians (Shortall et al., 2018); or between farmers and consumers, e.g. in the case of notions of good animal welfare on farms (Alonso et al., 2020). Furthermore, narratives around good and bad farming practices can be contested and conflicted. Muhammad et al. (2022), in their reflection on the social construction of narratives and arguments of animal welfare, suggest five distinct narratives: the animal rights/power-based, the farming as a business narrative, the research, legislative and political based narrative, the 'religion-based' and the higher welfare narratives. These narratives shape perceptions of a good or bad life for farmed animals and, as a result, what is perceived individually and collectively by stakeholders as being good farming and being a good farmer. Multiple academic papers have been written about defining what a 'good' farmer is, more recently through the lens of the environmental and nature-focussed issues (McGuire et al., 2013; Saunders, 2016; Cusworth & Dodsworth, 2021; Westerink et al. 2021); exotic livestock diseases (Naylor et al., 2018); and sustainable crop management practices (Lavoie & Wardropper, 2021).

GAP has also been repositioned more recently in terms of smart farming or smart agriculture. Smart farming is the integration of technology (such as sensors or artificial intelligence, data collation and data analysis) into machinery and equipment used within farming systems (Pivoto et al., 2018). Agriculture 4.0 is proposed as having the ability to increase productivity, allocate resources more effectively, avoid food loss and waste and enable adaption to climate change (Zhai et al., 2020). Some questions arise here. How does the evolution of farming systems reposition the framing of GAP, and what is perceived as good farming? Who are the actors who will be part of the discourse about what good farming is as a concept and how it is delivered in practice? This chapter will position how GAP has evolved over the last 30 years and will provide insight into how GAP practices could develop further in the future.

2 The evolution of GAP as a concept

Food safety, social, environmental and quality management controls within a farm assurance programme implemented on farms are derived firstly from the need to comply with national legislation and potentially with legislation in the country to which the food may be exported, and secondly to align with the standards contained in private market schemes (Manning & Luning, 2018).

One of the main drivers of farm assurance in the UK was a series of high-profile public health crises such as Bovine Spongiform Encephalopathy (BSE) in cattle and new variant Creutzfeldt–Jakob disease (CJD) in humans (Manning & Baines, 2004). The UK Food Safety Act (1990) also drove the development of farm assurance with its legal defence that required businesses to demonstrate that they had taken preventive measures and 'exercised due diligence', including by suitable approval of direct suppliers and others in the supply chain that might affect the safety of a business's products (Kirk-Wilson, 2002; Manning & Baines, 2004; Nabeshima et al., 2015), sometimes described as upstream assurance (Sharma & Modgil, 2013).

Due diligence, as a concept, is described as the process through which organisations identify, assess, mitigate and prevent the adverse impacts of their activities (OECD-FAO, 2016). The need to demonstrate due diligence has led to the complex development of standards and requirements throughout the supply chain, especially at the farm level (Manning & Luning, 2018). This implies that the definition of acceptable farming practices, such as standards for animal welfare or environmental protection, is shaped by social and cultural contexts. Additionally, it is influenced economically by what the market is willing to pay for these practices.

3 The development of GAP standards

Between 1995 and 2000, a number of international farm-focused standards emerged. These included EUREPGAP (Euro-Retailer Produce Working Group GAP), developed initially by European food retailers, and then in further versions by working group members that included multi-national corporations and retailers (Van der Grijp et al., 2005). EUREPGAP was renamed GLOBALGAP in 2007 and has been progressively developed since then (Manning et al., 2019). Farm assurance standards were developed and then routinely revised through a structured, iterative process where new requirements were added, and existing requirements changed over time (Manning & Luning, 2018). In 1996, when the European retailers came together, the aim was to harmonise a set of minimum GAP standards that encompassed food safety, environmental protection and worker welfare, the protocol being introduced in 2001 (Martinez & Poole, 2004; Van der Grijp et al., 2005).

In fresh produce (fruit and vegetables), a number of generic farm assurance schemes were developed in the United Kingdom, including the Assured Produce Scheme (APS) in 1996. Once established as a scheme, compliance with APS standards, demonstrated through TPC, became a pre-requisite to supply, i.e. a pre-requisite to market access. Compliance with elements of GAP standards can be evaluated via a checklist approach (da Cruz et al., 2006), although others have raised concerns about checklist approaches leading to shallow audits or evaluation myopia (Martz, 2010; Manning, 2018). The ability to comply with APS standards and further iterations, which then became more complex, has not been 'scale-neutral', i.e. larger organisations with greater access to information and technical resources were more able to comply (Loader & Hobbs, 1999; Martinez & Poole, 2004).

From the transactional viewpoint, GAP has been said to be a reference point for agricultural activities such as animal husbandry, soil management, water use, nutrient and crop protection product use to minimise the risk of pollution (Piorr, 2003). Transactional aspects of embedding GAP include considering the costs of compliance and certification borne by the farmers, and this has led to producers grouping together to form 'clusters' to create scale in terms of investing in post-harvest infrastructure and marketing and also in integrating, monitoring and verifying management systems and delivering socio-economic gains (Nain et al., 2015). Transactional aspects include the development of food safety management systems (FSMS), embedding traceability systems and documented or smart information management, which add cost, but are essential aspects of GAP (Abebe et al., 2017).

Specific UK retailers also developed their own schemes, such as Tesco's Natures Choice (which evolved to the Tesco Nurture Scheme) which included additional requirements to protect the environment and promote more sustainable practices (Martinez & Poole, 2004). Harmonisation of standards within farm assurance schemes coincided with activities to drive global harmonisation of food standards and regulations (Baines et al., 2000), e.g. across the European Union, or the harmonisation of organic standards through the activities of IFOAM (Winickoff & Klein, 2011). Thus, the development of

EUREPGAP certification meant that existing standards such as Assured Produce, Tesco Nature's Choice (later Tesco Nurture) or others could be assessed (benchmarked) to identify their level of equivalence, or divergence, with the benchmark standard (Martinez & Poole, 2004).

The Red Tractor Scheme was launched in the UK in 2000 by the National Farmers Union (NFU) and other stakeholders under the control of the scheme owner, Assured Farm Standards, into which Assured Produce was subsumed (Manning & Luning, 2018). The Red Tractor Scheme is often described as a baseline standard. Other standards may be described as higher level because their standards are deemed to be 'higher' with regard to aspects such as animal welfare or environmental protection. Higher farmer assurance standards are typically focussed on securing access to niche markets and premium prices from retail and tertiary supply businesses (Van der Grijp et al., 2005).

In April 2000, a larger group of international retailers came together to form the Global Food Safety Initiative (GFSI) with a focus on food safety, improving consumer confidence and consumer protection and improving the cost effectiveness of assurance processes from field to fork (Martinez & Poole, 2004). GLOBALG.A.P. has continued to develop as a standard with a generic central standard and then with a range of modules or 'bolt-on' products (Hobbs, 2001), developed by individual retailers or co-created by stakeholders to address specific issues such as worker or animal welfare. As well as generic GAP standards being developed as both baseline and higher-level determinants of good practice, local GAP standards have also emerged.

Many Asian and Southern Hemisphere countries have introduced local GAP standards, e.g. ChileGAP, ChinaGAP, VietGAP, JapanGAP and Thai GAP (Nabeshima et al., 2015; Manning et al., 2019; Tran & Goto, 2019). The key driver for the development of these standards has been national governments and their use of GAP standards to standardise and optimise food safety procedures and systems within their countries. For example, in Japan, the Ministry of Agriculture, Fishery and Forestry (MAFF) initially promoted voluntary adoption of GAP with self-certification. In 2006, JGAP was developed as a national scheme benchmarked to GLOBALG.A.P initially, and then to GFSI (Nabeshima et al., 2015).

VietGAP was issued by the Vietnamese Ministry of Agriculture and Rural Development (MARD) in 2008 as a set of minimal standards and guidelines from production to harvest for safe food from farming and aquaculture (Nabeshima et al., 2015). The drivers for GAP in Vietnam have historically been government (public-led) rather than privately led, and adoption has been hampered by the very small average size of farming businesses at around 0.25 ha (Nabeshima et al., 2015). In the Vietnamese aquaculture sector, good aquaculture practices have a pull dynamic as they are demanded by export markets. Whilst the cost of adoption in one study was seen to be high, it was outpaced by the increase in price for assured product with factors such as farm size, production system and level of education mediating farmer engagement (Quyen et al., 2022). In the tea sector in Vietnam, VietGAP certification has been linked to UTZ certification, which focuses on worker welfare and health (Tran & Goto, 2019). The UTZ certification programme comes from the Guatemalan term 'UTZ kapeh' meaning 'good coffee' (Rainforest Alliance, 2023).

4 Challenges facing GAP standards and implementation

GFSI as an organisation has evolved from a benchmarking organisation for third-party private certification standards developed by retailers and manufacturers aiming to harmonise food safety audits of food safety management systems and practices to an organisation with the aim of delivering strategic regulatory outcomes, facilitating regulatory compliance for organisations and facilitating improved understanding and application of CODEX standards (Gerardi, 2023). This means that GFSI has a prominent role in food safety governance, what Daugbjerg (2022) describes as a 'meta-governing' role (Verbruggen & Havinga, 2019).

Van der Grijp et al. (2005) raise questions about the transparency of governance of private standards. Marques (2019, p. 157) argues that the creation of intermediary organisations, such as GFSI, is "driven by global retailers' reliance on standardisation, low transaction costs, and regulatory harmonisation across all aspects of their operations [and] ... the harmonisation platforms are designed to leverage global retailers' market power and... intermediaries [transform] into de facto regulators that supplant existing private regulators." Demand-side procurement of this kind, often by international retailers, can provide opportunities, yet the food safety and quality standards they set can have significant impacts on supply chains, especially in developing countries (Moran, 2019).

One example is how GFSI benchmarks global standards from field to fork while, as Moran (2019) describes, GLOBALGAP certifies national GAP programmes such as KenyaGAP or ChileGAP, in the case of production destined for export. The development of these GAP programmes can catalyse inward investment and the provision of work opportunities in countries where employment opportunities are limited. Indeed, in countries such as Kenya, foreign investment in horticulture, especially flower production, can lead to 'transient secondary towns' emerging through the translocation of workers seeking employment (Kuiper & Greiner, 2021). This means that governance activities by organisations need to extend to worker welfare standards and wider aspects of social governance. Dolan and Humphrey (2000) argue that the type of governance these large companies develop and 'exercise in practice' has socio-economic consequences for indigenous businesses not only in terms of the factors that create inclusion and exclusion criteria, e.g. positioning of market access standards, but also the degree of inclusivity in the process for upgrading of organisations in terms of the degree of sophistication of their management and verification functions.

Creating supply chains with external multi-national food horticultural corporations in a given country can expose local organisations to the pressure of international markets (Heher & Steenbergen, 2021) and cause marginalisation of those organisations and workers who do not have the same market access. Indeed, voluntary, non-governmental social governance tools can cause the marginalisation of workers who are unable to influence governance processes, and the verification of these governance tools can lack structured and transparent monitoring (Sparks et al., 2022).

How can the development of GAP standards embed inclusive growth and avoid marginalisation? In embedding GAP in food supply chains, the leading actors (e.g. retailers, multinational corporations) need to consider the decision-making process they use for market access, i.e. the inclusion and exclusion criteria they use to ensure that GAP standards are met both in onboarding and in terms of continuing compliance with contract requirements, especially as discrete audits only provide limited oversight (Dolan & Humphrey, 2000). Can inclusion criteria be applied so that the result is not just a concentration of export trade in the hands of the few?

Upgrading individual organisations, or indeed a whole value chain, to embed higher environmental or social standards requires changes in infrastructure, knowledge, skills, capacities and production standards (Nadvi, 1999). Upgrading can include requirements for improvements in local infrastructure, embedding of training programmes to upgrade skills, knowledge and capabilities in terms of farming practices (Moran, 2019); as well as promoting improvements to drive efficiency and performance and reduce waste (Kiswili et al., 2021). Upgrading is a complex process, especially where value generation may deliver value for some supply chain actors and not for others, e.g. for retailers and not for farmers. How can GAP standards support more inclusive upgrading processes? Inclusive product and process upgrading in food supply chains needs to involve the intentional implementation of mitigation strategies to prevent the exclusion of disadvantaged actors, especially in developing economies where farmers and farm workers operate in a socio-economic environment influenced by both local and global economies (Park & Gachukia, 2021). These themes are now considered in a case study on VietGAP.

5 Ensuring the effectiveness and inclusivity of GAP: the case of VietGAP

In 2008 Vietnam launched the Good Agricultural Practices (VietGAP) programme to improve the quality and safety of locally produced agricultural and aquaculture products. VietGAP strives to ensure food safety, worker welfare and environmental sustainability in producing and harvesting essential crops.[2] VietGAP standards have been developed for major crops like rice, coffee, tea, fruits and vegetables, livestock and fisheries. The Ministry of Agriculture and Rural Development (MARD) develops the technical parameters of the standard and manages the programme with guidelines and procedures developed by sub-departments. Voluntary technical standards (TCVNs) drafted by MARD are appraised and published by the Ministry of Science and Technology (MOST) sub-department Directorate for Standards, Metrology and Quality (STAMEQ) based on a 5-year plan and an annual plan for National Standards developed by STAMEQ (STAMEQ, n.d.). This case study provides an overview of VietGAP, its requirements across different focus areas, implementation mechanisms, and strengths and weaknesses compared to other sustainability standards.

5.1 Emergence of VietGAP

VietGAP was implemented when Vietnam's agricultural exports were booming but concerns about pesticide usage and food contamination were growing. Multiple shipments of Vietnamese fruits, vegetables and aquaculture were being rejected for safety reasons by trading partners. The hazardous use of agrochemicals by farmers was also highlighted domestically (Van Hoi et al., 2009). VietGAP was conceived as a quality assurance programme that could improve production methods and ensure appropriate use of pesticides and other chemical products on farms.

The design of VietGAP was heavily influenced by worldwide GAP models such as ASEAN GAP, GLOBALG.A.P and JGAP from Japan. VietGAP highlighted a flexible, step-by-step approach for Vietnam's agriculture sector, in which smallholders, both export-oriented and domestically focused producers, dominate. As part of the government's aim to restructure and modernise agriculture and aquaculture, this approach has become a priority for quality improvement and food safety. In 2018, a partnership was formed between the Aquaculture Stewardship Council (ASC), the World Wildlife Fund (WWF) Vietnam and the Vietnamese government to benchmark VietGAP for aquaculture and create a transition pathway for VietGAP-certified aquaculture firms to migrate to internationally recognised ASC certification (ASC, 2018).

2 https://www.quacert.gov.vn/en/good-agriculture-practice.nd185/vietgap-standard.i88.html

5.2 Key principles and scope

VietGAP integrates principles of economic, social and environmental sustainability with the objective to ensure safe and sustainable agriculture that protects farmers' incomes and productivity. The aims of the assurance programme are to:

- improve food safety by controlling chemical, biological and physical hazards;
- promote environmentally responsible production practices that conserve resources;
- provide decent work conditions and welfare provisions for farm workers; and
- enable product traceback (from retail shelf to farmer) through documented procedures and record keeping.

VietGAP encompasses the entire agricultural production cycle, from inputs and field methods to harvesting and post-harvest handling. It has specific protocols for the main commodities produced in Vietnam, such as rice, coffee, tea, vegetables, fruits and cashews. Aquaculture and livestock are covered in VietGAP, but the procedures are modified for these individual production systems. The standards define both fundamental and advanced requirements for compliance with associated certification activities.

VietGAP defines a comprehensive GAP programme spanning various aspects of the farm-based production system (National Standards, TCVN 11892-1:2017). These include the following:

- **Site selection and land management:** Suitable farm sites must be identified based on pollution risk assessment and history of chemical use. Measures to conserve soil and water resources governed by mandatory national standards (QCVNs) and biodiversity are encouraged.
- **Planting materials and inputs:** Quality seeds, saplings and permitted agrochemicals should be used with a guidance document issued once a year by MARD. Guidance is provided on balanced fertilisation and need-based integrated pest management.
- **Farming techniques:** Techniques like integrated pest management, efficient water use and biodiversity promotion are recommended for field management practices. Pesticide application protocols like dosage, timing, protective clothing and container disposal are specified.
- **Facilities and equipment:** Food-grade packaging materials must be used. Secure and safe storage facilities for tools, machinery and chemicals must be constructed appropriately. Cleaning and sanitation protocols are defined.

- **Harvest and post-harvest:** Trained workers, sanitised containers, storage pest control and avoidance of cross-contamination are stipulated – pre-harvest intervals and withdrawal periods after pesticide spray application are defined.
- **Record keeping:** Detailed records of farm inputs, practices and yields must be maintained. Internal audits, product sampling and testing are required periodically. Application of pesticides and other crop treatments must be recorded.
- **Traceability:** Production areas, definition of batches and delivery records must allow traceback of all products.
- **Worker health and welfare:** Periodic health checks, safety training, hand washing facilities, clean drinking water, rest areas, and sanitary toilets should be provided. Decent work conditions, working hours and leave provisions are outlined.

VietGAP is administered by accredited certification bodies and their auditors, who are trained to inspect and certify the production procedures of registered farmers. MARD provides training on the requirements of VietGAP. Once formally certified, farmers/cooperatives can display the VietGAP brand on their produce and marketing materials. VietGAP auditing relies heavily on annual self-assessments based on a compliance criteria checklist. To enable traceability, it is required that data related to production, farm inputs and sales are documented. Regular product sampling and laboratory testing for residues are mandated but not strictly enforced. After more than a decade of implementation, conducting an objective evaluation of the effects of VietGAP's adoption in Vietnam is essential. Ongoing monitoring and evaluation will help determine the strengths of the scheme and which shortcomings (non-compliances) should be addressed moving forward.

5.3 Strengths and weaknesses of VietGAP

The VietGAP programme has several strengths that promote food safety, environmental management and farming practices, but also some limitations in its current format and application. The key strengths are as follows:

- A holistic set of good agricultural practices that extend beyond food safety.
- A programme which is customised to local conditions with a stepwise approach to adoption.
- An approach which leverages the participatory involvement of farmers and experts.
- A strong alignment with comparable international GAP standards supporting access to export markets.

Major weaknesses of the programme are as follows:

- The voluntary nature of participation limits widespread adoption.
- There is a need for continued training and auditing support for smallholder farmers within the programme.
- There is a lack of consumer awareness of the safety benefits of VietGAP-certified produce, over and above non-certified produce in the home market.
- The programme is susceptible to fraud, which, over time, impacts consumer trust. There are recurring incidents of fake VietGAP claims, creating a lack of trust (Nabeshima et al., 2015; Truong et al., 2021).
- The VietGAP logo is not standardised, resulting in multiple different formats on the market that could confuse consumers.
- The minimum testing regime involves one test per crop per season, before or during harvest.
- The cost of audits and compliance is problematic for smallholders (Nguyen, 2021).
- Weak integration with domestic food safety regulations and enforcement.
- Oversight and external audits are, in some cases, inadequate.
- Overall, system management of certification and registrations lacks internationally acceptable systems for integrity, independence and oversight.

Despite making a promising start, the implementation of VietGAP faces multiple challenges in its path ahead. Firstly, participation in the programme is voluntary, and there is insufficient market demand from commercial buyers and consumers. In a 2017 consumer survey in Vietnam (n = 500), more than 40% of survey participants were not aware of VietGAP (Nguyen et al., 2017). On the supply side, compliance costs appeared significant for smallholder farmers dominating the Vietnamese agricultural sector (Nguyen-Viet et al., 2017). Additionally, the benefits of engagement with the VietGAP programme are not evident in the absence of consumer recognition, limited market price increases and buyer demand for VietGAP certification. Therefore, expanding VietGAP to cover a significant portion of national production is challenging, especially if the costs of delivering upgrading of farm practices does not gain a greater market value for the finished product.

In addition, the systems for auditing and oversight remain inadequate. Currently, sampling and laboratory testing requirements are limited and not strictly enforced. Poor control over the quality and consistency of GAP checklists, auditor competency, verification and validation processes for certified farms are all issues which continue to plague the programme. The limited number of independent external third-party audits also may not be sufficient to prevent non-compliance (and fraudulent claims) in group certification schemes. Improving the independence and frequency of audits can increase trustworthiness in individual businesses and the programme as a whole. Despite these limitations, with appropriate policy support, VietGAP retains strong potential for broader impact. Both public and private investment to scale up on-farm adoption are crucial, going forward.

The government can advance VietGAP adoption in various ways, including upgrading VietGAP to better align it with international best practices, training accredited auditors and certification bodies (CBs) and improving governance and management to ensure credible verification and validation. The Vietnamese government could create a trusted, transparent public portal accessible to overseas buyers and domestic consumers and integrate domestic food control authorities in the auditing and enforcement of VietGAP. Investing in regular training and awareness building for farmers on the technical aspects and the benefits of VietGAP can lower adoption barriers. The government could also offer financial incentives like input subsidies and credit access for certified VietGAP producers (Nguyen, 2021), thereby offsetting the cost of upgrading.

Strengthening auditing capacity and quality of audits through greater oversight, field verification, accredited auditors and digitisation will advance the programme's credibility further. An example of a digital assurance tool is the adoption of VerifiK8 in the shrimp supply chain in Vietnam (Lien et al., 2023; Verifik8, n.d.). The government could also standardise the VietGAP logo and introduce further controls on its use, make VietGAP mandatory for high-risk commodities, mostly consumed fresh in the domestic market, procure only certified produce for public food distribution programmes such as school meals and raise consumer awareness through public health campaigns, mandatory certified product labelling and branding.

The private sector can also play a role by offering price premiums, purchase guarantees or contract farming support to VietGAP-certified farmers, maintaining traceability systems to the farmer level across the entire supply chain, and investing in trusted product labelling, branding and retail promotion campaigns for certified produce. The private sector could also partner with farmer groups to support capacity building and effective auditing, contribute resources for randomised surveillance testing of products across the value chain and engage consumers through social media, advertising, food bloggers, and influencers on safety and quality benefits.

5.4 Impact assessment and future outlook

As incomes grow, Vietnamese consumers will demand higher quality, safer and more traceable food. This will provide a sound economic basis to invest in nationwide GAP implementation. Realising the full potential of VietGAP requires greater resources and improved cooperation between public authorities and businesses in Vietnam. While VietGAP has made a good start in modernising Vietnamese agricultural methods, most exporters and importers rely on GFSI benchmarked certifications to access the international supply chain. VietGAP must expand in scale, scope and credibility and gain wider acceptance in countries to which Vietnamese farmers and food processors would like to export. To translate modest gains to date into safer and more sustainable food production in Vietnam, broader adoption, rigorous auditing and promotion to consumers within Vietnam and more widely is crucial. VietGAP can achieve its objectives and improve the image of Vietnamese agriculture if it is tightly integrated with national food safety governance (World Bank, 2016).

6 Conclusion

This chapter has described GAP and GAP standards. GAP standards are often transactional, focusing on key processes to assure the food produced is safe, of the quality expected and that sustainable practices have been used. The chapter has also shown the variability in adoption of GAP standards across the world and the varying processes of design and further evolution. Notions of good farming have also been considered, and the aspects of discourse and narratives around what makes a good or bad farmer. The innate exclusivity of GAP standards used by major retailers to differentiate and add value to their products appears at odds with using GAP standards, such as VietGAP, to drive increased farmer participation and inclusive improvement in national farm practices for both local supply and exports. Bottema et al. (2021) state that assurance programmes vary in terms of their degrees of prescription and their external verification and, as a result, how well they are trusted by food purchasers. Inclusive supply chains reflect not only equal access to marketing channels but also sharing food supply chain risks more equitably and recognise that ecosystem-level goals cannot necessarily be addressed by individual organisations gaining certification (Bush et al., 2019).

Pauwelussen and Bush (2020) considered inclusive assurance models and the degree to which they promoted inclusiveness in terms of farmer organisation and compliance assessment in the Vietnamese shrimp sector, where inclusiveness reflects both access to, and participation in, the scheme. In embedding GAP in food supply chains, the leading actors (e.g. retailers and multinational corporations) need to consider the design, adoption and decision-making processes they use for market access, i.e. the inclusion and exclusion criteria they use to ensure that GAP standards are met both in onboarding and in terms of continuing compliance with contract requirements, especially as discrete audits only provide limited oversight (Dolan & Humphrey, 2000). Can inclusion criteria be applied to ensure the outcome is not a concentration of export trade in the hands of the few? Yes, criteria can be framed for more inclusive participation and a greater sharing of the value generated and the risks of participation.

The discourse about what good farming is, and creating a more holistic rather than binary definition of what makes good agricultural practices in given contexts, is essential as a wider range of sustainability standards are increasingly embedded within GAP standards. Ensuring that the upgrading of farm standards is an inclusive, participatory approach is also crucial. This wider discourse, what Else et al. (2022) describe as sustainability storylines, can co-exist with the dominant storyline of developing descriptive, normative GAP standards that can be verified. At the same time developing more subjective, holistic sustainability-related storylines that engage multiple stakeholders is still a challenge that needs to be adressed, especially as GAP standards extend beyond food safety and environmental protection to consider wider aspects of sustainability.

7 Acknowledgements

The authors would like to acknowledge the contributions of Dr. Brian Bedard, John Duffill and Rolf Schoenert for their review, comments and feedback on the VietGAP case study.

8 References

Abebe, G. K., Chalak, A. & Abiad, M. G. (2017). The effect of governance mechanisms on food safety in the supply chain: Evidence from the Lebanese dairy sector. *Journal of the Science of Food and Agriculture*, *97*(9), 2908–2918.

Alonso, M. E., González-Montaña, J. R. & Lomillos, J. M. (2020). Consumers' concerns and perceptions of farm animal welfare. *Animals*, *10*(3), 385.

ASC. (2018). Mew ASC-VietGAP benchmark provides guidance for Vietnamese farmers. Available at: https://asc-aqua.org/news/new-asc-vietgap-benchmark-provides-guidance-for-vietnamese-farmers/ [Accessed on 31 July 2023].

Baines, R. N., Davies, W. P. & Ryan, P. (2000, June). Reducing Risks in the Agri-Food Supply Chain–Co-recognition of Food Safety Systems or a Single Global Scheme. In *IFAMA Conference, Chicago*. June 24–28, 2000.

Bamunuarachchi, B. A. D. S., Hitihamu, S. & Lurdu, M. S. (2019). Good Agricultural Practices (GAP) in Sri Lanka: Status, Challenges and Policy. Research Report No. 227. December 2019. Available at: http://www.harti.gov.lk/images/download/reasearch_report/new1/report_no_227.pdf [Accessed 20 February 2023]

Bottema, M. J., Bush, S. R. & Oosterveer, P. (2021). Assuring aquaculture sustainability beyond the farm. *Marine Policy*, *132*, 104658.

Burrell, A. (2011). 'Good Agricultural Practices' in the Agri-Food Supply Chain. *Environmental Law Review*, *13*(4), 251–270.

Burton, R. J., Kuczera, C. & Schwarz, G. (2008). Exploring farmers' cultural resistance to voluntary agri-environmental schemes. *Sociologia ruralis*, *48*(1), 16–37.

Bush, S. -R., Oosterveer, P., Bottema, M., Meuwissen, M., de Mey, Y., Chamsai, S., Ho, L. H. & Chadag, M. (2019). Inclusive environmental performance through 'beyond-farm' aquaculture governance. *Current Opinion in Environmental Sustainability*, *41*, 49–55.

Cusworth, G. & Dodsworth, J. (2021). Using the 'good farmer'concept to explore agricultural attitudes to the provision of public goods: a case study of participants in an English agri-environment scheme. *Agriculture and Human Values*, *38*(4), 929–941.

da Cruz, A. G., Cenci, S. A. & Maia, M. C. A. (2006). Good agricultural practices in a Brazilian produce plant. *Food control*, *17*(10), 781–788.

Daugbjerg, C. (2022). Accountability and integrity in private food safety regulation: evidence from the Australian food sector. *Australian Journal of Public Administration*, *81*(1), 37–-52.

Directorate for Standards, Metrology and Quality (STAMEQ). (n.d.). Available at: https://tcvn.gov.vn/?lang=en [Accessed 31 July 2023]

Dolan, C. & Humphrey, J. (2000). Governance and trade in fresh vegetables: the impact of UK supermarkets on the African horticulture industry. *Journal of development studies*, *37*(2), 147–176.

El Bilali, H., Hassen, T. B., Bottalico, F., Berjan, S. & Capone, R. (2021). Acceptance and adoption of technologies in agriculture. *AGROFOR*, *6*(1), 135–150.

Else, T., Choudhary, S. & Genovese, A. (2022). Uncovering sustainability storylines from dairy supply chain discourse. *Journal of Business Research*, *142*, 858–874.

Gerardi, A. (2023). Global Food Safety Initiative (GFSI): underpinning the safety of the global food chain, facilitating regulatory compliance, trade, and consumer trust. In *Present Knowledge in Food Safety* (pp. 1089–1098). Academic Press. Cambridge. Mass USA ISBN 9780128194706, https://doi.org/10.1016/B978-0-12-819470-6.00058-5.

Grunert, K. G., Hieke, S. & Wills, J. (2014). Sustainability labels on food products: consumer motivation, understanding and use. *Food policy*, *44*, 177–189.

Haggerty, J., Campbell, H. & Morris, C. (2009). Keeping the stress off the sheep? Agricultural intensification, neoliberalism, and 'good' farming in New Zealand. *Geoforum*, *40*(5), 767–777.

Heher, U. & Steenbergen, V. (2021). Kenya–supplying to multinationals exposed local firms to international horticulture markets. In C. Z. Qiang, Y. Liu, Y. V. & Steenbergen (Eds.),. *An investment perspective on global value chains*. World Bank Publications.

Hobbs, J. E. (2003). Incentives for the Adoption of Good Agricultural Practices (GAPs). *Food and Agriculture Organization*, Background paper for the FAO consultation on Good Agricultural Practices, Rome, November 10-12 2003.

Hobbs, J. E. (2001). *Dimensions of Competitiveness: Lessons from the Danish Pork Industry* (No. 519-2016-37617, pp. 1-11).

Kiswili, N. E., Shale, I. N. & Osoro, A. (2021). Influence of supply chain leagility on performance of humanitarian aid organisations in Kenya. *Journal of Business and Economic Development*, *6*(1), 37.

Kirk-Wilson, R. (2002). *Review of Food Assurance Schemes for the Food Standards Agency*, Food Standards Agency, London.

Kuiper, G. & Greiner, C. (2021). Export horticulture and labour migration in Kenya: translocality and transiency in a secondary town. *Geoforum*, *122*, 1–10.

Lavoie, A. & Wardropper, C. B. (2021). Engagement with conservation tillage shaped by "good farmer" identity. *Agriculture and Human Values*, *38*(4), 975–985.

Lien, H. H., de Mey, Y., Meuwissen, M. P. & Bush, S. R. (2023). Information practices for improved sustainability assurance in Vietnamese shrimp aquaculture. *Journal of Rural Studies*, *100*, 103015.

Loader, R. & Hobbs, J. E. (1999). Strategic responses to food safety legislation. *Food Policy*, *24*(6), 685–706.

Manning, L. (2018). Triangulation: effective verification of food safety and quality management systems and associated organisational culture. *Worldwide Hospitality and Tourism Themes*, *10*(3), 297–312.

Manning, L., & Baines, R.N. (2004). Effective management of food safety and quality. *British Food Journal*, *106*(8), 598–606.

Manning, L. & Luning, P. (2018). Food safety in agriculture. In C. A. Wallace, W. H. Sperber, & S. E. Mortimore (Eds.), *Food safety for the 21st century: managing HACCP and food safety throughout the global supply chain*. John Wiley & Sons.

Manning, L., Luning, P. A. & Wallace, C. A. (2019). The evolution and cultural framing of food safety management systems–Where from and where next?. *Comprehensive Reviews in Food Science and Food Safety*, *18*(6), 1770–1792.

Marques, J. C. (2019). Private regulatory capture via harmonisation: an analysis of global retailer regulatory intermediaries. *Regulation & Governance*, *13*(2), 157–176.

Martinez, M. G. & Poole, N. (2004). The development of private fresh produce safety standards: implications for developing Mediterranean exporting countries. *Food Policy*, *29*(3), 229–255.

Martz, W. (2010). Validating an evaluation checklist using a mixed method design. *Evaluation and Program Planning*, *33*(3), 215–222.

McGuire, J., Morton, L. W. & Cast, A. D. (2013). Reconstructing the good farmer identity: shifts in farmer identities and farm management practices to improve water quality. *Agriculture and Human Values*, *30*, 57–69.

Moran, T. H. (2019). FDI and supply chains in horticulture: diversifying exports and reducing poverty in Africa, Latin America, and other Developing Economies. *Thailand and The World Economy*, *37*(3), 1–24.

Muhammad, M., Stokes, J. E., Morgans, L. & Manning, L. (2022). The social construction of narratives and arguments in animal welfare discourse and debate. *Animals*, *12*(19), 2582.

Nabeshima, K., Michida, E., Nam, V. H. & Suzuki, A. (2015). *Emergence of Asian GAPs and its relationship to global GAP*. Inst. of Developing Economies, Japan External Trade Organization.

Nadvi, K. (1999). Collective efficiency and collective failure: the response of the Sialkot surgical instrument cluster to global quality pressures. *World development*, *27*(9), 1605–1626.

Nain, M. S., Singh, R., Kumar, S. & Chahal, V. P. (2015). Farmers producer organisation in reducing transactional costs: A study of Tamil Nadu mango growers federation (TAMAFED). *Indian Journal of Agricultural Sciences*, *85*(10), 1303–1307.

National Standards TCVN 11892-1:2017. GOOD AGRICULTURAL PRACTICES (VietGAP) Part 1: Crop production (see STAMEQ website for additional versions of VietGAP for other commodities).

Naylor, R., Hamilton-Webb, A., Little, R. & Maye, D. (2018). The 'good farmer': farmer identities and the control of exotic livestock disease in England. *Sociologia Ruralis*, *58*(1), 3–19.

Nguyen Tan, D. (2021). Evaluating WTA (Willingness to Accept) a Subsidy for Getting a VietGAP Certificate in Shrimp Farming: A Case Study in Tra Vinh Province, Vietnam. *Egyptian Journal of Aquatic Biology and Fisheries*, *25*(2), 623–650.

Nguyen, N. D. M., Rutsaert, P., Van Loo, E. J. & Verbeke, W. (2017). Consumers' familiarity with and attitudes towards food quality certifications for rice and vegetables in Vietnam. *Food Control*, *82*, 74–82.

Nguyen-Viet, H., Tuyet-Hanh, T. T., Unger, F., Dang-Xuan, S. & Grace, D. (2017). Food safety in Vietnam: where we are at and what we can learn from international experiences. *Infectious Diseases of Poverty*, *6*(01), 83–88.

OECD-FAO. (2016), OECD-FAO Guidance for Responsible Agricultural Supply chains OECD 2016. Available at: http://www.oecd.org/daf/inv/investment-policy/rbc-agriculture-supply-chains.htm [Accessed 6 April 2023]

Pandit, U., Nain, M. S., Singh, R., Kumar, S. H. I. V. & Chahal, V. P. (2017). Adoption of good agricultural practices (GAPs) in Basmati (Scented) rice: a study of prospects and retrospect. *Indian Journal of Agricultural Sciences*, *87*(1), 36–41.

Park, E. & Gachukia, M. K. E. (2021). The role of the local innovation system for inclusive upgrading in the global value chain: The case of KenyaGAP in the Kenyan Horticultural Sector. *The European Journal of Development Research*, *33*, 578–603.

Pauwelussen, A. P. & Bush, S. R. (2020). *Inclusive assurance models in Vietnamese shrimp aquaculture: Deliverable no. 3.2*. European Commission. Available at: https://library.wur.nl/WebQuery/wurpubs/fulltext/537197 [Accessed 1 August 2023].

Piorr, H. P. (2003). Environmental policy, agri-environmental indicators and landscape indicators. *Agriculture, Ecosystems & Environment*, *98*(1–3), 17–33.

Pivoto, D., Waquil, P. D., Talamini, E., Finocchio, C. P. S., Dalla Corte, V. F. & de Vargas Mores, G. (2018). Scientific development of smart farming technologies and their application in Brazil. *Information Processing in Agriculture*, *5*(1), 21–32.

Quyen, N. T. K., Yen, T. T. B. & Riple, A. K. L. (2022). Adoption of Vietnamese Good Agricultural Practices (VietGAP) in aquaculture: evidence from small-scale shrimp farming. *Asian Fisheries Science*, *34*(4), 393–403.

Rainforest Alliance. (2023). UTZ certification (Now Part of the Rainforest Alliance). Available at: https://www.rainforest-alliance.org/utz/#:~:text=UTZ%20means%20good,Guatemalan%20Mayan%20language%20of%20Quiché. [Accessed 11 November 2023]

Rees, W., Tremma, O. & Manning, L. (2019). Sustainability cues on packaging: the influence of recognition on purchasing behavior. *Journal of Cleaner Production*, *235*, 841–853.

Saunders, F. P. (2016). Complex shades of green: gradually changing notions of the 'Good farmer'in a S wedish context. *Sociologia Ruralis*, *56*(3), 391–407.

Sharma, S. & Modgil, S. (2013). Supply chain efforts among downstream and upstream: a developed view. *International Journal of Engineering Management and Economics*, *4*(1), 54–71.

Sparks, J. L. D., Matthews, L., Cárdenas, D. & Williams, C. (2022). Worker-less social responsibility: how the proliferation of voluntary labour governance tools in seafood marginalise the workers they claim to protect. *Marine Policy*, *139*, 105044.

Shortall, O., Sutherland, L. A., Ruston, A. & Kaler, J. (2018). True cowmen and commercial farmers: exploring vets' and dairy farmers' contrasting views of 'good farming'in relation to biosecurity. *Sociologia Ruralis*, *58*(3), 583–603.

Speedy, A. W. (2004). FAO activities in relation to good agricultural practices. In *A farm-to-table approach for emerging and developed dairy countries. IDF/FAO international symposium on dairy safety and hygiene*. Cape Town 2-5 March 2004. Available at: https://www.fao.org/3/y6007e/y6007e00.pdf#page=13 [Accessed 5 April 2023]

The UK Food Safety Act 1990. Available at: https://www.legislation.gov.uk/ukpga/1990/16/contents.

Thomas, E., Riley, M. & Spees, J. (2019). Good farming beyond farmland–Riparian environments and the concept of the 'good farmer'. *Journal of Rural Studies*, *67*, 111–119.

Tran, D. & Goto, D. (2019). Impacts of sustainability certification on farm income: evidence from small-scale specialty green tea farmers in Vietnam. *Food Policy*, *83*, 70–82.

Truong, V. A., Conroy, D. M. & Lang, B. (2021). The trust paradox in food labelling: an exploration of consumers' perceptions of certified vegetables. *Food Quality and Preference*, *93*, 104280.

UN FAO (Food and Agriculture Organisation). (2016). A Scheme and Training Manual on Good Agricultural Practices (GAP) for Fruits and Vegetables. Volume 1. The scheme – standard and implementation structure. Available at: https://www.fao.org/3/i6677e/i6677e.pdf [Accessed 20 February 2023]

Van Der Grijp, N. M., Marsden, T. & Cavalcanti, J. S. B. (2005). European retailers as agents of change towards sustainability: The case of fruit production in Brazil. *Environmental Sciences*, *2*(4), 445–460.

Van Hoi, P., Mol, A. P. & Oosterveer, P. J. (2009). Market governance for safe food in developing countries: the case of low-pesticide vegetables in Vietnam. *Journal of environmental management*, *91*(2), 380–388.

Verbruggen, P. & Havinga, T. (2019). Transnational business governance interactions in food safety regulation: Exploring the promises and risks of enrolment. Stepan Wood, Rebecca Schmidt, Errol Meidinger, Burkard Eberlein, and Kenneth W. Abbott ;(Eds). In *Transnational Business Governance Interactions* (pp. 28–51). Edward Elgar Publishing, Cheltenham UK, ;https://doi.org/10.4337/9781788114738.

Verifik8. (n.d.). Available at: https://www.verifik8.com [Accessed on 1 August 2023].

Westerink, J., Pleijte, M., Schrijver, R., van Dam, R., de Krom, M. & de Boer, T. (2021). Can a 'good farmer'be nature-inclusive? Shifting cultural norms in farming in The Netherlands. *Journal of Rural Studies*, *88*, 60–70.

Winickoff, D. E. & Klein, K. (2011). Food labels and the environment: Towards harmonisation of EU and US organic standards. D. Vogel & J. Swinnen ;(Eds). In *Transatlantic Regulatory Cooperation*. The Shifting Roles of the EU, the US and California. Edward Elgar Publishing, Cheltenham UK. ISBN: 978 0 85793 242 6.

World Bank. (2016). *Vietnam Food Safety Risks Management Challenges and Opportunities. Technical working paper*. Hanoi, Vietnam: World Bank.

Zhai, Z., Martínez, J. F., Beltran, V. & Martínez, N. L. (2020). Decision support systems for agriculture 4.0: Survey and challenges. *Computers and Electronics in Agriculture*, *170*, 105256.

Chapter 10

Improving sustainable agricultural principles and practices: the Sustainable Agriculture Network (SAN)

Mona McCord and Oliver Bach, Sustainable Agriculture Network (SAN), USA

1 Introduction

The history of agriculture can be viewed as a series of key events dating back to around 10,000 years ago when tribes began collecting seeds from edible grasses and vegetation on their nomadic routes and replanting them in regions suitable for year-round cultivation. This was the start of the Neolithic Revolution, also called the First Agricultural Revolution, when human cultures transitioned from a lifestyle of hunting and gathering to one of agriculture and settlement. Since then, human culture has experienced other key events in the transformation of agriculture that have irrevocably altered the relationship between human societies and nature. The early exchange of agriculture knowledge and technology is attributed to a network of routes used by traders, including the

http://dx.doi.org/10.19103/AS.2024.0133.17
Published by Burleigh Dodds Science Publishing Limited, 2024.

"Silk Road" linking Asia and Europe, allowing agricultural and other products to spread from China to other countries from as early as 130 BCE[1].

From the late nineteenth century, the industrialisation of agriculture, or the Second Agricultural Revolution, changed the scale of farming, improving crop production through the development of the first synthetic fertilisers and pesticides as well as the beginning of mechanisation of operations such as ploughing, planting and harvesting. More recently, the Green Revolution, or the Third Agricultural Revolution, brought with it significant increases in grain yield through the introduction of new, high-yielding seed varieties of wheat and rice to address the threat of malnutrition in developing countries. Taken in conjunction with a dramatic increase in use of more advanced chemical fertilisers, herbicides and pesticides, together with a global increase in more intensive methods of cultivation, these new varieties (developed with new, more sophisticated breeding techniques) revolutionised food production all over the world (Sebby, 2010).

While the Third Agricultural Revolution has led to a huge increase in volumes of food production, relying heavily on more intensive, industrialised production methods across the agri-food supply chain, this transformation has come at a huge environmental cost. This includes dramatic declines in biodiversity, particularly in the tropics where forested areas are transformed into monocultures (which are effectively ecological deserts) to increase productivity. According to Our WorldinData.Org (Poore and Neecek, 2018), our current food system is responsible for 80% of tropical deforestation worldwide. Scientists modelling these changes say that current biodiversity loss could be one of the largest mass extinctions in the history of the Earth (Bellard et al., 2012).

Moreover, agriculture-related emissions, mainly from use of chemical fertilisers and large-scale, factory-style livestock production, make up more than 20% of global greenhouse gas emissions (GHGEs) annually ("Sources of Greenhouse Gas Emissions," 2020). GHGEs are driving changes in the climate which, in the long term, when combined with biodiversity loss, pose an existential threat to agricultural production. According to modelling by the World Bank, we can expect to see significant climate change-related yield decreases in some of the world's key crops by 2050, including wheat, rice, maize and millet (as well as field pea, sugar beet, sweet potato, soybean, groundnut, sunflower and rapeseed) (Müller et al., 2010).

To make matters worse, the United Nations forecasts that by 2050 the world's population will grow by more than two billion people compared to the 2020s. Half will be born in sub-Saharan Africa, and another 30% in South and Southeast Asia[2]. These trends suggest that we will need significantly more food

1 https://education.nationalgeographic.org/resource/silk-road.
2 https://www.un.org/development/desa/en/news/population/world-population-prospects-2019.html.

Published by Burleigh Dodds Science Publishing Limited, 2024.

while facing an increasingly hostile environment due to climate change and diminishing resources. In other words, the agriculture sector will need to grow, while simultaneously becoming more climate smart, to produce an estimated 70% more food if it is to feed a population of 9.7 billion people by 2050.

2 Towards a more sustainable future for agriculture: the Sustainable Agriculture Network

An alternative and more sustainable agricultural model is provided by agroecology. Agriculture, in this context, is viewed as an adapted form of ecosystem in which an understanding of fundamental biogeochemical flows/cycles and functional interactions between organisms can be used to optimise its ability to deliver a range of ecosystem services, including food production. Examples of practices promoting agroecosystem function include optimising the physical, chemical and biological health of soils to promote crop growth and health (reducing the need for external inputs such as fertilisers and pesticides), actively promoting biodiversity (e.g. incorporating a range of plants and landscape features to encourage beneficial species such as pollinators or predators of crop pests), and ensuring diversity in agricultural landscapes (to optimise functional interactions between species). In a functional agroecosystem, healthy soil microbes provide nutrients to plants, the root systems of which hold the soil in place. Plants provide food and habitat to beneficial insects and birds that pollinate them and manage pests. Livestock can recycle leftover parts of crops and provide natural fertiliser to fields and pasture through manure.

The International Assessment of Agricultural Knowledge, Science and Technology for Development (IAASTD) (IAASTD, 2009) has concluded that the future of agriculture lies in biodiverse, agroecological-based farming that can meet social, economic and environmental goals as well as maintain and increase productivity. Agroecosystems fundamentally depend on diversity to stay in balance.[3] As an example, a recent meta-analysis showed that the establishment of hedgerows on croplands leads to a significant increase in soil organic carbon (SOC) stocks – on average 32% higher than in the adjacent cropland (Drexler et al., 2021). Similarly, research from California (Long et al., 2017) on establishing hedgerows has highlighted pollination services provided by native bees and pest control by natural enemies. Among other results, they found bird predation resulted in 35% reduction in codling moth pests, reduced weed abundance (10–100%) and diversity (35%), reduced sediment run-off (60% or more) and with 20% greater carbon sequestration in hedgerows than surrounding fields.

3 https://foodprint.org/issues/biodiversity-and-agriculture/.

For many decades, multinational companies have sought to improve safety and quality of agri-food supply chains to meet regulatory or customer requirements through individual or collective voluntary sustainability standards (VSS) and certification labels. However, there are concerns that this approach is not sufficient to achieve the major changes required to improve environmental and social sustainability, particularly in managing the transition from deeply embedded 'productivist' approaches to more sustainable 'climate-smart', agroecological and regenerative approaches to agricultural production.

In this respect, non-governmental organisations (NGOs) have a key role to play, both in advocating for change and in providing expertise and solutions that can help key players in agri-food supply chains, from multi-national corporations to smallholder farmers, manage the transition. The Sustainable Agriculture Network (SAN)[4] is a global collaborative network of 33 NGOs with a mission to transform agriculture to secure a sustainable future for food, nature and rural communities. It is paving the way for a sustainable agricultural revolution that recognises the inherent interconnectedness of people and planet with solutions that promote sustainable communities that are equitable, environmentally resilient and economically viable across the globe. SAN's work began in 1998 with development of the SAN Certification Standard that has enabled more than one million small, medium and large-scale producers to embark on and then continually progress their journey towards more sustainable farming.

Based on this experience of developing and implementing its own certification scheme, SAN's remit has grown to provide guidance to help others develop and improve their own standards and certification initiatives. A particular focus is performance measurement. Traditionally sustainability standards have been practice-based standards. Achieving certification required implementing specified production practices or adopting particular management systems. The underlying assumption was that these practices would deliver the desired social and environmental outcomes. However, this also brought the risk that implementing particular practices might become a mere box-ticking exercise with little real impact if the relevant motivation, knowledge and resources were lacking. More recently, there has been more of a focus on standards with outcome or performance-based measurements better suited to measuring actual impacts (Komives and Jackson, 2014).

3 The Sustainable Agriculture Framework

The Sustainable Agriculture Framework (SAF) is an outcome-based sustainability indicator framework designed to guide organisations towards more effective standard-setting processes and ultimately deliver much more

4 https://www.sustainableagriculture.eco/sustainable-agriculture-framework-2021.

Published by Burleigh Dodds Science Publishing Limited, 2024.

positive changes for landscapes, farmers and communities. Launched in 2018, SAF is rooted in the 2017 Sustainable Agriculture Network (SAN) Standard which built on SAN's extensive experience in standard development. The 2017 SAN Standard was developed after a 3-year process including five global stakeholder consultation exercises, two rounds of field testing on farms in Africa, Asia and Latin America, and peer reviews by a number of expert committees. The undertaking of this rigorous process ensured its compliance with the ISEAL Code of Good Practice for Setting Social and Environmental Standards. The 2017 SAN Standard has become a respected model for global standard-setting and certification systems by both the private sector and NGOs.

While the 2017 SAN Standard was designed as a VSS for supply chain certification and assurance, it was not suitable for organizations looking to develop their own, outcome-based solutions. This led SAN to convert the 2017 SAN Standard into the outcome-based SAF technical framework. SAN applied a number of steps to develop SAF:

- It left intact the core technical elements of the 2017 SAN Standard previously validated by regular stakeholder consultation; it did not change elements that were already outcome-based but also left practice-based elements where these were best suited to achieving a particular outcome.
- It restructured the technical contents of the Standard into ten impact areas and established indicators within different outcomes categories of each impact area.
- It added new impact areas on climate change mitigation and adaptation and biodiversity outcomes for pollinator-friendly agriculture.
- It compiled and synthesised the large number of questions and observations received from its internal and external stakeholders about the 2017 SAN Standard.

It further improved SAF by:

- Eliminating unpractical or ineffective elements and replacing them with technically robust solutions with proven efficacy based on SAN's more than 20 years of experience. For example, while SAN fully supports the six high conservation values (HCV), its long experience in sustainable agriculture, public consultation and field-testing has led it to develop a specific HCV implementation approach that can work better for smallholder groups and medium-sized plantations.
- Substituting non-outcome focused criteria for outcome-based indicators. For example, the existence of an integrated pest management (IPM) plan does not actually mean that IPM is practiced effectively; better measures

include the presence of predator species and measurable declines in pest populations.
- Filling in content gaps with new outcome-related indicators. These new elements were developed by experienced SAN technicians and its member network.
- Checking that outcome-based indicators were practical, credible and meaningful, and relevant to a variety of crop or cattle production contexts, as well as big plantations, medium farms and smallholder groups.
- Removing the continuous improvement stepwise-system originally designed for certification to enable more innovation and flexibility in demonstrating continuous improvement.

These five steps have given SAF both rigour and flexibility.

Further, SAF measures progress based on ten attributes:

- Sustainable management of agricultural and livestock operations
- Conservation and management of biodiversity
- Conservation and management of natural resources
- IPM
- Protection of workers' rights
- Occupational health and safety
- Wellbeing of workers and their families
- Wellbeing of rural communities
- Sustainable livestock production
- Climate change mitigation and adaptation.

The ten attributes of the SAF are shown in Fig. 1.

Each attribute includes various sustainability goals, which are composed of expected outcomes and performance indicators to demonstrate progress over time. The modular design of SAF is flexible enough to allow tailored solutions for diverse farming systems and regions around the world. The interactions of impact areas, attributes, sustainability goals, outcomes and performance indicators within the SAF structure are shown in Fig. 2.

The result of adopting this process of identifying impact areas, goals, outcomes and performance indicators supports the effective transition of a farm to a production system with reduced GHG emissions, improved biodiversity and soil health, and improved livelihoods for farmers and farm workers. SAF also provides a technical framework for SAN's own sustainable agriculture project planning and implementation.

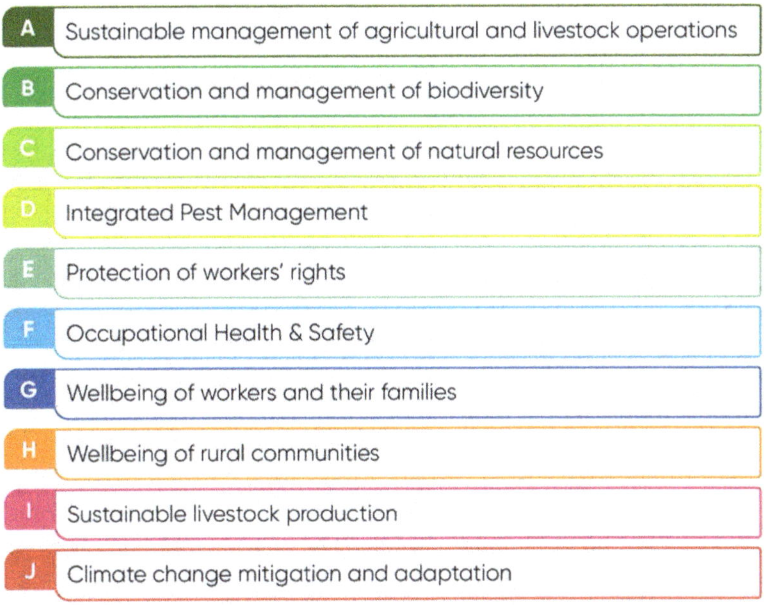

Figure 1 Attributes of SAN SAF.

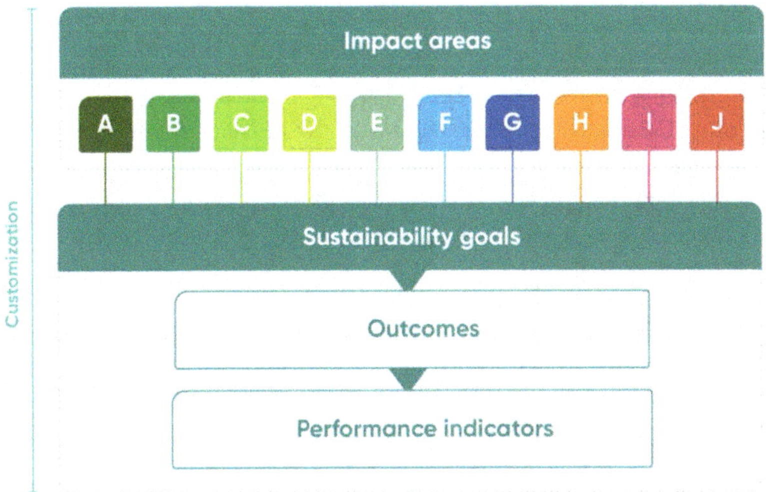

Figure 2 SAF structure.

4 Examples of sustainability programmes guided by SAF: the Integrated Landscape Approach (TerraViva)

SAF has guided SAN to create several signature programmes that, through collaboration, coordination and planning with a range of stakeholders, facilitates agriculture transformation with a focus on biodiversity and conservation while ensuring sustainable livelihoods for rural communities.

One of SAN's signature programmes is based on an integrated landscape approach (ILA) that allows for holistic management of land use and stakeholder needs within a region. It is a governance strategy that acknowledges the interdependencies of human and natural systems and seeks to optimise synergies and minimise tradeoffs to harmonise wellbeing of rural communities within their environment.

SAN's ILA, called TerraViva, incorporates several attributes from SAF including:

- conservation and management of natural resources and biodiversity
- climate change mitigation and adaptation

For multi-national brands focusing on producers within particular regions in their supply chains, TerraViva provides an opportunity for a collective approach aimed at improving the sustainability of entire sourcing regions. It also has the potential to guide local activities, and inform policy and investment decisions. TerraViva provides stakeholders with a holistic view of landscape sustainability, credible data to inform decision makers and consistent information to support adaptive management.

Currently TerraViva is being piloted in a coffee-growing region in Colombia (South America). It includes:

- using GIS mapping and climatic analysis to understand current and historic land use, and
- collecting information from local stakeholders through surveys and interviews.

The combination of data acquisition and participatory dialogues with stakeholders helps to foster a common vision and purpose developed from the perspective of local stakeholders that can lead to collective solutions to meet both production and environmental goals. Particular features of TerraViva include:

- *A stakeholder identification and engagement plan* to build trust, legitimacy and optimum participation.
- *A common territorial agenda* to establish common concerns shared by multiple stakeholders as a basis for co-developing actionable strategies to address for example, climate change adaptation, forest loss or food security.

- *A multi-stakeholder platform* to provide space for negotiations of land-use trade-offs that will pave the way for systemic interventions so that regeneration is not isolated on single farms but is integrated across an entire region.
- *Business models* to recognise ecosystem services and their longer-term returns for rural communities, including sustainable commodity sourcing and credible carbon credits, among others.
- *Monitoring and evaluation systems* to measure progress towards relevant socio-economic, environmental, and governance objectives.

While ILAs such as TerraViva hold great promise to establish more sustainable and equitable multifunctional landscapes, ILAs are a relatively new approach and more evidence of effectiveness is needed to bridge knowledge gaps, particularly as these need to be long-term projects to deliver durable changes across regions. A study, which was part of the CGIAR Research Programme on Forests, Trees, and Agroforestry (CRP-FTA), and led by James Reed from the Centre for International Forestry Research and University of Cambridge Conservation Research Institute (Reed et al., 2020), has taken stock of ILAs in the tropics. The study found that, despite considerable enthusiasm for landscape approaches, the evidence base within the scientific literature remains poorly developed. Given that there are several elements of ILAs that still lack a robust empirical evidence base, the research community can play an important role in further development of this approach within complex tropical landscapes.

5 Examples of sustainability programmes guided by SAF: on-farm biodiversity and integrated pest management (Ecoasis)

One of the attributes for SAF is IPM. Fostering communities of beneficial wild animals and insects is a key part of IPM. IPM treats pesticides as the option of last resort for protecting crops and focuses instead on using beneficial relationships between insects and plants as a first line of defense (Altieri, 1993). Additionally, wildlife habitats contribute to reduced pollution, improved human and animal health, aesthetic and recreational value.

To capitalise on the beneficial relationships between insects and plants, SAN developed a programme called Ecoasis that simultaneously improves the resilience and sustainability of both farmer livelihoods and natural ecosystems. Enhancing natural ecosystem services through improved habitat design and farm management strategies are at the heart of the programme. Ecoasis involves co-designing multi-functional areas of native flowers and shrubs with local producers to provide important habitats for pollinators, natural pest enemies and other beneficial insects. Increases in species diversity, habitat complexity and resource availability provide a rich substrate to provide the beneficial interactions necessary to optimise the ecological benefits and economic productivity of agroecosystems.

Key features of Ecoasis include:

- Equipping farmers to re-establish farmland biodiversity and improve production.
- Ensuring adequate agricultural production by recommending viable short-term and long-term crops.
- Helping farmers acquire knowledge and tools to support diversified farming systems, and working with local government and the private sector to facilitate markets.
- Implementing community-led approaches to foster socially and environmentally sustainable farming.

Currently, Ecoasis is being trialled with palm oil growing communities in Malaysia to demonstrate and evaluate its potential. Producers are trained to establish and monitor insect habitats and apply alternative pest and disease controls. In doing so, they are empowered to become local biodiversity conservation and IPM champions. SAN anticipates expanding the programme to include other tropical crop systems and regions in the future, including incorporating hedgerows, windbreaks, live fences and better soil cover that can also act as a carbon sink and provide multiple ecosystem services.

6 Examples of sustainability programmes guided by SAF: multifunctional field margins (Operation Pollinator)

SAN works closely with corporate clients to bring tangible returns for their sustainability investments and drive measurable action towards zero-net emissions, enhanced livelihoods and environmental impact in the communities that are directly connected to their business activities. SAN has adapted Scott Poynton's VT-TV approach (Poynton, 2017) to bring transformation to product supply chains. SAN's adapted approach is:

- **Value**: understand the outcomes that a company wants to achieve and work backwards to co-create a 'climate transition programme' with the company that includes actionable strategies to achieve the desired outcomes.
- **Transparency**: map a company's supply chain and understand the main challenges. SAN, together with its local partners/network members,

conduct desk assessments, field visits, key stakeholder interviews and root-cause analysis to identify the main areas for improvement and potential reputational risks associated with a specific supply chain.

- **Transformation**: Once a supply chain is mapped using SAF, local partners and/or SAN members provide guidance to suppliers/processers in the development, implementation, monitoring and evaluation of interventions at the farm-level that will help transition their supply chain to regenerative agriculture practices
- **Verification**: by providing independent verification, SAN can not only support companies in managing their supply chain responsibly but also provide opportunities for market recognition.

This approach has been applied in the case of multifunctional field margins (MFMs). MFMs can serve as a key tool in IPM and fostering communities of beneficial wild animals. Planting strips of beneficial flowers, incorporating perennials, establishing hedgerows (rows of trees or shrubs separating fields) and leaving areas of land uncultivated are methods of increasing diversity on non-cropped land. Benefits include:

- Planting crops that support natural enemies or directly inhibit insect attack helps to stabilise insect communities.
- Spatially and temporally diverse plantings ensure that natural pest predator populations are provided with a continuous availability of resources.
- Beneficial insects can also be provided food and habitat by including areas of adjoining, uncultivated land, and wild vegetation.

When developed using participatory approaches, local stakeholders are empowered to become local biodiversity conservation and IPM champions.

In 2018 Syngenta, Arcadis and Bioversity International joined forces to review how agricultural biodiversity can be improved using MFMs to produce natural, social and economic benefits and suggested principles and protocols to establish and manage them effectively (Syngenta, Arcadis and Bioversity International, 2018). The review included contributions from scientific researchers, advice from practitioners, interviews with selected project stakeholders across nine countries, and a literature review of 74 scientific papers. The study documented 20 natural and environmental benefits along with 15 social capital benefits (shown in Table 1).

In addition, six of these benefits (pollination services provision, soil erosion prevention, water filtration and storage, recreation and aesthetics, carbon sequestration and climate regulation) were selected for monetary valuation. The study estimated the average economic values (US$ per hectare per year) that particular MFM practices could bring:

- Establishment of MFMs with flowers, grasses and shrubs margins: US$1600/ha/year.
- Establishment of MFMs with tree rows: US$1200/ha/year.
- Restoration of connectivity in tropical forest: US$4,100/ha/year.

Other main conclusions of the study were that:

- the more the range of ecological features and biodiversity richness in MFMs, the more benefits they provide, and
- improvements in agro-biodiversity are optimum when a minimum of 3% of farmland is dedicated to MFFMs.

However, challenges in wide-spread uptake of MFMs include optimising their cost-effectiveness, and continuous documentation of the positive impacts to convince the agriculture sector to adopt these practices. Establishment costs for MFMs can be improved through the testing of seed mixes adapted to local crop and biophysical conditions as well as the active management of existing understory vegetation, focusing on the most beneficial host plants for insect pollinators and predators.

Syngenta International has developed a programme called Operation Pollinator that boosts the number of pollinating insects on commercial farms

Table 1 Natural, environmental and social capital benefits.

Natural and environmental benefits	Social capital benefits
Reintroduction of local species	Employment
Pollinators	Yield
Ecosystem disturbance moderation	Gender equality
Waste treatment	Recreational activities
Nutrient cycling	Tourism and aesthetics
Nursery service	Raw material provision
Ornamental resources	Spiritual
Natural pest predators	Medicinal resources
Earthworm populations and activity	Genetic resources
Food sources and nesting sites	Air quality regulation
Migration corridors	Knowledge acquisition and education
Pest control	Local community benefits
Soil quality and erosion prevention	Landscape aesthetics
Water filtration and storage	Cultural and historical heritage
Water quality	Product branding/market access
Flood control	
Wind breaks	
Genetic diversity	
Climate regulation	
Carbon sequestration	

using MFMs. Operation Pollinator has been active for more than 15 years in creating habitats and resources for local wildlife in 22 countries, creating five million hectares of habitat in field margins or on fallow land on commercial farms or golf courses. The habitat provides nesting and food resources for bees, pollinators, beneficial insects, as well as for small mammals and farmland birds, enhancing overall biodiversity and ecosystem services that improve yields, secure sustainable farming and environmental balance.

In collaboration with Syngenta, SAN coordinated Operation Pollinator on berry farms in the State of Jalisco in Mexico for 3 years (2019–2021), focusing not only on promoting pollinators but also natural pest enemies (parasites and predators). Working with farmers in a participatory and outcome-oriented approach, SAN and its project partners (ABC Mexico, ANAC and Zrum Agroecología) created more than 1800 m^2 of native plants on six pilot farms with a total area of 1200 hectares. Great care was taken to optimise flower strip design with participating farmers in terms of diversity, costs and potential impact of beneficial insects according to their flight range. A diversity of 20 different plant species with a variety of flower shapes and colours were planted to attract a maximum variety of insects.

Lead farmers were trained in insect identification and their model farms and experiences were used for on-boarding of new producers in each region. Data collection on beneficial insect diversity and abundance continues and has initially demonstrated a 56% increase in insect diversity in Jalisco and 300% in the Puebla Municipality. The frequency and intensity of pest and beneficial insect monitoring can be increased via community insect monitoring conducted by farmer and rural community members with guidance from regional and local entomologists and botanists. These local 'para-taxonomists' can classify

Figure 3 Results from Jalisco.

Figure 4 Results from Puebla.

insects at functional groups – the necessary level to associate the diversity of beneficial insects with insect pest abundance and crop productivity. Figure 3 shows results for Jalisco and Fig. 4 shows for results for Puebla.

Other types of spatial design can be more appropriate for other crops. In Malaysia, SAN is currently working with Ferrero, Wild Asia and Universiti Putra Malaysia (UPM) entomologists and botanists in developing a square block design of multiple natural vegetation plots within oil palm production areas that can deliver ecological benefits while not affecting necessary oil palm production management activities. The results are promising as documented in a recently published study in Cogent Food & Agriculture (Norhisham et al., 2024).

7 Future trends

Looking ahead, SAN recognises that more collaboration is needed between practitioners and researchers to bridge the gap between science and practice. In particular, SAN is interested in research that contributes to sustainable agriculture in a number of areas, including the following.

7.1 Biochar

Biochar is a lightweight black residue, composed of carbon and ashes, left after pyrolysis of biomass (used to produce green energy or reduce waste). Biochar is defined by the International Biochar Initiative as 'the solid material obtained from the thermochemical conversion of biomass in an oxygen-limited environment' ('Standardized production definition and product testing guidelines for biochar that is used in soil', 2015). While the biochar production process releases CO_2, the carbon in the final product is extremely stable (Woolf et al., 2010). Biochar in soil can therefore improve soil carbon sequestration, as well as increase soil fertility, improve water quality and raise agricultural productivity, thus reducing pressure on natural forests (Laird, 2008). Currently agribusiness Barry Callebaut is upcycling its cocoa shells into biochar to prodce green energy and reduce carbon emissions in their operations.[5]

7.2 Crop wild relatives

Crop wild relatives (CWR) contain a multitude of genes of potential value for plant breeding. Among these are many traits that are relevant for climate change adaptation. Crop wild relative species are distributed across a wide range of habitats, including mountains, deserts, grasslands, salt marshes and rainforests. They have evolved many different strategies for surviving in these diverse climatic conditions. The genetic traits that allow CWR species to thrive in different, sometimes, extreme habitats, represent a valuable resource for plant breeding in the context of climate change.[6] As an example, Michael Kantar of the University of Hawaii (USA) is using techniques from genetics, genomics, plant breeding, ecology, geography and agroecology to develop new crops and improve old ones for a changing environment.[7]

7.3 Orphan crops

Orphan crops are those crops which are typically not traded internationally but which can play an important role in regional food security.[8] Despite their huge importance for present and future agriculture, orphan crops have generally received little attention by the global scientific community. A lack of research means they often produce inferior yields in terms of both quantity and quality. The major bottlenecks affecting the productivity of these crops include little or no selection for improved genetic traits, challenges in management (often by smallholders with limited knowledge and resources) as well as unfavorable government policies to promote more investment and resources. Concerted efforts are urgently needed to advance the research and development of orphan crops so that food security and the livelihoods of rural population can

5 https://www.foodingredientsfirst.com/news/converting-cocoa-shells-into-green-energy-barry-callebaut-utilizes-agricultural-left-overs-to-create-valuable-biochar.html.
6 https://www.cwrdiversity.org/project/cwr-and-climate-change-adaptation/.
7 https://mbkantarlab.weebly.com/about.html.
8 https://fse.fsi.stanford.edu/research/orphan_crops#:~:text=Orphan%20(or%20minor)%20crops%20are,wishing%20to%20improve%20their%20productivity.

be improved (Tadele, 2019). Currently, the African Orphan Crops Consortium (ACC) is assembling, sequencing and annotating the genomes of 101 traditional African food crops to improve their nutritional content. The resulting information will be put in the public domain with the endorsement of the African Union.[9]

8 Conclusion

Agriculture and those who depend on it for their livelihoods face increasing challenges in such areas as climate change (with more extreme and variable weather), rising input costs and regulatory uncertainty. Climate change and biodiversity loss are closely linked to food production and trade. To address these existential threats, companies in the agri-food supply chain need to use their global footprint and influence to transform our global food system. There is a greater recognition that current certification and farmer training initiatives are no longer enough. SAN has developed SAF as a guide towards a more sustainable agriculture model. SAF provides the measurement criteria for an ecosystem-based approach to farming that aims to improve farmer resilience, yields and produce quality by improving soil health, enhancing biodiversity and effective use of IPM. The aim of SAF is to help producers and supporting organisations re-imagine farming and create a pathway to progress.

9 References

Altieri, Miguel A. (1993). Ethnoscience and biodiversity: key elements in the design of sustainable pest management systems for small farmers in developing countries. *Agriculture, Ecosystems & Environment*, 46(1–4), 257–272. https://doi.org/10.1016/0167-8809(93)90029-o.

Bellard, C., Bertelsmeier, C., Leadley, P., Thuiller, W. and Courchamp, F. (2012). Impacts of climate change on the future of biodiversity. *Ecology Letters*, 15(4), 365–377. https://doi.org/10.1111/j.1461-0248.2011.01736.x.

Drexler, S., Gensior A. and Don, A. (2021). Carbon sequestration in hedgerows biomass and soil in the temperate climate zone. *Regional Environmental Change* 21, Article No. 74. https://link.springer.com/article/10.1007/s10113-021-01798-8.

IAASTD. (2009). *Agriculture at a crossroads. International assessment of agricultural knowledge, science and technology for development*. Island Press, Washington, DC. http://www.agassessment.org.

Komives, K. and Jackson, A. (2014). Introduction to voluntary sustainability standard systems. In: C. Schmitz-Hoffmann, M. Schmidt, B. Hansmann and D. Palekhov (eds.). *Voluntary standard systems – a contribution to sustainable development*. Springer, Berlin.

Laird, David A. (2008). The charcoal vision: a win–win–win scenario for simultaneously producing bioenergy, permanently sequestering carbon, while improving soil

9 https://www.worldagroforestry.org/output/african-orphan-crops-consortium.

and water quality. *Agronomy Journal*, 100, 178–181. https://doi.org/10.2134/agronj2007.0161.

Long, R., Garbach, K. and Morandin, L. (2017). Hedgerow benefits align with food production and sustainability goals. *California Agriculture*, 71(3), 117–119. https://doi.org/10.3733/ca.2017a0020.

Müller, C., Bondeau, A., Popp, A., Waha, K. and Potsdam, M. F. (2010). *Institute for Climate Impact Research (PIK)*. Germany.

Norhisham, A. R., Yahya, M. S., Nur Atikah, S., Jamian, S., Bach, O., McCord, M., Howes, J. and Azhar, B. (2024). Non-crop plant beds can improve arthropod diversity including beneficial insects in chemical-free oil palm agroecosystems. *Cogent Food & Agriculture, 10*(1). https://doi.org/10.1080/23311932.2024.2367383.

Poore, Joseph and Neecek, Thomas. (2018). *Reducing Food's Environmental Impacts through Producers and Consumers*. Published in Science.

Poynton, Scott. (2017). *Beyond Certification*. Routledge. https://www.earthworm.org/uploads/files/Beyond-Certification.pdf.

Reed, J., Ickowitz, A., Chervier, C., Djoudi, H., Moombe, K., Ros-Tonen, M., Yanou, M., Yuliani, L. and Sunderland, T. (2020). Integrated landscape approaches in the tropics: a brief stock-take. *Land Use Policy*, 99, 104822.

Sebby, K. (2010). The green revolution of the 1960's and its impact on small farmers in India. *Environmental Studies Undergraduate Student Theses*, 10. https://digitalcommons.unl.edu/envstudtheses/10.

"Sources of Greenhouse Gas Emissions." *EPA*, Environmental Protection Agency, 4 Dec. 2020. www.epa.gov/ghgemissions/sources-greenhouse-gas-emissions.

"Standardized production definition and product testing guidelines for biochar that is used in soil" (PDF). 2015. Archived (PDF) from the original on 25 February 2019. Retrieved 23 November 2015.

Syngenta, Arcadis and Bioversity International. (2018, April). *'Multifunctional field margins' Assessing the benefits for nature, society, and business; a position paper*.

Tadele, Z. (2019). Orphan crops: their importance and the urgency of improvement. *Planta*, 250, 677–694. https://doi.org/10.1007/s00425-019-03210-6.

Woolf, D., Amonette, J. E., Street-Perrott, F. A., Lehmann, J. and Joseph, S. (2010, August 10). Sustainable biochar to mitigate global climate change. *Nature Communications*, 1(5), 56. https://doi.org/10.1038/ncomms1053. ISSN 2041-1723. PMC 2964457. PMID 20975722.

Chapter 11

Fulfilling the promise of sustainability certification: reflections from the Rainforest Alliance[1]

Deanna Newsom, Rainforest Alliance, USA; Henk Gilhuis, Rainforest Alliance, The Netherlands; Rens Rutten, PUM Netherlands, The Netherlands; and Talia Sechley, Rainforest Alliance, USA

1 Introduction

In 1987, the Rainforest Alliance was founded by a handful of young people who were concerned about rainforest destruction and wondered whether rewarding responsible forestry companies in the tropics, through third-party certification, might be a better approach to forest conservation than boycotting all tropical producers. Twelve years later, in 1999, a Belgian-Guatemalan coffee grower and a Dutch roaster applied the same principle to the sustainable coffee sector and founded the Utz Kapeh certification programme.[2]

The initial premise of both the Rainforest Alliance and UTZ was simple: consumer demand for sustainable products, combined with the desire of retailers and other buyers to bolster their green credentials, would allow voluntary sustainability certification to harness market forces to benefit both the environment and the people whose livelihoods depend on these natural

1 Views and opinions expressed in this chapter are solely those of the authors and do not necessarily reflect the position of the Rainforest Alliance. Please note the chapter reflects the situation at the point the text was written in September 2022.
2 The Rainforest Alliance and UTZ formally merged in 2018, continuing their operations under the Rainforest Alliance name.

http://dx.doi.org/10.19103/AS.2023.0133.19

resources. The logic was that rising consumer demand for sustainable products – and potentially even a willingness to pay more for them – would enable companies to reward producers who farmed sustainably and conserved forests. It was anticipated that these rewards, such as better market access or a price premium, would offset the costs incurred by producers as they improved their practices. Moreover, the hope was that producers would realise intrinsic benefits from adopting sustainable practices, improving the productivity, efficiency, and profitability of their farms and strengthening the business case for certification. And those difficult issues like child labour? Ideally, the threat of de-certification and losing access to certified markets would motivate farmers to stamp them out; at the very least certification would provide a space to raise awareness and discuss solutions.

Now, 35 years after the founding of the Rainforest Alliance and 23 years since the founding of UTZ, we see that many aspects of the two organisations' founding visions have, in fact, materialised, and the initial fear that corporate demand for certified products would be weak seems to be unwarranted. Today, virtually all the largest global companies have adopted sustainability commitments and set carbon reduction targets (Winston, 2021), and governments and investors are increasingly focusing attention on the environmental and social impact of business through voluntary and mandatory environmental, social, and governance reporting and due diligence legislation. Some governments have directly adopted sustainability standards, such as the Forest Stewardship Council (FSC), as a proxy for regulation to help them track progress on sustainability policies and meet global commitments (FSC, 2020). In addition, collaborations between standard-setting organisations and other non-governmental organisations (NGOs) to tackle tough issues such as high-conservation value forests and the living wage gap have led to the establishment of new global norms.

This demand has caused the market share of certified products to grow steadily over the years, especially in export-oriented commodity markets. Currently, Rainforest Alliance Certified products make up approximately 11% of global coffee production, 7% of cocoa production, and 22% of tea production, with the certification programme reaching over 400 000 coffee farmers, 890 000 cocoa farmers, 950 000 tea farmers, and 2100 banana farmers. The number of hired workers covered by the standard stands at 3.3 million[3] and currently, over 6 million hectares are certified to the Rainforest Alliance standard, spanning 70 countries.

Independent research shows that, in many contexts, the positive outcomes envisioned by the founders of the Rainforest Alliance and UTZ have been realised. Implementing the certification standards has boosted farm yields

3 This figure includes 185 000 banana workers, 790 000 tea workers, 33 000 cocoa workers, and 660 000 coffee workers.

(Dompreh et al., 2021; Brako et al., 2021; Bini et al., 2016), market prices (Haggar et al., 2017; Mitiku et al., 2017), and income (Dompreh et al., 2021; Iddrisu et al., 2020; Mitiku et al., 2017; Akoyi and Maertens, 2018) for certified farmers, improved living and working conditions for hired labour (Premaratne et al., 2018; Ochieng et al., 2013), and led to better implementation of environmental best management practices on certified farms (Rueda and Lambin, 2013; Haggar et al., 2012), increasing ecosystem quality (Haggar et al., 2017; Takahashi and Todo, 2017; Hardt et al., 2015).

Despite these successes in programme growth and in certain outcome areas, many Rainforest Alliance Certified farmers or farm workers have seen less economic progress as a result of engaging with the certification programme than was envisioned all those years ago. Over 90% of certified cocoa farmer families in Côte d'Ivoire still do not earn a living income (IDH, 2021). In all sectors and in most countries, considerable gaps continue to exist between earned wages and the living wage benchmarks established for those locations. This highlights the observation – pointed out by critics and supporters of certification alike – that many of the most critical social and environmental challenges cannot be addressed by voluntary, market-based regulatory mechanisms like certification alone (Grabs, 2020; LeBaron, 2020; Ponte, 2019). In addition, certification is inherently confined to the area within a farm's boundaries, while many urgent environmental issues require landscape-level cooperation. And even within the farm's boundaries, societal norms and the policy environment can make it difficult for some farmers to change practices, despite the incentive of certification.

Recognising these realities has motivated the Rainforest Alliance to make significant changes to its programmes and to forge new partnerships to better address these challenges. Introducing the Shared Responsibility concept – which requires buyers to share the costs and benefits of certification more equally – is one such example. Shifting away from a pass/fail approach to a due diligence approach to identifying and remediating human rights violations is another. The 2020 Rainforest Alliance Standard requires farmers to develop systems to assess and address the risks of child labour, forced labour, discrimination, and workplace violence and harassment. This approach reduces farmers' incentives to hide practices such as child labour and encourages them to address problems proactively.

The purpose of this chapter is to describe in more detail the challenges outlined earlier and why we remain optimistic about the solutions being implemented by the Rainforest Alliance. Just as importantly, we conclude by voicing support for a complementary longer-term solution that, we think, would directly address one of the market failures inherent in the certification model. We believe that by including *all* social and environmental costs in the prices of tea, coffee, and any product in the marketplace, sustainable products would finally compete against unsustainable ones on an even playing field – and would become the default choice for consumers.

2 Origins

The Rainforest Alliance was founded in 1987, and by 1989, it opened its first office in Costa Rica. Shortly thereafter, the Rainforest Alliance established the first principles and criteria of sustainable agriculture, motivated by the human rights violations and toxic agrochemical use that were rampant in the banana sector. Though the concept of sustainability certification was rejected by the vast majority of banana growers at the time, the companies Chiquita and Favorita decided to implement the standard, which was a bold move at the time that provided critical momentum for the certification concept (Taylor and Scharlin, 2004).

In response to growing awareness of the deforestation crisis and demand for sustainably produced wood products, the Rainforest Alliance co-founded the FSC in 1993. The Rainforest Alliance's auditing unit, SmartWood, began auditing companies against the FSC standard, the first independent global forestry certification system. Building on this effort to connect responsible production with burgeoning markets for sustainably sourced goods, the Rainforest Alliance co-founded the Sustainable Agriculture Network (SAN) to develop a sustainable agriculture standard that aimed to balance environmental conservation, farm economics, and the well-being of workers and farming communities.

In the second half of the 1990s, two important events occurred in Guatemala. Firstly, the Rainforest Alliance certified its first coffee farm and began working with partners in the Maya Biosphere Reserve to implement a community forestry model there. Soon after, the Rainforest Alliance began working with a SAN member organisation in Ecuador to certify a cocoa farm, and with an NGO in Colombia on flowers, as the concept of sustainability certification quickly gained momentum. Around the same time, in 1999, the Utz Kapeh programme was launched in Guatemala, with the certification of a coffee farm. The programme was founded partially as a response to the observation by one of its founders that the fairtrade concept, which stipulated a minimum price and fixed premiums, had not gained traction among mainstream coffee roasters, brands, and retailers in the early 2000s, since the concept was difficult to apply to large coffee farms. In 2009, the programme, now called UTZ Certified, expanded into the cocoa sector, followed by tea and rooibos (2010) and hazelnuts (2014).

As two well-established certification programmes, the Rainforest Alliance and UTZ soon realised that the impact of their work could be significantly increased by joining forces. The two organisations began collaborating on efforts such as the founding of the Global Living Wage Coalition, banning hazardous pesticide use on farms, and joint impact studies in collaboration with ISEAL. Although the Rainforest Alliance focused more on tea and banana and the US and UK markets, and UTZ was more oriented toward coffee and cocoa and the EU market, the two standards had many similarities. There was a growing recognition too, that the proliferation of standards was causing hardships for those farmers who were asked to certify to multiple standards and pay for multiple audits. So, to increase the scale and impact of their work, and minimise costs for producers, the Rainforest Alliance and UTZ decided to merge, doing so officially in 2018 and continuing their operations under the Rainforest Alliance's name.

One of the first tasks of the new organisation was to create a joint Sustainable Agriculture Standard (discussed in more detail in the next section). The Rainforest Alliance had taken full ownership of the shared Rainforest Alliance/SAN certification in 2017 and had sold its certification unit to the independent auditor NEPCon (now called Preferred by Nature), which allowed the organisation to focus on driving impact more directly through developing and implementing the standard. Now, in 2022, the UTZ logo is about to phase out from the marketplace to be replaced by the green frog as certificate holders of the pre-merger programmes are transitioning to the new standard.

As will be explored further in this chapter, in recent years, the Rainforest Alliance's scope and impact has broadened far beyond its certification programme. The organisation's Integrated Landscape Management Programmes, which aim to drive change beyond individual farms or forests by facilitating dynamic landscape partnerships that include all land users, covered over 14 million hectares in 25 different projects by the end of 2021. In its far-reaching advocacy work, the Rainforest Alliance partners with governments, companies, and civil society organisations to shape policies and practices that support sustainable agriculture, forestry, and business. The Rainforest Alliance also offers customised Supply Chain Services to help companies define and achieve sustainability commitments; to date, the organisation has helped over 8000 different suppliers in 80 countries navigate complex sustainability challenges within their supply chains. Finally, the Rainforest Alliance effects change through its leading role in key partnerships such as the Accountability Framework, 1000 Landscapes for 1 Billion People, and LandScale.

3 How Rainforest Alliance certification works

3.1 The standard

Published in 2020 and implemented from July 2021 onwards, the new Rainforest Alliance Sustainable Agriculture Standard consists of both Farm Requirements and Supply Chain Requirements (formerly called Chain of

Custody). As a full member of ISEAL,[4] the Rainforest Alliance developed the standard in accordance with ISEAL's Standard-Setting Code of Good Practice. Two public consultations of the standard were held between December 2018 and August 2019; these consultations gathered input from more than 1000 people in nearly 50 countries, representing farmers, companies, NGOs, governments, and research institutions. This input was incorporated into the process not only by Rainforest Alliance staff but also by the independent, 10- to 15-member Standards Committee, composed of experts in agriculture, certification, and standards from farmer associations, academia, certification bodies (CBs), NGOs, and the corporate sector. At least half of the Committee members had to be from countries where Rainforest Alliance Certified products were cultivated. The Committee was responsible for discussing, weighing, and deciding on the feedback received during the multi-stakeholder consultations, and for approving draft versions of the Standard.

The 2020 Standard is divided into six thematic chapters that work in parallel to drive social, environmental, and economic sustainability at the farm level: management, traceability, income and shared responsibility, farming, social, and environment. The new Farm Requirements offer more options for farmers to manage their own sustainability performance. While the base of the Standard is the core requirements, which must always be complied with to achieve or maintain certified status, the Standard also includes improvement requirements, which are designed to promote and measure progress in steps,[5] and self-selected criteria, which farms can choose to implement based on their own sustainability aspirations. The Supply Chain Requirements, which contain the same thematic chapters as the Farm Requirements (minus the farming chapter), go above and beyond traditional traceability rules by, for example, including strengthened risk assessment requirements.

To support producers who were certified under the previous Rainforest Alliance and UTZ systems in their transition to the 2020 Standard, the Rainforest Alliance has implemented a 2-year 'transition period' during which producers need to comply only with the core requirements of the new Standard. The transition period has also allowed the organisation to learn from the experience of producers implementing the new Standard and amend the requirements as needed. In particular, the Rainforest Alliance has been working closely with a subset of 15 'early implementer' farmers since the beginning of 2020, who are providing invaluable feedback on the feasibility, clarity, and usefulness of the new Standard. It has also established a 'sounding board' involving a wider set of certificate holders who will provide feedback on implementation challenges

4 ISEAL (isealalliance.org) is an independent organisation that sets Codes of Good Practice to support sustainability systems to improve how they operate and deliver greater impact. ISEAL defines credible practices based on emerging global consensus; convenes forums for collaboration and sharing of experience; delivers expertise, advice, and training; and facilitates innovation to strengthen sustainability practice.

5 'Level 1' criteria must be complied with after 3 years of certification while 'level 2' must be complied with after 6 years.

and proposed amendments. Initial feedback reveals that farmers and farm groups are eager for the data-driven insights that they will gain access to but are concerned about the high number of requirements, tools, and guidance documents that are included in the new Standard and the complexity and cost of implementation.

As a result of this feedback, the Rainforest Alliance has taken steps to make the 2020 Standard and its related documentation more user-friendly and less complex. The large number of annexes providing additional details that govern the implementation of standard requirements has been reduced to one annexe for each chapter of the Standard, and the contents have been shortened and simplified. The Rainforest Alliance has also reviewed the content and structure of the requirements to make them easier to understand and implement.

3.2 Assurance

The Rainforest Alliance's assurance system has multiple elements to ensure the quality of the third-party auditing of its standards by CBs. The assurance system includes rules that set out how auditors must evaluate compliance with the farm and supply chain requirements and a grievance mechanism that allows stakeholders to submit complaints about CBs or any other element of the assurance system, which the Rainforest Alliance then investigates (see Fig. 1).

As in the standard itself, the assurance system is moving toward a risk-based approach, which requires CBs to assess the risks for compliance before each audit and take special measures for farms at high risk for issues such as child labour, freedom of association, deforestation, or encroachment into protected areas. Risk assessments are conducted at several stages of the certification process, for both farms and supply chains, using data from the Rainforest Alliance and external sources, along with country- and sector-specific risk information. If risks are detected, auditors will investigate those specific issues more carefully. For example, if satellite imagery shows that farm groups have members located in areas of recent tree cover loss (i.e. deforestation), the auditor should verify that these members have been excluded from the group or that sufficient evidence has been gathered to show that deforestation did not occur within the farm boundaries.

To ensure that CBs audit consistently across all regions and crops, our assurance system provides detailed guidance on auditing for specific topics, such as deforestation, child and forced labour, and traceability. The Rainforest Alliance monitors CB performance through shadow audits (where Rainforest Alliance staff observe auditors (shadowing) in the field) and through parallel audits (where Rainforest Alliance staff conduct separate audits and compare the findings with those from CBs). It has implemented a rigorous process to

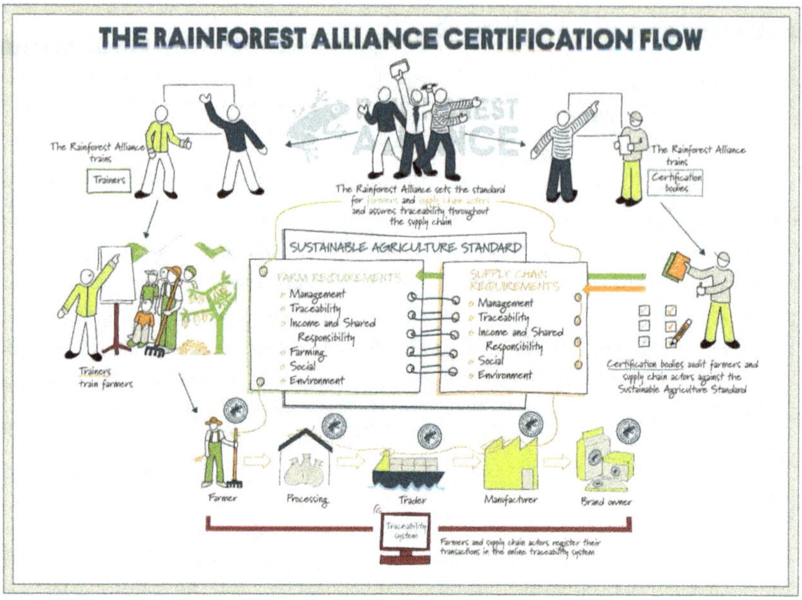

Figure 1 The Rainforest Alliance certification process.

authorise CBs to conduct audits for the new certification programme and to ensure that there is adequate CB capacity in each country with certified producers. The Rainforest Alliance is also actively innovating in its assurance system, including developing processes for enabling and verifying the quality of remote audits, which were frequently used during the COVID-19 pandemic when in-person audits were often not possible.

3.3 Traceability, fees, labelling, and claims

Producers and buyers have to register their commercial transactions involving certified commodities in the Rainforest Alliance's traceability platform. This platform enables the organisation to trace produced and traded certified volumes from origin to manufacturer. The ability to trace a product back to the farm of origin depends on the traceability level. 'Identity preserved' traceability means that the certified product from a specific farm is kept separate from products from any other farm throughout the supply chain. 'Segregated' traceability means that all certified products have been kept separate from non-certified products throughout the supply chain, although the certified product may come from different farms. Some commodities use a 'mass balance' traceability system, in which a manufacturer can mix certified and non-certified products in the final product as long as the amount of certified product purchased is equivalent to the amount used in manufacturing the product. This system provides flexibility in the supply chain and is commonly used in the cocoa sector, as well as in orange juice and coconut oil.

A volume-based royalty is paid to the Rainforest Alliance, once in each supply chain. The Rainforest Alliance License Agreement General Terms & Conditions[6] outlines the royalty amount per certified product, how it is collected, and at which point in the supply chain it is invoiced. Royalties are applicable whenever companies claim that a crop is purchased from a Rainforest Alliance Certified farm or sell it as 'certified'. This happens when the claim is on a product, on packaging, in a contract or invoice, or on promotional materials (on- or offline), regardless of whether the seal or other marks are used. These royalties generated 62% of Rainforest Alliance's revenues in 2021 and are used to support the development of the Standard and to finance producer training on-site in countries of origin, the communication and marketing teams, as well as organisational maintenance costs.[7]

A Supply Chain Risk Assessment is conducted once per certification cycle. Supply chain companies assessed as 'very low' risk do not need an external audit. All others must hire (and pay for) an external audit conducted by an approved CB of their choice. The Rainforest Alliance does not regulate audit costs, which vary considerably depending on the location(s), size, and complexity of the client's operations. Supply chain actors are also responsible for direct financial support to producers through the Sustainability Differential (SD) and Sustainability Investment (SI), which are new requirements in the 2020 Standard and are described in detail in the next section.

Brands and manufacturers that source Rainforest Alliance Certified products can use the Rainforest Alliance trademark seal on their products (Fig. 2). In 2022, thousands of products carried the Rainforest Alliance frog seal, which signals to consumers that the products they are buying are manufactured with ingredients produced according to its Standard. Each on-pack claim undergoes revision and approval by Rainforest Alliance staff. The traceability platform enables the Rainforest Alliance to monitor and verify the on-pack claims on final products. This builds consumer trust in the Rainforest Alliance name and sustainability claims and enhances the value of brands and products bearing the seal. The Rainforest Alliance logo, on the other hand, can be used by organisations that wish to denote a cause-related marketing relationship or to represent another partnership with the Rainforest Alliance. The logo can be used off-pack only.

6 https://www.rainforest-alliance.org/resource-item/rainforest-alliance-license-agreement-general-terms-and-conditions/.

7 For details on the Rainforest Alliance's revenue sources, see the annual report https://www.rainforest-alliance.org/annual_report/2021/.

Figure 2 The Rainforest Alliance certification seal (left) and logo (right).

3.4 Monitoring implementation and evaluating change

The cornerstone of the Rainforest Alliance's monitoring and evaluation system is the theory of change that underpins the certification programme. It explains how certification, supported by other interventions such as training and advocacy, contributes to improving farming practices and livelihoods as well as the social and environmental conditions of producers, their families, and hired workers. The Rainforest Alliance collects many types of monitoring and evaluation data, which are prioritised based on the results of risk assessments, existing knowledge gaps, the urgency of critical issues, and demand from internal and external actors. Some variables are recorded for all certificate holders and supply chain actors that are part of the certification system, irrespective of their geographical location and commodity. Other variables are assessed for a sample of certificate holders and supply chain actors; these are often pilot projects or learning studies to answer a specific question about how the programme is working. Lastly, the Rainforest Alliance commissions impact evaluations to test specific hypotheses about its programme's impact on key topics. The choice of countries, crops, and topics for impact evaluations is determined by their priority level and geographic spread.

The monitoring and evaluation system contains indicators that examine both intended and unintended effects. Of course, not every outcome can be monitored, nor can every causal relationship in the theory of change be evaluated. Some causal links are well understood and need not be confirmed time and again through primary data collection and reporting. Other change pathways of the certification programme have a weaker evidence base. This may be due to a dearth of research on specific topics, as well as to the fact that certification outcomes have been shown to be very context-dependent, with mixed and modest impacts (Oya et al., 2018; Ingram et al., 2018).

Like other sustainability organisations, the Rainforest Alliance is broadening its use of data beyond the monitoring of programme reach and outcomes, to expand the range of services offered to producers and supply chain partners. With the release of the 2020 Standard, our certification programme has become more data-driven: initiatives like the Farm Intelligence App and Smart Meters put farm data into the hands of producers and group managers, allowing them to track and analyse metrics about their own performance to support decision-making and adaptive management. While this is an exciting prospect, the data will only be useful if the Rainforest Alliance can ensure that it is of sufficient quality and has the analytic capacity to distinguish vital signals from noise.

Other challenges related to monitoring and evaluation of the certification programme include some age-old problems that many certification systems face: the increased difficulty of finding an evaluation counterfactual as compliance with the Standard becomes more mainstream; the difficulty in maintaining a stable treatment group as participants self-select into the programme and are free to leave the programme as they wish; and, of course, the near-impossibility of randomly assigning different groups to different interventions. These issues create obstacles to the use of experimental and quasi-experimental evaluation designs. Evaluating certification in this context requires medium- to long-term planning, which is often at odds with the need for adaptive management.

4 Challenges and future directions

In this section, we would like to take a closer look at three challenges that have affected the Rainforest Alliance's ability to reconcile farming, environmental, and social sustainability through certification alone and have driven the organisation to continue innovating to increase its impact. The key challenges are as follows:

- The mismatch between the financial costs and benefits of certification for many farmers, particularly smallholder farmers;
- The limitations of the traditional high-stakes pass/fail certification approach, which does not sufficiently support remediation and behaviour change to achieve intended results; and
- The limited scope of farm-level certification, which is not able to address larger systemic issues that affect wider communities, such as governance systems, attitudes, and market dynamics.

We will describe herein some of the Rainforest Alliance's recent efforts to address these challenges, by developing new approaches to drive value to producers both on and off certified farms, working creatively to equalise supply chain power imbalances, shifting from a binary pass/fail to a due diligence approach to human rights issues, and forging new partnerships and approaches that support systemic change on the wider landscape beyond the scale of the individual farm.

We conclude by highlighting a novel approach that we believe, in time and with the right kind of support from governments and companies, could even up the playing field between certified and non-certified production: true price. By increasing transparency on social and environmental externalities and factoring these externalities into the actual price of products, the true price concept would dramatically increase the costs of unsustainable production and reward sustainable practices and the farmers and workers who implement them.

4.1 Challenge: Often, farmers bear the costs of sustainable agricultural production while the rewards accrue elsewhere

The vast majority of the world's poor live in rural areas and depend largely on agriculture for their livelihoods (UNEP, 2013). Many of these people are smallholder farmers. An estimated 2.5 billion people worldwide are supported by smallholder farms through full or part-time work (UNEP, 2013). Unsurprisingly, research suggests that improving sustainable livelihood opportunities for farming and forest communities is the most effective way to reduce poverty in rural areas (Townsend, 2015). Despite this reality, smallholders in developing countries remain marginalised in global economic and development policy due to a lack of access to technological and institutional innovation, making it difficult to achieve this goal (Gatzweiler and von Braun, 2016; UNEP, 2013).

While sustainability certification programmes are working hard to address this challenge, farming in accordance with the Rainforest Alliance Standard and selling products into the certified supply chain often is not enough to lift smallholders out of poverty. Research shows that certified farmers, including those in the Rainforest Alliance programme, can often achieve higher yields, prices, and incomes than non-certified farmers but that these benefits do not adequately reduce the likelihood of living in poverty or improve food security (Dompreh et al., 2021; Iddrissu et al., 2020; Brako et al., 2021; Dietz et al., 2019; Ingram et al., 2017, 2018). Despite the positive impacts of certification on intermediate economic outcomes, the systemic inequalities and structural challenges present in many agricultural supply chains make it impossible to lift farming communities out of poverty with certification alone.

This situation persists against a backdrop of many public, well-funded corporate commitments to source from sustainable suppliers, driven by investors, public opinion, and more recently, governments and consumers. The question the Rainforest Alliance faces is how to ensure that the resources that companies and others put behind their sustainability commitments, which number in the billions of US dollars, adequately benefit producers and compensate them for the costs of adopting sustainable farming practices.

Sustainable food production and farming practices do cost more. Integrated pest management takes more time and labour than controlling pests through pesticides; leaving on-farm natural ecosystems intact can be a lost opportunity to plant more crops; paying a living wage to workers increases the payroll. In addition, there is the cost of the audit, associated recordkeeping, internal inspections, training, and setting up committees, though ideally the cost of many of these activities will decrease over time as knowledge increases and systems are improved. While some practices required by the standard, such as crop renovation, pruning, soil assessments, and mulching, can increase yields and help offset the costs of certification (Akoyi and Maertens, 2018; Bini et al., 2016; Waarts et al., 2012; Krain et al., 2011), for most producers, there is highly likely to be a net cost to compliance.

And yet, despite the corporate commitments and the basic logic of supply and demand, producers of sustainable products are not able to recoup these costs by simply raising their prices. The main reasons for this inability are classic market failures: purchasing power is concentrated in the hands of a relatively small number of buyers, and current market mechanisms do not incorporate the costs of environmental and social externalities into prices. And while companies' sustainability departments are usually genuinely committed to supporting producers, procurement departments tend to be driven by the lowest price. Certification organisations like the Rainforest Alliance have also not adequately contributed to analysing and communicating the real costs of sustainable production.

The ultimate cost of implementing a sustainability standard obviously depends on the acceptable level of practice, or 'bar', at which the standard is set, and how far from reaching that bar a given producer is. The process of setting any sustainability standard is typically, to some degree, a negotiation between the interests of different stakeholders. Buyers of certified products, who want to minimise their supply chain risk, may push for the requirements within the standard to be more demanding. Producers, on the other hand, may push for less demanding requirements to lower their costs of compliance. The Rainforest Alliance is therefore constantly striving to strike a balance that allows it to maintain the rigor and credibility of its certification programme while minimising the burden on producers as they work toward the social and environmental goals that are central to the integrity of the system.

We have introduced mechanisms that are aimed to address supply chain imbalances and ultimately improve the economic situation of certified farmers. One of these is the Shared Responsibility requirements that provide two pathways for increasing the payments from companies to farmers; another is a system that enhances farmer control over certification targets and data collection. These two pathways are now explained.

4.1.1 Shared responsibility

In recent years, the Rainforest Alliance has introduced specific mechanisms to lessen the financial burden of certification on smallholder farmers. These initiatives, discussed below, fall under the umbrella of 'Shared Responsibility', which contends that both the risks and value of sustainable production must be distributed more equitably between farmers and other supply chain actors.

As noted in a previous section, a key element of shared responsibility is the Sustainability Differential (SD) (similar to the concept of a price premium). The SD is a per-volume payment awarded to certified farmers above the market price for their product. Although the SD is new in the 2020 Rainforest Alliance Standard, fixed premiums and/or minimum prices have been used by Fairtrade for decades. UTZ's previous standard also required first buyers to pay a premium for certified products to the certificate holder, who could be either an individual (typically large) farmer or a group of smallholder farmers managed by its members or by a trader as a producer group. The UTZ premium could be used to cover the management costs of the producer group, including internal (self) and external audits. In contrast to the former UTZ requirement, the new SD must be transferred (and traced) by the certificate holder to the individual farmers. It must also be paid in cash; in-kind payments are not allowed. For most sectors, there is no set level of SD that must be paid; the amount is determined through negotiations between certified producers and first buyers. In the cocoa sector, however, the Rainforest Alliance has set a minimum SD of US$70 per metric tonne (MT) of cocoa based on an analysis of the price premiums paid to cocoa farmers in the previous Rainforest Alliance and UTZ certification systems.

In addition to the SD, we also introduced the concept of 'SI' in the 2020 Rainforest Alliance Standard. SI are a means through which supply chain actors can financially support producers to implement sustainability measures required by the Standard. Companies have to make cash or in-kind contributions (investments) which can be earmarked for specific improvements (such as wages or conservation activities) based on the needs identified by producers. Like the SD, the investment amount is not set by the Rainforest Alliance, with the exception of the banana sector. In other sectors, it is up to companies and certified farmers to negotiate the amount.

The intended benefits of these elements of shared responsibility are, of course, first and foremost ensuring a better livelihood for farmers and supporting their ability to adopt sustainable practices that comply with the Rainforest Alliance Standard. But other benefits are anticipated as well. There is evidence that establishing direct communication and an active partnership between buyers and producers leads companies to better understand the challenges faced by farmers and creates more resilient supply chains. An additional benefit for buyers is the ability to report to stakeholders on specific improvements and tell detailed stories about the projects that they have supported.

The Rainforest Alliance is closely monitoring the implementation and outcomes of the SD and SI to assess how these requirements are translated to practice across sectors. Are farmers and workers indeed receiving additional payments? Are the amounts commensurate to the needs outlined in the SI plans? Are value chain relations showing signals of shared responsibility?

4.1.2 Enhanced farmer control of certification targets and data collected

The new Rainforest Alliance Standard aims to develop a culture of continuous improvement that both benefits the environment and empowers farmers. To do this, the Rainforest Alliance has introduced the concept of 'Smart Meters', which are quantitative targets for certain practices, which are set by farmers and allow them to track their own performance over time. Both the targets and the actions needed to achieve them are determined by farmers, who set their own sustainability goals based on what they consider to be most beneficial and feasible. Some Smart Meters are mandatory while others are self-selected by farmers who want to focus on these issues. Our approach aims to ensure that farmers can accurately monitor their progress against targets and build on it year over year, adapting their practices as needed.

While many certified farmers and farmer producer groups already have sophisticated, data-driven approaches to management, the majority of certified farmers do not, and the hope is that Smart Meter data will improve the efficiency, yield and bottom line for these farms. Smart Meter data can also help farmers develop their investment plans for SI, described earlier. The data generated through the new certification approach will allow other supply chain actors to see where further support is required. The sharing of data enables farmers to better display their efforts and quantify the benefits of more sustainable practices, which contributes to better market access and prices for their goods. Supply chain actors can ask farmers to use specific Smart Meter data for their own reporting and aggregated Smart Meter data can be used for advocacy purposes; for instance, to show the living wage gap in certain countries.

The collection of Smart Meter data will become mandatory in the second half of 2023 but is currently being piloted by a group of certified producer volunteers. Preliminary findings show that many producers need additional support to build and implement a system for data collection and analysis, but that the insights generated through the data are very valuable for farmers. For example, data on water and energy usage have enabled farmers to improve their farms' efficiency and decrease production costs. Additionally, an increase in data collection and use means increased attention to privacy and data management concerns; the Rainforest Alliance is developing systems and safeguards to ensure that Smart Meter data are handled ethically and responsibly.

4.2 Challenge: the high-stakes environment created by pass/fail approaches alone can be counterproductive to promoting behaviour change and undermine impact

One challenge observed during audits is that the traditional pass/fail approach to certification creates a high-stakes environment that can provide an incentive for farmers to hide their shortcomings or infractions. Although some elements of sustainable practice are quite straightforward to observe, such as the availability of personal protective equipment, quality of worker housing, or the density of shade trees, some practices, such as the use of child labour, are easier to conceal when an auditor is visiting.

4.2.1 Moving beyond pass-fail to a due diligence approach for human rights

Previous standards took a zero-tolerance approach to key topics such as child labour. Infractions usually meant losing certified status and the license to sell to certified markets. As noted earlier, the Rainforest Alliance found that this approach can drive farms to attempt to hide bad practices. The 2020 Standard takes a due diligence approach, called 'assess-and-address'.[8] This method focuses on prevention, engagement, improvement, and incentivising farms to tackle challenging issues such as human rights violations, rather than conceal them. In the case of child labour, farms (or groups of farms) are required to set up internal committees to prevent and monitor for any child labour and to remediate any cases that might be found. Farmers and group leaders receive training on this approach and must follow a remediation protocol developed by the Rainforest Alliance.

The Rainforest Alliance has also developed risk maps based on credible third-party data to identify countries and sectors where the risks of child labour and forced labour are high, and therefore where producers must implement stricter risk mitigation measures. During the audit, an auditor evaluates the processes that the farm has established, in addition to looking for actual signs of child labour. What the Rainforest Alliance has learned thus far is that producers find the assess-and-address system a good tool for tackling child labour and other human rights issues as it provides a practical approach to remediation.

8 https://www.rainforest-alliance.org/resource-item/whats-in-our-2020-certification-program-assess-and-address/.

4.3 Challenge: certification systems are often expected to singlehandedly solve systemic sustainability issues

Sometimes implementation of sustainable practices is hindered by problems that are not within the power of a certification programme to solve. These include, for example, situations where there are attitudes and norms that run counter to the norms of the certification standard, or where a government will or cannot provide sufficient permits or proof of land ownership. Child labour is an apt example of this. In some places, child labour is viewed as an acceptable alternative to school and as a way for children to learn the family business. Encroachment into protected areas is another example. In the absence of other livelihood opportunities and government enforcement of laws, farmers who have created their farms illegally within a protected area will likely not be deterred by being removed from the Rainforest Alliance certification programme.

In addition, the fact that certification works at the level of the individual farm means that challenges that span a broader geographic scope or require coordination among many regional actors are inherently difficult to solve. While it is true that many farms or groups of farms are large enough to have measurable impacts on landscape-level issues like habitat connectivity or water quality, for the majority of farms, there is simply a mismatch between the farm boundaries and the scale at which many important issues must be addressed.

4.3.1 Joining with partners to change norms and create an enabling environment

Positive impacts of sustainability certification will be limited if programmes are implemented in an environment that has legal or normative barriers to the adoption of the required sustainable practices. Many such barriers are deeply entrenched and addressing them requires sector-wide cooperation, well beyond the limits of certification alone. In these contexts, the Rainforest Alliance works with partners who have the skills and mandate to address these issues. An example is the organisation's work with Kyagalanyi Coffee Limited, a Rainforest Alliance Certified coffee trader, to bring the Child Labour Free Zone concept to the Nebbi District in West Nile, Uganda.[9] Here, a coalition of local and international partners sought to tackle the root causes of child labour by raising awareness of the detriments of child labour and recognition of the line between child labour and age-appropriate work, not just among coffee farmers, but within the whole community, including local government and schools. This effort showed that certification on its own was not an optimal way to address

9 https://www.rainforest-alliance.org/wp-content/uploads/2021/07/eliminating-child-labor-uganda.pdf.

the root causes; collaboration with local organisations and communities, and buy-in from local governments, were key to successfully preventing and remediating child labour.

In efforts to improve the livelihoods and living situation of farming communities, the Rainforest Alliance plays an important role in linking partners and fostering collaboration between supply chain actors, civil society, and governments. For example, in Cote d'Ivoire, the Rainforest Alliance helped establish a Land Management Board (LMB) of local farmers and other stakeholders to improve the advocacy capacity of farming communities at the edge of Taï National Park.[10] In collaboration with local NGOs and government, the Rainforest Alliance worked to ensure that the national forestry code now allows cocoa farmers at the edge of the park to own the trees on their property, increasing farmers' willingness to plant agroforestry systems that can bring them extra income while also offering important ecological benefits such as shading, drought protection, and soil stabilisation. Through the LMB, farmers have become a central player in local forest management and are better equipped to advocate for their needs.

4.3.2 Bringing stakeholders together to tackle larger-scale problems

To expand its impact beyond the level of individual farms and forests, the Rainforest Alliance works with communities, companies, NGOs, and local governments to design and implement landscape-level interventions. These landscape projects are designed to tackle complex and interconnected challenges such as climate change and ecosystem restoration that require action across a wider geographical area and with multiple actors. Projects often include activities such as training in management best practices and production diversification, improving access to finance, stimulating demand for sustainably produced goods, connecting communities to new markets, and strengthening local and regional cooperatives and trade unions. Through enhancing our understanding of sustainability challenges and solutions in different contexts, these landscape projects complement and reinforce farm-level certification.

Evidence from many long-term projects shows that bringing stakeholders together at the landscape level is effective: in Guatemala's Maya Biosphere Reserve, for example, communities supported by the Rainforest Alliance (and others) have maintained near-zero deforestation rates for almost 20 years while supporting thousands of sustainable forestry jobs and generating millions in annual revenue. In Ghana's Juabeso-Bia landscape, the Rainforest Alliance's

10 https://www.rainforest-alliance.org/in-the-field/cocoa-farmers-help-protect-last-primary-rainforest-in-cote-d-ivoire/.

support helped 36 cocoa farming communities establish an LMB to spearhead reforestation efforts on formerly degraded lands and cultivate shade trees on cocoa farms. In Peru's San Martín region, the Rainforest Alliance has worked with indigenous communities to diversify their incomes through helping them bring a range of sustainable products to market while conserving 26 000 hectares of forest (Rainforest Alliance, 2021).

Stakeholders can also be engaged around a specific topic, rather than a specific landscape. Voluntary sustainability standards, including the Rainforest Alliance, joined together with many other organisations to develop the High Conservation Value approach, which has been adopted by over 160 companies and financial institutions, helping to protect millions of hectares of old-growth forest and other important natural habitats and cultural sites. Another example of such cooperation is the Global Living Wage Coalition, which has helped to develop living wage benchmarks and other methodologies in several sectors. The Rainforest Alliance, as part of a diverse coalition of organisations, also helped establish the Accountability Framework Initiative, which provides a standardised approach for assessing and communicating sustainability performance in order to improve accountability and accelerate progress on company commitments to protect natural ecosystems and human rights.

5 The next frontier: a level playing field

The previous sections described the efforts the Rainforest Alliance is making firstly to lower the financial and technical burden that currently falls on the shoulders of certified farmers; secondly to make our certification programme better suited to addressing tough topics by shifting away from a strict pass/fail to a due diligence approach; and lastly to work through partnership to address systemic landscape-level issues and norms that currently hinder the adoption of sustainable practices.

We would like to conclude by highlighting a promising approach that has the potential to remedy many of the problems described earlier by making the hidden costs of production transparent. In current markets, social and environmental externalities are not paid for by companies or consumers; instead, these costs fall on taxpayers, future generations, and individuals throughout the value chain who bear the brunt of low wages, pollution, and other damages. The goal of the 'true price' approach is to include in a product's final price all costs incurred during production, including underpayment of workers, pollution, biodiversity loss, carbon emissions, and more. In theory, a pricing/taxing system that would internalise the externalities would make prices and markets radically transparent and would convey sustainability information via the price mechanism. Under that scenario, sustainable products would beat non-sustainable products on price alone.

Shifting social and environmental externalities onto supply chain actors would alleviate much of the financial burden on producers. As unsustainable practices such as poverty and pollution have a greater economic, environmental, and social cost, businesses could be incentivised to reduce carbon emissions, minimise pollution, and ensure that workers and farmers in their supply chains are well-paid and healthy. Doing so would lower the market cost of their goods. The money that is generated from better resource management could be invested into remediation efforts to address these externalities, benefiting farmers and workers. Further, the true price concept would theoretically account for environmental and social costs that extend beyond farm boundaries, such as climate change – issues that certification alone can't address. Needless to say, accounting for and monetising these costs would be complex and challenging.

Reducing the true price gap – that is, working to include hidden costs in market prices – would even the playing field between certified and conventional products. But how do we get there? Creating transparency on true costs is a first step. This must be coupled with credible communication so that consumers can make informed choices about purchases and be supported by efforts to remediate harmful externalities (True Price, 2019). In turn, these practices would reduce the price of certified products relative to conventional ones, since certification systems are already addressing many of these externalities through their programmes and interventions.

Realistically, governments would have to play a central role in mainstreaming the true price concept. In particular, they would have to enact regulations that require disclosure of true prices, in support of increasing transparency. They can support the development of standardised methods for assessing the true price of goods and services, in collaboration with initiatives such as True Cost Accounting, True Price, and True Value who are working to develop these practices. Finally, governments can use information about external costs to adjust tax and subsidy policies and to better align incentives with the true costs of production.

Although the Rainforest Alliance's system does not explicitly embrace the 'true price' concept, the costs of externalities are already partially addressed through our Shared Responsibility approach and the SD and SI paid to farmers and workers. The SD, for example, is paid over and above the market price and is an effort to shift some of the responsibility for addressing structural inequalities and externalities – including persistently low wages and incomes – onto buyers. Nonetheless, the SD is not designed to capture all the externalities of production.

If the true price concept continues to gain support and become more widespread, the Rainforest Alliance could play an important role in helping to answer key questions such as: Which external costs should be included? And how can these be quantified and monetised (True Price, 2019)? Much of the research that we regularly conduct addresses questions like these. Assessing living wage gaps, analysing the costs, benefits, and impacts of sustainable production, and implementing monitoring systems for agricultural pollutants, all of which are part of our current work, are critical steps in the process of determining the true price of agricultural commodities.

6 Conclusion

During the authors' combined 40 years of experience monitoring and evaluating the certification programmes of UTZ and Rainforest Alliance, we have drafted and challenged theories of change; evaluated outcomes; consulted with stakeholders; analysed data; engaged with academic researchers; and learned from colleagues who understand markets, producers, and companies. This has given us insights into the potential and limitations of a market-based, voluntary standard and certification programme. Despite this vantage point, we can only make an educated guess about what lies ahead. Past performance does not guarantee future results, as they say.

There is certainly much about our past to celebrate. What started out 35 years ago as a sustainability standard for one banana farm now spans multiple major agricultural sectors, thousands of businesses along the value chain, and a significant percentage of global trade in these commodities. Certification has gone mainstream and has played an important role in making production and trade of export crops more sustainable. It has provided a platform on which diverse stakeholders can agree and collaborate on what 'sustainably produced' means in practical terms; it has boosted the traceability of global commodities; and it was one of the key drivers for due diligence policies adopted by companies and laws passed by several European governments.

Nonetheless, the purpose of this chapter is not to laud the successes of our certification standard and programme but rather to acknowledge a few hard truths about the challenges of using this market-based tool to tackle entrenched social and environmental issues and explain how the Rainforest Alliance has evolved to address them.

One challenge is that some issues, particularly human rights violations, are difficult to identify and remediate using a high-stakes pass/fail approach to infractions. In our new standard, therefore, we have shifted to a due-diligence or 'assess and address' approach that encourages farmers to address potential problems proactively. A second challenge is that the scope of many systemic sustainability issues is either culturally or geographically broader than the scale at which certification works. In response, we have developed landscape-level projects and community-based partnerships to address the norms, systems, and policies that hinder sustainable production.

But the most fundamental challenge is that the logic of the market, on its own, has not consistently transferred value from consumer to producer. Left to markets, certified products tend to cost the same as non-certified equivalents, and certification becomes a condition for market access instead of a force to drive monetary rewards to farmers who grow food sustainably. It is not realistic or ethical to expect farmers to adopt sustainable practices if it forces them to operate at a loss and deplete their natural and social capital to stay in business. The Rainforest Alliance's 2020 Standard acknowledges that sustainable production requires better terms of trade for producers. The introduction of the Shared Responsibility concept (with its two main components, the SD and SI) marks a departure from our previous assumption that markets would self-regulate and somehow arrive at a fair price for certified products.

We have continued this line of reasoning and asked ourselves under which conditions markets could deliver sustainable outcomes that incentivise companies and consumers to choose sustainable options, while simultaneously enabling producers to earn a decent living from sustainable farming. We believe that, in the long term, true pricing – in which all social and environmental externalities are included in a product's market price – holds promise. While the drive to make the true price concept a reality would require a herculean, creative, and committed effort on the part of governments, companies, and others, we see it as a potential long-term solution to the failures that have hindered the ability of markets to reward sustainable production. With truly transparent pricing, products that enhance society and the environment would be less expensive than those that harm them, and the elusive business case for sustainability would emerge.

7 References

Akoyi, K. T. and Maertens, M. (2018). Walk the talk: private sustainability standards in the Ugandan coffee sector, *The Journal of Development Studies* 54(10), 1792–1818.

Bini, D., Pinto, L. F. G., Vian, C., et al. (2016). Socioenvironmental certification of farms is economically advantageous, *Sustentabilidade em Debate* 3, 25–33.

Brako, D. E., Richard, A. and Alexandros, G. (2021). Do voluntary certification standards improve yields and wellbeing? Evidence from oil palm and cocoa smallholders in Ghana, *International Journal of Agricultural Sustainability* 19(1), 16–39.

Dietz, T., Chong, A. E., Grabs, J. and Kilian, B. (2019). How effective is multiple certification in improving the economic conditions of smallholder farmers? Evidence from an impact evaluation in Colombia's coffee belt, *The Journal of Development Studies* 6, 1141–1160.

Dompreh, E. B., Asare, R. and Gasparatos, A. (2021). Sustainable but hungry? Food security outcomes of certification for cocoa and oil palm smallholders in Ghana, *Environmental Research Letters* 16(5), 055001.

Forest Stewardship Council (FSC) (2020). FSC in Gabon: starting the journey towards 100% certification, Available at: https://fsc.org/en/newscentre/fsc-in-gabon-starting-the-journey-towards-100-certification, accessed September 13, 2022.

Gatzweiler, F. W. and von Braun, J. (2016). Innovation for marginalized smallholder farmers and development: an overview and implications for policy and research. In: Gatzweiler, F. W. and von Braun, J. (Eds). *Technological and Institutional Innovations for Marginalized Smallholders in Agricultural Development*, Springer, New York.

Grabs, J. (2020). *Selling Sustainability Short? The Private Governance of Labor and the Environment in the Coffee Sector*, Cambridge University Press, Cambridge UK.

Haggar, J., Jerez, R., Cuadra, L., Alvarado, U. and Soto, G. (2012). Environmental and economic costs and benefits from sustainable certification of coffee in Nicaragua, *Food Chain* 2(1), 24–41.

Haggar, J., Soto, G., Casanoves, F. and de Melo Virginio, E. (2017). Environmental-economic benefits and trade-offs on sustainably certified coffee farms, *Ecological Indicators* 79, 330–337.

Hardt, E., Borgomeo, E., dos Santos, R. F., Pinto, L. F. G., Metzger, J. P. and Sparovek, G. (2015). Does certification improve biodiversity conservation in Brazilian coffee farms?, *Forest Ecology and Management* 357, 181–194.

Iddrisu, M., Aidoo, R. and Wongnaa, C. A. (2020). Participation in UTZ-RA voluntary cocoa certification scheme and its impact on smallholder welfare: evidence from Ghana, *World Development Perspectives* 20, 100244.

IDH (The Sustainable Trade Initiative) (2021). *New Insights on Reaching Living Income: Impact Analysis*, IDH, Utrecht, The Netherlands.

Ingram, V., van Rijn, F., Waarts, Y., et al. (2017). Towards sustainable coca in Cote d'Ivoire. The impacts and contribution of UTZ certification combined with services provided by companies. Report. 041, Wageningen Economic Research, Wageningen (vol. 2018).

Krain, E., Millard, E., Konan, E. and Servat, E. (2011). *Trade and Pro-poor Growth: Introducing Rainforest Alliance Certification to Cocoa Production in Cote d'Ivoire*, Deutsche Gesellschaft für Internationale Zumsammenarbeit. Eschborn, Germany.

LeBaron, G. (2020). *Combatting Modern Slavery: Why Labour Governance Is Failing and What We Can Do About It*, Polity Press, Cambridge, UK.

Mitiku, F., de Mey, Y., Nyssen, J. and Maertens, M. (2017). Do private sustainability standards contribute to income growth and poverty alleviation? A comparison of different coffee certification schemes in Ethiopia, *Sustainability* 9(2), 246.

Ochieng, B. O., Hughey, K. F. D. and Bigsby, H. (2013). Rainforest Alliance certification of Kenyan tea farms: A contribution to sustainability or tokenism?, *Journal of Cleaner Production* 39, 285–293.

Oya, C., Schaefer, F. and Skalidou, D. (2018). The effectiveness of agricultural certification in developing countries: a systematic review, *World Development* 112, 282–312.

Ponte, S. (2019). *Business, Power and Sustainability in a World of Global Value Chains*, Zed Books, London.

Premaratne, S. P., Priyanath, H. M. S., Yoosuf, A. and Maurice, D. (2018). Technical efficiency for tea smallholder farmers under UTZ certification system in Sri Lanka: A stochastic frontier approach, *SEISENSE Journal of Management* 1(2), 1–19.

Price, T. (2019). A roadmap for true pricing. Vision paper, consultation draft, Available at: https://trueprice.org/wp-content/uploads/2021/12/True-Price-A-roadmap-for-true-pricing-v1.0.pdf, accessed August 29, 2022.

Rainforest Alliance (2021). Thriving communities, healthy landscapes, Available at: https://www.rainforest-alliance.org/in-the-field/thriving-communities-healthy-landscapes/, accessed September 9, 2022.

Rueda, X. and Lambin, E. F. (2013). Responding to globalization: impacts of certification on Colombian small-scale coffee growers, *Ecology and Society* 18(3), 21.

Takahashi, R. and Todo, Y. (2017). Coffee certification and forest quality: evidence from a wild coffee forest in Ethiopia, *World Development* 92, 158–166.

Taylor, J. G. and Scharlin, P. J. (2004). *Smart Alliance: How a Global Corporation and Environmental Activists Transformed a Tarnished Brand*, Yale University Press, London.

Townsend, R. F. (2015). *Ending Poverty and Hunger by 2030: an Agenda for the Global Food System*, World Bank Group, Washington, DC.

United Nations Environment Programme (UNEP) (2013). Smallholder farmers key to lifting over one billion people out of poverty, Available at: https://www.unep.org/news-and-stories/press-release/smallholder-farmers-key-lifting-over-one-billion-people-out-poverty, accessed August 31, 2022.

Waarts, Y., Ge, L., Ton, G. and Jansen, D. (2012). *Sustainable Tea Production in Kenya: Impact Assessment of Rainforest Alliance and Farmer Field School Training, LEI Report*, Lei, Wageningen University and Research Center, The Hague, 2012–2043.

Ingram, V. J., Van Rijn, F., Waarts, Y. R. and Gilhuis, H. (2018). The impacts of cocoa sustainability initiatives in West Africa, *Sustainability* 10(11), 4249.

Winston, A. (2021). Sustainable business went mainstream in 2021, *Harvard Business Review*, Available at: https://hbr.org/2021/12/sustainable-business-went-mainstream-in-2021, accessed September 12, 2022.

Chapter 12

Forest certification – an answer to the challenge of sustainable forest management?

Anna Thorning, Mid Sweden University, Sweden and Cecilia Mark-Herbert, Swedish University of Agricultural Sciences, Sweden

1 Introduction

Forest certification was developed because of a growing global concern for issues related to the sustainable management of forest resources and deforestation. It can be seen as a form of governance without government, a way to regulate global markets in the absence of a single authoritative body. Today it is one of the most common tools for sustainable forest management and is essential for considering the multiple values of forests (Malek & Abdul Rahim, 2022). Starting as a global environmental concern, forest certification has developed to become an organisational governance tool to differentiate. It is based on the assumption that consumers are willing to pay a premium price for certified products (products with an eco-label) much the same way as for food products.

Standards for forest certification include three main types of standards that cover different parts of the value system: forest management certification (FM), chain of custody certification (CoC), and trademark licenses. This chapter focuses on forest management certification because it serves as a condition for other value system certifications.

http://dx.doi.org/10.19103/AS.2024.0133.22

As with most standards (Brunsson & Jacobsson, 2000), forest management certification is voluntary. It is a market-driven tool that considers environmental, social and economic dimensions of value creation, and as such, it has effectively raised awareness of sustainable forest management (Rametsteiner & Simula, 2003). It enables responsible forest owners to consider the nature of corporate social responsibility (CSR) in their resource management choices (Uggla, 2017). Sustainable forest management is described as a practice to ensure that forests are able to provide current and future generations with forest goods and services (FAO, 2023). However, many forest stakeholders view the definition of sustainability and sustainable forest management as too broad and prefer the term responsible forest management (FSC, 2023). In this context, the main forest stakeholders refer to actors in the forest sector such as different types of forest owners, entrepreneurs that provide services, forest owner organisations processing industries and customers (industrial and end consumers), as well as environmental organisations, local and indigenous communities, forest workers, and outdoors and tourism organisations (Baumgartner, 2019).

Forest certification as a concept and forest certification programmes have, over the years, been criticised for both how the system is built around continuous improvement and specific issues regarding credibility (Röstlund, 2022). The biggest challenge for forest certification schemes has been attaining legitimacy. It refers to establishing in practice, how forest certification influences sustainability concerning, in particular, environmental, social, and financial dimensions of value production. Despite developed systems for value-based indicators for each of these dimensions, measurements in practice prove to be challenging.

In Sweden and other countries, many small private forest owners combine their forest ownership with agricultural land and the associated practices. This group of forest owners is often described as a traditional type of forest owners, that value their autonomy, with larger holdings and dependency on forestry income. Therefore, motivating small private forest owners to adopt forest certification requires awareness of what certification implies for them.

This chapter focuses on two well-established (Malek & Abdul Rahim, 2022) forest certification schemes, the Forest Stewardship Council (FSC) and the Programme for the Endorsement of Forest Certification (PEFC). A short presentation of each of these systems is provided, followed by a case study and a synthesis of the context of the forestry certification.

2 How forest certification works and different forest standards

Forest certification can be described as a social contract established by a range of forest stakeholders (Lehtonen *et al.*, 2021). It aims to deliver sustainable (or responsible) forest management by validating, through third-party audits, that the forest management in place fulfils standard requirements, in a production process that may take between 50 and 100 years. The forest certification systems involve several complex governance processes, including setting standards, auditing, certification and accrediting.

Governance of forest certification schemes occurs on a global, national and sometimes regional level. Internationally governance is carried out through a board, nomination committee and general assembly for international members. Many forest certification schemes are member organisations where members are part of the governance system and can influence objectives, strategies and motion proposals at the general assembly (Meidinger *et al.*, 2003). The international governance level can also create the framework for national standards, for example, in the case of FSC's criteria and indicators.

The *standard-setting process* is mainly carried out nationally and aims to define specific requirements for sustainable forest management. The standard-setting process for forest certification is often described as a negotiation process between members. Members can be part of different working groups for specific standard requirements, where the members negotiate standard rules, requirements and policies applied to national or regional forest conditions. Proposed standards are also open for more comprehensive consultations, during which stakeholders can comment on the standard draft. The standard-setting process can be more or less complicated depending on the number and heterogeneity of members and stakeholders participating (Tröster & Hiete, 2018).

The *certification implementation process* consists of sub-processes such as certification implementation, auditing, labelling and administration (Meidinger *et al.*, 2003). The forest owner that wants to become certified often starts with choosing and contacting an accredited certification body for approval for certification and audits. Certification bodies are accredited by an accreditation body appointed by the certification scheme. It is also not uncommon for actors desiring certification to use the service of a certification consultant to assist in modifying resource governance and account for the forestry management. The implementation process of forest certification described by Higman *et al.* (2005) often starts with a pre-assessment before closing out gaps and stakeholder consultation. When these tasks are completed, the main assessment can take place. During the main assessment, documentation is examined, and field visits and interviews with employees and stakeholders are conducted. Certified forest owners must pass the main assessment and annual audits by the independent (often third-party) certification body. Non-compliance is listed as either a minor or major corrective action request (CAR). Minor CARs arise as a result of the forest owner demonstrating only partial compliance or a failure that can be considered as minor and not systemic. Major CARs arise during an audit as a result of the identification of complete non-compliance and systemic failures to meet criteria with an element(s) of the standards. Major CARs hinder the implementation of forest certification and need to be closed out before proceeding with the certification process. After a successful main assessment, independent specialists often peer-review the certification report based on their regional and forest ecosystem expertise. If no new issues are raised, the final certification decision can be made by a panel appointed by the certification body. Routine annual audits occur to assess continuous compliance by following up and closing any outstanding CARs or stakeholder complaints, visiting new sites and ensuring that any changes to the standard that have been made are addressed by the forest owner and any non-compliance that could result is categorised as minor or major CARs, with the resultant corrective action being required.

The products from certified forests can then be assured and labelled through Chain of Custody, CoC and trademark standards for eco-labelling, ensuring the certified products' traceability and credibility. Actors not required to be certified often at the end of the value chain can apply for a trademark license to use the eco-label in marketing certified products. The next section provides a closer analysis of each of the certification schemes, FSC and PEFC.

2.1 Forest Stewardship Council and Programme for Endorsement of Forest Certification

FSC is an international membership organisation founded in 1993 in Toronto by representatives from various forest industry and environmental organisations, aiming for globally responsible forest management (Overdevest, 2009). FSC responsible forestry is based on economic, social, and environmental value dimensions. This is reflected in the organisation of FSC. It is founded on three-member chambers: the social, the economic and the environmental chamber. Each chamber has equal decision-making rights, as decisions are primarily made based on consensus between the chambers. The system also considers that each chamber consists of actors from the global economic north respectively south to avoid unjustness due to resource allocation. Members negotiate the standard for forest certification through participation in different working groups and committees. Agreed-upon national standards then need to be approved by FSC International. FSC can be considered the most studied scheme in forest certification research (Wolff & Schweinle, 2022). FSC relies on Assurance Services International (ASI) as accreditors for certification bodies.

The FSC standard-setting process for forest management standards is based on ten principles and criteria that function as a framework for national standards (FSC, 2023) (see Table 1).

Table 1 Overview of principles for Forest Steward Certification and Programme for Endorsement of Forest Certification

FSC certification	PEFC certification
Compliance with laws	Legal framework
Workers right and employment conditions	Workers' rights, local employment
Indigenous peoples' rights	Indigenous peoples' rights
Community relations	
Benefits from the forest	Ecosystem services
Environmental values and impacts	
Management and planning	
Monitoring and assessment	
High conservation values	Biodiversity
Implementation of management activities	
	Natural alternatives

The Pan European Forest Certification Council (PEFC) was founded in 1999 as an alternative certification scheme for European smallholders or here referred to as small private forest owners. PEFCs member system is based on chambers for forest management, forest industry and forest stakeholders. PEFC later came to approve national forest certification standards globally and, therefore, in 2004, changed its full name to Programme for the Endorsement of Forest Certification. PEFC has often referred to the slogan 'act locally, think globally' and functions as an umbrella organisation that endorses national and regional forest certification standards (PEFC, 2023). Sustainable Forestry Initiative (SFI) and Malaysian Timber Certification Council (MTCC) are members today, and their standards are endorsed by PEFC (Gutierrez Garzon *et al.*, 2020). PEFC national requirements for sustainable forest management are guided by a common framework (PEFC, 2023). PEFC uses national accreditation bodies that are members of the International Accreditation Forum (IAF). A PEFC Sustainable Forest Management certification encompasses the following aspects of forestry management (Forest Eco Certification, 2023) (see Table 1).

These principles are part of the PEFC Sustainability Benchmark, further developed with over 300 criteria that serve as a basis for national certification systems assessed in a PEFC endorsement audit.

3 Delivering environmental, social and financial value

Forest certification is an independent assessment process of the environmental and social impact of forestry (Gullison, 2003). The idea of forest certifications as a market-driven tool is based on their capacity to effectively deliver value for different stakeholders. Numerous stakeholders' needs are negotiated in a certification scheme. Forest certification should, for example, reduce deforestation and improve the lives of local communities and the conditions for forest workers while also increasing the financial viability for forest owners. Ensuring the forest certification sustainability impact is essential for the credibility of the certification system. Research on forest certification is often done to evaluate issues related to the impact of certification (Malek & Abdul Rahim, 2022). Several studies have aimed to assess the outcomes of forest management certification and how well it delivers on environmental, social, financial and policy aspects. However, studying and assessing the impact of forest certification can often be a complex and costly process (Romero *et al.*, 2017). The level of impact of forest certification also depends on governance and policy on a national and regional level and not only the applied certification system. Governmental support for forest certification is essential for a well-functioning system (Niedziałkowski & Shkaruba, 2018). The three dimensions of sustainable value related to forest certification are now considered more in detail.

3.1 Environmental impact

The history of forest certification emphasises the conflict between environmental and financial values in forests. Therefore, the ability of forest certification to provide documented environmental benefits in terms of reduced deforestation and biodiversity conservation as well as informing environmental legislation and models of forest ownership is an essential topic for research (Table 2).

Table 2 points to several positive environmental effects from forest certifications. However, environmental aspects and the associated value have also been found to be an understudied area (Wolff & Schweinle, 2022). Forest certification has been found to support several positive environmental outcomes regarding biodiversity at comparable levels as forests with a minimum of activities (Campos-Cerqueira *et al.*, 2019; Polisar *et al.*, 2017). Forest certification tends to benefit the environment (reduce deforestation rate), but more studies are needed to understand the impact evaluation (Burivalova *et al.*, 2017). Other studies support the notion of an inconclusive impact from forest certification's with regard to deforestation (Rana & Sills, 2018; Romero & Putz, 2018).

Villalobos *et al.* (2018) and Johansson and Lidestav (2011) question the environmental benefits of forest certification provided by small private forest owners. Although, this is done without considering that certification also affects the whole forest industry and impacts uncertified forests (Lehtonen *et al.*, 2021). However, the certification system based on price premiums might attract already active forest owners. Market tools that promote premiums depending on the forest products sold are related to challenges in the level of extraction in certified areas. For example, certification in intact forest landscapes could

Table 2 Research pointing to environmental impacts of forestry certification, presented in alphabetical order

Environmental factor	Reported impact in certified forests	Author, year
Biodiversity conservation	Higher biodiversity. Forest certification can provide protection for threatened species. Set-asides in forest certification can have positive impact on biodiversity but need to be complemented with other policy tools.	Campos-Cerqueira *et al.*, 2019; Polisar *et al.*, 2017; Alveira *et al.*, 2023; Elbakidze *et al.*, 2011
Deforestation	Reduced rate of de-forestation. Geographic location had greater impact on hindering de-forestation in tropical forests than certification. First certification reduced de-forestation. Later forest certification was found to increase de-forestation. Certified plantation had a higher degree of native forest cover regeneration. The practice of clear-cuts undermines ecological effectiveness of forest certification in boreal forests.	Burivalova *et al.*, 2017; Panlasigui *et al.*, 2018; Rana & Sills, 2018; da Silva *et al.*, 2019; Blümröder *et al.*, 2020
Environmental consideration among small private forest owners	Minor positive impact. However, differences between active and non-active forest ownership. Forest certification did not have any positive environmental impact.	Johansson & Lidestav, 2011; Villalobos *et al.*, 2018
Environmental legislation	Certification schemes may influence the development of and enforcement of legislation.	Haywood & Henriot, 2019; Sundstrom & Henry, 2017

lead to that other areas close to certified forest land that are uncertified being logged more intensively (Kleinschroth *et al.*, 2019). Other studies also question the environmental benefits of forest certification altogether and propose changes in forest operations practices such as clear-cuts (Blumröder *et al.*, 2020). Forest standards contribute to reducing deforestation, as a complement to national legislation (Haywood & Henriot, 2019). The practice of certification can impact state forestry policy (Sundstrom & Henry, 2017). Non-conformities are most frequently found, by certification bodies during audits concerning the environmental impact of FSC (Trishkin *et al.*, 2019) and tend to increase in frequency with the size of forest management units (Dagnaisser *et al.*, 2022). However, this also relates to the country's economic development, as non-conformities were found to be more likely in countries with less developed forestry practices (Halalisan *et al.*, 2016).

3.2 Financial and socioeconomic impact

Financial and socioeconomic impacts from forest certification are often reduced to monetary dimensions, in terms of benefits or costs. Critical aspects in these assessments relate to which system boundaries, stakeholders and time dimensions are considered. Research findings may appear contradictory partly because different system boundaries, stakeholder focus and time perspectives have been selected (Table 3).

Table 3 illustrates a number of financial factors that are influenced by forestry certification. Assuming a consumer's willingness to pay a price premium and market access were the main driver for certification (Perera *et al.*, 2022). Certification can enable price margins and negatively affect quantitative margins, essential aspects of a country's export growth (Zhang *et al.*, 2022). Forest certification can increase sales prices of hardwood in auctions but is often related to the increased cost of certification (Deniz, 2023). However, premiums are often limited and might only occur on higher-quality logs (Durusoy & Özdemir, 2021). Furthermore, certification cost is one of the most significant barriers to becoming certified (Sugiura & Oki, 2018).

Managers of certified forests perceive improved forest management and changes in stakeholder communication due to certification (Halalisan *et al.*, 2018). Certified forest owners recognise that certification contributes to demonstrating environmental consideration and improving public relations (Paluš *et al.*, 2018). However, certification standards often include restrictions on the use of chemicals, which can lead to lower forest productivity and yield (Mendell *et al.*, 2015). A critical motive for companies to adopt forest certification has been as a signal to the market and end-consumers that they are responsible actors (Galati *et al.*, 2017). Forest certification is an essential communication tool for sustainable forest management utilised by forest actors (Lombardo *et al.*, 2021). A forest certification may provide market opportunities and an improved corporate image which may lead to positive competitiveness and improved stakeholder relations (Perera *et al.*, 2022; So & Lafortezza, 2022). Also, the future of digitalisation tools and remote sensing in forest certification offers lowered costs, and higher transparency (Lopatin *et al.*, 2016) associated with higher credibility.

The cost of certification is one of the biggest barriers to becoming certified and having a small forest ownership (Sugiura & Oki, 2018). Therefore, it is also important that global standards are adapted to local contexts (Lemes *et al.*, 2017). Forest certification is found to impact increased stakeholder

Table 3 Research pointing to financial and socioeconomic impacts of forestry certification

Financial or socioeconomic factor	Reported impact in certified forests	Author, year
Consumer behaviour	Environmental attitudes play an important role in willingness to pay for certified products.	Panico *et al.*, 2022
Contextual interpretations	Global standards need to be adapted to local conditions.	Lemes *et al.*, 2017
Costs	Cost and sales price increase with forest certification.	Deniz, 2023
Legitimacy and image	Digital tools, remote sensing – offers transparency. The main reason to adopt forest certification is as signalling mechanism. Grounds to communicate.	Lopatin *et al.*, 2016; Galati *et al.*, 2017; Lombardo *et al.*, 2021
Management skills	Managers grounds for communication is enhanced.	Halalisan *et al.*, 2018; Paluš *et al.*, 2018
Market access	Forest certification aids in accessing new markets, keep market share, and sell products in existing markets.	Perera *et al.*, 2022
Price premium	Limited price premiums for forest certification. Cost of certification must be balanced with a price premium.	Durusoy & Özdemir, 2021; Sugiura & Oki, 2018
Profit	Influence price margins and imports. Companies that perform well financially have predisposition for forest certification.	Zhang *et al.*, 2022; Zubizarreta *et al.*, 2023
Smallholders	Subsidies for forest certification are important for small private forest owners adopting certification. Long-term forest certification can lead to economic resilience. Administration can hinder smallholders from selling certified wood as labelled.	Wibowo *et al.*, 2018; Ehrenberg-Azcárate & Peña-Claros, 2020; Hermudananto & Supriatno, 2020
Stakeholder engagement	Increased stakeholder participation	Lescuyer *et al.*, 2021
Trade	Forest certification has become a trade barrier for developing countries trading with developed countries.	Chen *et al.*, 2020
Yield and productivity	Restricted use of chemicals may lower productivity.	Lemes *et al.*, 2021; Mendell *et al.*, 2015

participation. However, this does not affect national governance or levels of corruption or increased transparency (Lescuyer *et al.*, 2021).

3.3 Social impact

Forestry is, for the most part, carried out in rural areas. The rural localisation influences social aspects, such as the need for transport, limitations in labour recruitment, a potential for labour abuses, and a need for stakeholders to keep a very long financial time perspective. Social, generational perspectives on forestry also mean locking up much financial capital for a long time (50–100+ years) between harvests means that the social impacts are key and intimately connected to the financial aspects of forestry certification.

As in most CSR research fields, the social dimension is less researched. Key findings are synthesised in Table 4.

Forest certification has positively influenced workers' rights, health and safety, and forest companies' relations to the community (Pezdevšek Malovrh *et al.*, 2019). Although living and working conditions improved for forest

Table 4 Research pointing to the social impacts of forestry certification

Social factor	Reported impact in certified forests	Author, year
Democracy	Transparency promotes democratic traditions, fair distribution and gender equality. A challenge to include social sustainability in forest certification systems.	Loveridge *et al.*, 2021; Boström, M, 2011
Free, Prior and Informed Consent (FPIC)	Risk of transferring responsibility and duties to actors with fewer resources. Possibility to achieve mutual understanding and mitigate conflicts. Social norms and stakeholders affect the implementation and effectiveness of certification tools.	Dobrynin *et al.*, 2020; Tysiachniouk *et al.*, 2021
Human rights	Positive effects in areas of workers' rights, health and safety, living conditions and community relations.	Cerutti *et al.*, 2017; Pezdevšek Malovrh *et al.*, 2019
Indigenous communities	Forest certification can involve indigenous communities and strengthen their rights. Forest certification as a learning tool for forest companies to consider indigenous communities. Forest certification activities can impact indigenous communities negatively.	Tikina *et al.*, 2010; Teitelbaum & Wyatt, 2013; Doremus, 2019
Local community development	Forest certification tends to favour larger actors. Benefits local communities but cannot resolve deeply rooted social conflicts.	McDermott *et al.*, 2015; Tricallotis *et al.*, 2018

workers, forest certification also meant considering a forest management plan that sets the rules for how the certified forest area can be utilised, thereby limiting local access to forest resources and customary rights (Cerutti *et al.*, 2017).

Forest certification and associated principles of democracy and transparency promote democratic decision-making, fair distribution and gender equality (Loveridge *et al.*, 2021). Although, this requirement comes with an increased administrative burden for community forests and smallholders. Therefore, it is necessary to support capacity building among local communities to ensure these principles are met. Forest certification activities have also created disadvantages for local and indigenous communities as they tend to favour larger actors with the infrastructure and economic ability to engage with certification (Doremus, 2019; McDermott *et al.*, 2015). Forest certification systems could greatly benefit from focusing on social sustainability that enables inclusion and acceptance among social stakeholders (Boström, 2011; Tikina *et al.*, 2010).

Forest certification programmes have adopted tools for comprehensively including and considering indigenous communities. This is accomplished through a consultation and participation process called Free, Prior, and Informed Consent (FPIC). However, these are often complex issues to resolve where forest companies need to develop capacities (Teitelbaum & Wyatt, 2013). FPIC can be considered as a tool for mutual consent that mitigates potential conflicts. There is also the risk of transferring responsibility from state governance to private actors and narrowing the scope of included stakeholders (Dobrynin *et al.*, 2020; Tysiachniouk *et al.*, 2021). The next case study explores some of these themes described in the previous sections.

4 Case study: small private forest owners' motives for forest certification

The perceived and proven impact of forest certification affects the adoption of forest certification, which is especially true for small private forest owners. A study of a diverse group of private forest owners points to the forest certification process as being complex. As suggested by Weiss *et al.* (2019), their motives for making a forest certification decision depends on their specific objectives, attitudes and behaviour. A qualitative approach was conducted to understand further small private forest owners' motives and objectives for adopting forest certification in Sweden (Thorning & Mark-Herbert, 2022). Their lived experience of forest certification (n = 14) was explored through interviews with forest owners. Through a process called laddering, important forest certification attributes that give rise to perceived benefits and disadvantages that, in turn, trigger emotions and are essential for understanding the motives, objectives and values behind the reason for adopting forest certification were developed. The results showed a range of experiences and motives for forest certification. These groups were mainly women and men, and owners with forestry as their primary occupation compared to those with another main income.

For both men and women, forest certification was important for the credibility and legitimacy of the forest industry and, for some, showing solidarity with their forest owner association. Women's important forest certification attributes were set-asides, rules and regulations, eco-labelling and price premiums. For women, price premiums were both a financial benefit and an acknowledgement of doing the right thing. Forest certification was also seen as an instrument for fulfilling environmental objectives. Social and environmental considerations through forest certification are essential aspects for female forest owners and provide a sense of pride and being a knowledgeable and accomplished forest owner. This was also a result of understanding the forest certification standard that most female forest owners stated that they had no problems complying with.

Attributes for men were the same as for women, with the addition of a certified forest management plan that functioned as an important forest management tool resulting in feeling pride in being a good forest owner and forest steward. For male forest owners, set-asides not only resulted in environmental and social considerations (as for women) but also increased costs. The male interviewees differed in their opinion of forest certification as being either easy to follow the rules or difficult to follow. The group of men who found it difficult also questioned forest certification's environmental impact.

Small private forest owners with forestry as a primary occupation frequently combined this with agricultural practices. These forest owners then often chose to apply for certification for forest management and agricultural practices, resulting in high costs. For these forest owners, attributes such as set-asides, standard rules and regulations, labelling and price premiums were important attributes. For them, price premiums needed to result in financial benefits, which could make up for increased costs and loss in timber production that occurred through set-asides. Complying with forest certification was also considered as a decrease in autonomy. Environmental considerations in forest certification fulfilled environmental interests and objectives, resulting in credibility and legitimacy. Forest certification also included having a long-term perspective on forestry which was connected to taking pride in forest ownership and having it as a lifestyle. Forest certification could also strengthen the feeling of being knowledgeable and accomplished forest owners through gaining forestry and certification knowledge.

Increased forestry and certification knowledge and the feeling of being an accomplished forest owner were even more important for forest owners with a main income other than forestry. This group shared the same view on important attributes as those with forestry as the main occupation but also included the benefit of a forest management plan that could fulfil environmental interests and objectives. The objective of environmental consideration was also achieved by environmental considerations through set-asides and having a long-term perspective. Set-asides also resulted in social and environmental considerations as well as increased costs. However, the group of forest owners with other main income than forestry perceived no problems with certification compliance, which allowed them to take pride in their forest ownership.

5 Forest certification for small private forest owners

Forest certification began as a market-driven policy instrument developed under pressure by NGOs and companies that invited stakeholders to the standard-setting process. Early on, small private forest owners (often referred to as smallholders in forest certification) considered forest certification a threat to autonomy and forest income, costly and with uncertain benefits (Bensel, 2001; Lindström *et al.*, 1999; Rickenbach, 2002). Problems that arise for smallholders often relate to understanding and complying with requirements and the cost of certification. When PEFC was founded, it was seen as an alternative to forest certification offered to small private forest owners. FSC later developed a standard for small- or low-intensity managed forests (SLIMF) to cater to the needs of smallholders in the certification process and auditing and lessen the cost for certification. Here forest management can be classified as small, if 100 hectares or less, but can also be considered for management sizes up to 1000 hectares under certain conditions. Low-intensity forest management is considered when harvesting is less than 20% of the mean annual increment and the annual harvest is less than 5000 cubic metres. The forest can also be eligible for SLIMF if it is managed for non-timber forest products, not including plantations.

Arguments for offering forest certification are that small private forest owners contribute to the UN Sustainable Development Goals and that this is not solely a responsibility of the state or companies (Danley, 2018). However, forest certification rates among small private forest owners or smallholders are at a different rate than among forest companies. First, to increase in uptake of certification when the forest owners' associations started to offer certification, including a price premium, to its members (Johansson & Lidestav, 2011). Then a need for financial incentives or cost compensation was suggested to encourage small private forest owners to consider conservation or more sustainable forest practices (Kilgore *et al.*, 2008; Langpap, 2006). An ongoing policy and media debate encourages small private forest owners to consider environmental objectives (Lindahl *et al.*, 2017). Forest conservation is also expected to increase among small private forest owners (Karppinen *et al.*,

2019). Therefore these forest owners are now experiencing increased pressure from society to consider sustainable development (Cashore *et al.*, 2007).

The financial benefits of forest certification are often highlighted and described as price premiums and market access. Johansson and Lidestav (2011) noted that certified forest owners are more likely to increase forest management activities. They proposed that certified forest owners are generally more active or become more active when forest management plans following certification are in place. A forest management plan refers to a detailed plan for each local forest area that gives information about how it will be managed over the next ten or more years. A green forest management plan includes forest certification.

Administrative burden acted as a barrier for smallholders to sell certified wood as labelled, and often it was sold into uncertified markets (Hermudananto & Supriatno, 2020). Challenges in certification for small private forest owners can be addressed through group certification managed by forest owners' associations, forest companies or certification organisations. Enrolment by association is encouraged by offers of green forest management plans that include forest certification. For small private forest owners who are not members of an association, the route to becoming certified either includes a self-initiative or a process as suggested by timber-procuring companies. The motives for becoming certified vary from one forest owner to another. Forest owners' background, values and own forest management experience serve as background to understand differences in motives. Therefore, the intermediary organisation facilitating and administrating group certificates are essential in lowering the barrier threshold to certification and creating value for participating forest owners (Boakye-Danquah & Reed, 2019).

Small private forest owners are transforming from a relatively homogenous group (males living close to their forestland) to a more heterogeneous group as they are increasingly urban, a growing proportion of female owners and less financially dependent on their forest land (Follo *et al.*, 2016; Keskitalo, 2017). Socio-demographic characteristics, such as gender and residency of the forest owner include categories, e.g. female and urban forest owners as a new type of forest owner. Previous studies show that female and non-resident forest owners have other objectives than solely timber production and are inclined to assess forest values linked to environmental and social aspects higher than the traditional forest owner (Berlin *et al.*, 2006; Umaerus *et al.*, 2019). The socioeconomic shift also means forest owners are increasingly highly educated and less financially dependent on their forestland (Karppinen *et al.*, 2019). The ongoing change in ownership structure and demographics indicates new types of forest owners with goals other than primarily financial, which in turn influences forest management and policy and sustainability goals (Weiss *et al.*, 2019). These changes in forest owner types are predicted to create challenges for the forest industry and policy-makers due to their expected non-traditional forest views and objectives (Follo *et al.*, 2016). These forest owners require the forest industry to develop new types of service and view value creation from multiple uses of forests (Kurttila *et al.*, 2019). Sustainable development in the form of forest certification can be viewed as a source of value creation and a possibility to innovate and develop services specific to forest owners (Toppinen *et al.*, 2019).

6 Conclusion and future trends

A transition to sustainable forest management is guided by standards such as FSC and PEFC. Critical perspectives would argue that landowners pay for the enforcement of rules, in this case, compliance with these standards. In contrast, positive perspectives point to the opportunities to differentiate forest management approaches, signaling responsible management for which they can receive a price premium for products or services, and a feeling of 'doing the right thing'.

Forestry certification differs from certification of agricultural production in that the production process has long-time dimensions, between 50 and 100 years in Scandinavia (and shorter production spans in a warmer climate where the production cycle is shorter). It means that a decision made by one generation of forest owners will influence the management of the forest resource for the next generation as well as the current generation.

A similar logic can be applied to understand the landowner's motives for certification. Forestry owners are a diverse group. Their lands may differ in size and location and thereby be more or less suitable for certification, but what is more, the importance of the income from forest management varies. For a small private forest owner who makes a prime living from the forest, the financial aspects (added value from certification and the use of an eco-label) might be crucial to attaining a price that supports a livelihood. On the other hand, a forest owner that is not dependent on the income from forest management may have another job on the side, which explains other expectations from a forest certification, and other needs to contract management support. As the profile of forest owners gradually changes over time, from a male-dominated domain, where a forest owner lives close to the forest property and manages his or her own forest, a transition is taking place where a larger proportion of forest owners are women and/or living in urban areas.

From a market development perspective, industrial market incentives are crucial. A solution to management challenges for small forest owners is seen in the option of group certification, to relieve the administrative burdens. Service and value co-creation are becoming essential to the timber procuring process to attract and build relationships with new types of forest owners (Andersson & Keskitalo, 2019). Nevertheless, path dependence in the forest industry is often a barrier to developing new value creation and service offerings. Their primary focus on keeping their market position hinders more minor actors from entering the market (Mattila *et al.*, 2013).

Future development of certification schemes is influenced by societal development, in particular, awareness of sustainability aspects of forest management. The continued development of carbon markets is of particular interest as part of eco-system services provided in forest contexts. From a consumer perspective, it may continue to fill an essential educational role for fibre-based products. Compared to food products, where eco-labels are part of everyday consumption decisions for consumers, the eco-labels for fast-moving consumer fibre goods (hygiene products, packaging, paper in books etc.) are not given the same attention.

Forestry management and standards appear on the sustainability agenda as a concern for resource management focusing on biodiversity, carbon sequestration, ecosystem services, social services and cultural heritages. Just like FSC and PEFC, future standards need to engage a wide set of stakeholders in the dialogue of managing resources to avoid conflicts (Dare *et al.*, 2011). The democratic traditions of forest certification may serve as well-grounded principles for managing forestry resources for sustainable development.

7 Where to look for further information

- Cashore, B., Auld, G. and Newsom, D. (2004). *Governing through markets: Forest certification and the emergence of non-state authority.* Yale University Press. doi:10.12987/9780300133110
- Forest Stewardship Council (FSC): https://fsc.org/en.
- Nussbaum, R. and Simula, M. (2005). *The forest certification handbook.* 2nd edition. London. Earthscan.
- Malek, E. J. and Abdul Rahim, A. R. (2022). A thematic review of forest certification publications from 2017 to 2021: analysis of pattern and trends for future studies. *Trees, Forests and People, 10.* doi:10.1016/j.tfp.2022.100331
- Program for Endorsement of Forest Certification (PEFC): https://pefc.org.

Journals

- *Forest Policy and Economics: https://www.sciencedirect.com/journal/forest-policy-and-economics.*
- *Forests: https://www.mdpi.com/journal/forests.*

8 References

Alveira, M., de Bustos, S. and Tálamo, A. (2023). Certified forest management and conservation of medium and large-sized mammals: biodiversity, trampling, browsing and seed removal in the Argentine Austral Yungas. *Forest Ecology and Management, 533*. doi:10.1016/j.foreco.2023.120860

Andersson, E. and Keskitalo, E. C. H. (2019). Service logics and strategies of Swedish forestry in the structural shifts of forest ownership challenging the old and shaping the new. *Scandinavian Journal of Forest Research, 34*, 508–520. doi:10.1080/02827581.2019.1604990

Baumgartner, R. J. (2019). Sustainable development goals and the forest sector–a complex relationship. *Forests, 10*(2). doi:10.3390/f10020152

Bensel, T. (2001). Promoting certified sustainable forestry on private woodlots in Northwestern Pennsylvania: challenges and opportunities. *Local Environment, 6*(3), 257–278. doi:10.1080/13549830120073266

Berlin, C., Lidestav, G. and Holm, S. (2006). Value placed on forest property benefits by Swedish NIPF owners: difference between members in forest owners associations and non-members. *Small-scale Forest Economics, Management and Policy, 5*, 83–96.

Blumröder, J. S., Hoffmann, M. T., Ilina, O., Winter, S., Hobson, P. R. and Ibisch, P. L. (2020). Clearcuts and related secondary dieback undermine the ecological effectiveness of FSC certification in a boreal forest. *Ecological Processes, 9*(1). doi:10.1186/s13717-020-0214-4

Boakye-Danquah, J. and Reed, M. G. (2019). The participation of non-industrial private forest owners in forest certification programs: the role and effectiveness of intermediary organisations. *Forest Policy and Economics, 100*, 154–163. doi:10.1016/j.forpol.2018.12.006

Boström, M. (2011). The problematic social dimension of sustainable development: the case of the Forest Stewardship Council. *International Journal of Sustainable Development & World Ecology, 19*(1), 3–15. doi:10.1080/13504509.2011.582891

Brunsson, N. and Jacobsson, B. (2000). *A world of standards* (N. Brunsson & B. Jacobsson, Eds.). Oxford: Oxford University Press.

Burivalova, Z., Hua, F., Koh, L. P., Garcia, C. and Putz, F. (2017). A critical comparison of conventional, certified, and community management of tropical forests for timber in terms of environmental, economic, and social variables. *Conservation Letters, 10*(1), 4–14. doi:10.1111/conl.12244

Campos-Cerqueira, M., Mena, J. L., Tejeda-Gómez, V., Aguilar-Amuchastegui, N., Gutierrez, N., Aide, T. M., Pettorelli, N. and Lecours, V. (2019). How does FSC forest certification affect the acoustically active fauna in Madre de Dios, Peru? *Remote Sensing in Ecology and Conservation, 6*(3), 274–285. doi:10.1002/rse2.120

Cashore, B., Egan, E., Auld, G. and Newsom, D. (2007). Revising theories of nonstate market-driven (NSMD) governance: lessons from the Finnish forest certification experience. *Global Environmental Politics, 7*, 1–44. doi:10.1162/glep.2007.7.1.1

Cerutti, P., Lescuyer, G., Tacconi, L., Eba'a Atyi, R., Essiane, E., Nasi, R., Tabi Eckebil, P. P. and Tsanga, R. (2017). Social impacts of the Forest Stewardship Council certifi- cation in the Congo basin. *International Forestry Review 19*, 50–63.

Chen, J., Wang, L., Li, L., Magalhães, J., Song, W., Lu, W., Xiong, L., Chang, W.-Y. and Sun, Y. (2020). Effect of forest certification on international trade in forest products. *Forests, 11*(12). doi:10.3390/f11121270 da Silva, R. F. B., Batistella, M., Palmieri, R., Dou, Y. and Millington, J. D. A. (2019). Eco-certification protocols as mechanisms to foster sustainable environmental practices in telecoupled systems. *Forest Policy and Economics, 105*, 52–63. doi:10.1016/j.forpol.2019.05.016

Dagnaisser, L. S., Chaves Cardoso, J., Ribeiro, L. O. M. M., Monte, M. A., Silva, E. V. and Basso, V. M. (2022). Forest stewardship council certification of forest plantation management in the Southern Cone of America. *Small-scale Forestry, 21*(4), 601–619. doi:10.1007/s11842-022-09515-x

Danley, B. (2018). Skepticism of state action in forest certification and voluntary set-asides: a Swedish example with two environmental offsetting options [Article]. *Scandinavian Journal of Forest Research, 33*(7), 695–707. doi:10.1080/02827581.2018.1479442

Dare, M., Schirmer, J. and Vanclay, F. (2011). Does forest certification enhance community engagement in Australian plantation management? [Article]. *Forest Policy and Economics, 13*(5), 328–337. doi:10.1016/j.forpol.2011.03.011

Deniz, T. (2023). The effect of forest certification on log sale prices: a case study in Northwestern Turkey. *Forests, 14*(3). doi:10.3390/f14030596

Dobrynin, D., Smirennikova, E. and Mustalahti, I. (2020). Non-state forest governance and 'responsibilization': the prospects for FPIC under FSC certification in Northwest Russia. *Forest Policy and Economics, 115*. doi:10.1016/j.forpol.2020.102142

Doremus, J. (2019). Unintended impacts from forest certification: evidence from indigenous Aka households in Congo. *Ecological Economics, 166*. doi:10.1016/j.ecolecon.2019.106378

Durusoy, İ. and Özdemir, O. (2021). Price premiums for certified roundwood: evidence from auction sales in Turkey. *Journal of Forest Research, 26*(6), 395–399. doi:10.1080/13416979.2021.1940664

Ehrenberg-Azcárate, F. and Peña-Claros, M. (2020). Twenty years of forest management certification in the tropics: major trends through time and among continents. *Forest Policy and Economics, 111*. doi:10.1016/j.forpol.2019.102050

Elbakidze, M., Angelstam, P., Andersson, K., Nordberg, M. and Pautov, Y. (2011). How does forest certification contribute to boreal biodiversity conservation? Standards and outcomes in Sweden and NW Russia. *Forest Ecology and Management, 262*(11), 1983–1995. doi:10.1016/j.foreco.2011.08.040

FAO. (2023). *Sustainable forest management*. Retrieved from https://www.fao.org/sustainable-forests-management/en/

Follo, G., Lidestav, G., Ludvig, A., Vilkriste, L., Hujala, T., Karppinen, H., Didolot, H. and Mizaraite, D. (2016). Gender in European forest ownership and management: reflections on women as "new forest owners". *Scandinavian Journal of Forest Research, 32*(2), 174–184. doi:10.1080/02827581.2016.1195866

Forest Eco Certification. (2023). *Forest Eco certification*. Retrieved from https://forestecocertification.com/pefc-requirements-and-criteria/

FSC. (2023). *The 10 FSC principles*. Retrieved from https://fsc.org/en/fsc-standards

Galati, A., Gianguzzi, G., Tinervia, S., Crescimanno, M. and La Mela Veca, D. S. (2017). Motivations, adoption and impact of voluntary environmental certification in the Italian Forest based industry: the case of the FSC standard. *Forest Policy and Economics, 83*, 169–176. doi:10.1016/j.forpol.2017.08.002

Gullison, R. E. (2003). Does forest certification conserve biodiversity? *Oryx, 37*(2), 153–165. doi:10.1017/s0030605303000346

Gutierrez Garzon, A. R., Bettinger, P., Siry, J., Abrams, J., Cieszewski, C., Boston, K., Mei, B., Zengin, H. and Yeşil, A. (2020). A comparative analysis of five forest certification programs. *Forests, 11*(8). doi:10.3390/f11080863

Halalisan, A., Abrudan, I. and Popa, B. (2018). Forest management certification in Romania: motivations and perceptions. *Forests, 9*(7). doi:10.3390/f9070425

Halalisan, A., Ioras, F., Korjus, H., Avdibegovic, M., Maric, B., Pezdvesek Malovrh, S. and Abrudan, I. V. (2016). An analysis of forest management non-conformities to FSC standards in different European countries. *Notulae Botanicae Horti Agrobotanici Cluj-Napoca, 44*(2), 634–639. doi:10.15835/nbha44210263

Haywood, C. and Henriot, C. (2019). Protecting forests from conversion: the essential role of supply-side national laws. *Frontiers in Forests and Global Change, 2*. doi:10.3389/ffgc.2019.00035

Hermudananto, H. and Supriatno, N. (2020). Evaluation of timber harvesting planning in Indonesian certified small and low-intensity managed forests. *Journal of Tropical Forest Science, 32*(3), 283–288. doi:10.26525/jtfs32.3.283

Higman, S., Mayers, J., Bass, S., Judd, N. and Nussbaum, R. (2005). *The sustainable forestry handbook* (2nd ed.). London: Eartschan.

Johansson, J. and Lidestav, G. (2011). Can voluntary standards regulate forestry?– Assessing the environmental impacts of forest certification in Sweden. *Forest Policy and Economics, 13*, 191–198. doi:10.1016/j.forpol.2010.11.004

Karppinen, H., Ní Dhubháin, Á. and Butler, B. J. (2019). Family forest owners' changing values and other service-demand drivers. In *Services in family forestry* (pp. 83–102). doi:10.1007/978-3-030-28999-7_5

Keskitalo, E. C. H. (2017). *Globalisation and change in forest ownership and forest use: natural resource management in transition*. London: Palgrave Macmillan.

Kilgore, A., M, Snyder, S., Taff, S. and Schertz, J. (2008). Family forest stewardship: do owners need a financial incentive? *Journal of Forestry, 106*, 357–362.

Kleinschroth, F., Rayden, T. and Ghazoul, J. (2019). The dilemma of maintaining intact forest through certification. *Frontiers in Forests and Global Change, 2*. doi:10.3389/ffgc.2019.00072

Kurttila, M., Butler, B. J., Hujala, T., Miina, J., Vanhanen, H. and Vacik, H. (2019). The diversifying use of family forests–opportunities for new services. In T. Hujala, A. Toppinen and B. J. Butler (Eds.), *Services in family forestry* (pp. 67–81). Berlin: Springer.

Langpap, C. (2006). Conservation of endangered species: can incentives work for private landowners? *Ecological Economics, 57*(4), 558–572. doi:10.1016/j.ecolecon.2005.05.007

Lehtonen, E., Gustafsson, L., Lohmus, A. and von Stedingk, H. (2021). What does FSC forest certification contribute to biodiversity conservation in relation to national legislation? *Journal of Environmental Management, 299*, 113606. doi:10.1016/j.jenvman.2021.113606

Lemes, G., P., Zanuncio, J. C., Jacovine, L. A. G., Wilcken, C. F. and Lawson, S. A. (2021). Forest stewardship council and responsible wood certification in the integrated pest management in Australian forest plantations. *Forest Policy and Economics, 131*. doi:10.1016/j.forpol.2021.102541

Lemes, G., P., Zanuncio, J., C., Serrao, J., E. and Lawson, S., A. (2017). Forest Stewardship Council (FSC) pesticide policy and integrated pest management in certified tropical plantations. *Environmental Science and Pollution Research International, 24*(2), 1283–1295. doi:10.1007/s11356-016-7729-3

Lescuyer, G., Tsanga, R., Nziengui, S., Forni, E. and Romero, C. (2021). Influence of FSC certification on the governance of the logging sector in the Congo basin. *Natural Resources Forum, 45*(3), 289–304. doi:10.1111/1477-8947.12231

Lindahl, K. B., Sténs, A., Sandström, C., Johansson, J., Lidskog, R., Ranius, T. and Roberge, J.-M. (2017). The Swedish forestry model: more of everything? *Forest Policy and Economics, 77*, 44–55. doi:10.1016/j.forpol.2015.10.012

Lindström, T., Hansen, E. and Juslin, H. (1999). Forest certification: the view from Europes NIPFs. *Journal of Forestry, 97*(3), 25–30.

Lombardo, E., Crescimanno, M., Vrontis, D. and Galati, A. (2021). Driving forces affecting the adoption of certifications in the forest-based industry: a systematic literature review. *Jōurnal of Global Business Advancement, 14*, 453–476.

Lopatin, E., Trishkin, M. and Gavrilova, O. (2016). Assessment of compliance with PEFC forest certification indicators with remote sensing. *Forests, 7*(12). doi:10.3390/f7040085

Loveridge, R., Sallu, S. M., Pfeifer, M., Oldekop, J. A., Mgaya, M., da Silva, D. A., Latham, J., Platts, P. J. and Marshall, A. R. (2021). Certified community forests positively impact human wellbeing and conservation effectiveness and improve the performance of nearby national protected areas. *Conservation Letters, 14*(6). doi:10.1111/conl.12831

Malek, E. J. and Abdul Rahim, A. R. (2022). A thematic review of forest certification publications from 2017 to 2021: analysis of pattern and trends for future studies. *Trees, Forests and People, 10*. doi:10.1016/j.tfp.2022.100331

Mattila, O., Toppinen, A., Tervo, M. and Berghäll, S. (2013). Non-industrial private forestry service markets in a flux: results from a qualitative analysis on Finland. *Small-Scale Forestry 12*, 559–578. doi:10.1007/s11842-012-9231-1

McDermott, C. L., Irland, L. C. and Pacheco, P. (2015). Forest certification and legality initiatives in the Brazilian Amazon: lessons for effective and equitable forest governance. *Forest Policy and Economics, 50*, 134–142. doi:10.1016/j.forpol.2014.05.011

Meidinger, E., Elliott, C. and Oesten, G. (2003). The fundamentals of forest certification. In E. Meidinger, C. Elliott, & G. Oesten (Eds.), *Social and political dimensions of forest certification 3*. Forstbuch Verlag. Retrieved from https://open.fsc.org/handle/resource/996

Mendell, B. C., Lang, A. H., Caldwell, W. and Garrett, D. L. (2015). Chemical use and forest certification: productivity and economic implications. *Journal of Forestry, 113*(4), 367–371. doi:10.5849/jof.14-121

Niedziałkowski, K. and Shkaruba, A. (2018). Governance and legitimacy of the forest stewardship council certification in the national contexts – a comparative study of Belarus and Poland. *Forest Policy and Economics, 97*, 180–188. doi:10.1016/j.forpol.2018.10.005

Overdevest, C. (2009). Comparing forest certification schemes: the case of ratcheting standards in the forest sector. *Socio-Economic Review, 8*(1), 47–76. doi:10.1093/ser/mwp028

Paluš, H., Parobek, J., Šulek, R., Lichý, J. and Šálka, J. (2018). Understanding sustainable forest management certification in Slovakia: forest owners' perception of expectations, benefits and problems. *Sustainability (Switzerland), 10*(7). doi:10.3390/su10072470

Panico, T., Caracciolo, F. and Furno, M. (2022). Analysing the consumer purchasing behaviour for certified wood products in Italy. *Forest Policy and Economics, 136*. doi:10.1016/j.forpol.2021.102670

Panlasigui, S., Rico-Straffon, J., Pfaff, A., Swenson, J. and Loucks, C. (2018). Impacts of certification, uncertified concessions, and protected areas on forest loss in Cameroon, 2000 to 2013. *Biological Conservation, 227*, 160–166. doi:10.1016/j.biocon.2018.09.013

PEFC. (2023). *What is sustainable forest management?* Retrieved from https://pefc.org/what-we-do/our-approach/what-is-sustainable-forest-management

Perera, P., Rupasinghe, R. L., Weerasekera, D., Vlosky, R. and Bandara, R. (2022). Revisiting forest certification in Sri Lanka: the forest management and export wood-based manufacturing sector perspectives. *Forests, 13*(2). doi:10.3390/f13020179

Pezdevšek Malovrh, Š., Bećirović, D., Marić, B., Nedeljković, J., Posavec, S., Petrović, N. and Avdibegović, M. (2019). Contribution of forest stewardship council certification to sustainable forest management of state forests in selected Southeast European Countries. *Forests, 10*(8). doi:10.3390/f10080648

Polisar, J., de Thoisy, B., Rumiz, D. I., Santos, F. D., McNab, R. B., Garcia-Anleu, R., Ponce-Santizo, G., Arispe, R. and Venegas, C. (2017). Using certified timber extraction to benefit jaguar and ecosystem conservation. *Ambio, 46*(5), 588–603. doi:10.1007/s13280-016-0853-y

Rametsteiner, E. and Simula, M. (2003). Forest certification–an instrument to promote sustainable forest management? *Journal of Environmental Management, 67*(1), 87–98. doi:10.1016/s0301-4797(02)00191-3

Rana, P. and Sills, E. (2018). Does certification change the trajectory of tree cover in working forests in the tropics? An application of the synthetic control method of impact evaluation. *Forests, 9*(3). doi:10.3390/f9030098

Rickenbach, G. M. (2002). Forest certification of small ownership: some practical challenges. *Journal of Forestry, 100*(6), 43–47.

Romero, C. and Putz, F. (2018). Theory-of-change development for the evaluation of forest stewardship council certification of sustained timber yields from natural forests in Indonesia. *Forests, 9*(9). doi:10.3390/f9090547

Romero, C., Sills, E., Guariguata, M., Cerutti, P., Lescuyer, G. and Putz, F. (2017). Evaluation of the impacts of Forest Stewardship Council (FSC) certification of natural forest management in the tropics: a rigorous approach to assessment of a complex conservation intervention. *International Forestry Review, 19*, 36–49.

Röstlund, L. (2022). Frihet under ansvar. Om certifieringar som lovar hållbarhet *Skogslandet*. Stockholm: Bokförlaget Forum.

So, H. W. and Lafortezza, R. (2022). Reviewing the impacts of eco-labelling of forest products on different dimensions of sustainability in Europe. *Forest Policy and Economics, 145*. doi:10.1016/j.forpol.2022.102851

Sugiura, K. and Oki, Y. (2018). Reasons for choosing Forest Stewardship Council (FSC) and Sustainable Green Ecosystem Council (SGEC) schemes and the effects of certification acquisition by forestry enterprises in Japan. *Forests, 9*(4). doi:10.3390/f9040173

Sundstrom, L. and Henry, L. (2017). Private forest governance, public policy impacts: the forest stewardship council in Russia and Brazil. *Forests, 8*(11). doi:10.3390/f8110445

Teitelbaum, S. and Wyatt, S. (2013). Is forest certification delivering on First Nation issues? The effectiveness of the FSC standard in advancing First Nations' rights in the boreal forests of Ontario and Quebec, Canada. *Forest Policy and Economics, 27*, 23–33. doi:10.1016/j.forpol.2012.09.014

Thorning, A. and Mark-Herbert, C. (2022). Motives for sustainability certification–private certified forest owners' perspectives. *Forests, 13*. doi:10.3390/f13050790

Tikina, A., Innes, J. L., Trosper, L., R. and Larson, C., B. (2010). Aboriginal peoples and forest certification: a review of the canadian situation. *Ecology and Society, 15*, 33.

Toppinen, A., Mikkilä, M., Tuppura, A. and de Vries, G. (2019). Sustainability as a driver in forestry-related services Hujala, T., Toppinen, A. & Butler., B. J. (Eds.), *Services in family forestry* (pp. 289–306). Cham, Switzerland, Springer.

Tricallotis, M., Gunningham, N. and Kanowski, P. (2018). The impacts of forest certification for Chilean forestry businesses. *Forest Policy and Economics, 92*, 82–91. doi:10.1016/j.forpol.2018.03.007

Trishkin, M., Karjalainen, T. and Kangas, J. (2019). An analysis of the non-conformities of certified companies operating in Northwestern Russia. *Forests, 10*(12). doi:10.3390/f10121061

Tröster, R. and Hiete, M. (2018). Success of voluntary sustainability certification schemes – a comprehensive review. *Journal of Cleaner Production, 196*, 1034–1043. doi:10.1016/j.jclepro.2018.05.240

Tysiachniouk, M. S., McDermott, C. L., Kulyasova, A. A., Teitelbaum, S. and Elbakidze, M. (2021). The politics of scale in global governance: do more stringent international forest certification standards protect local rights in Russia? *Forest Policy and Economics, 125*. doi:10.1016/j.forpol.2021.102407

Uggla, Y. (2017). Negotiating responsible forestry: forest owners' understanding of responsibility for multiple forest values. *Environmental Sociology*. doi:10.1080/23251042.2017.1414659

Umaerus, P., Högvall Nordin, M. and Lidestav, G. (2019). Do female forest owners think and act "greener"? *Forest Policy and Economics, 99*, 52–58. doi:10.1016/j.forpol.2017.12.001

Villalobos, L., Coria, J. and Nordén, A. (2018). Has forest certification reduced forest degradation in Sweden? *Land Economics, 94*, 220–238.

Weiss, G., Lawrence, A., Hujala, T., Lidestav, G., Nichiforel, L., Nybakk, E., Quiroga, S., Sarvašová, Z., Suarez, C. and Živojinović, I. (2019). Forest ownership changes in Europe: state of knowledge and conceptual foundations. *Forest Policy and Economics, 99*, 9–20. doi:10.1016/j.forpol.2018.03.003

Wibowo, A., Pratiwi, S. and Giessen, L. (2018). Comparing management schemes for forest certification and timber-legality verification: complementary or competitive in Indonesia? *Journal of Sustainable Forestry, 38*(1), 68–84. doi:10.1080/10549811.2018.1498359

Wolff, S. and Schweinle, J. (2022). Effectiveness and economic viability of forest certification: a systematic review. *Forests, 13*(5). doi:10.3390/f13050798

Zhang, J., Zhao, Z., Zhao, W., Tao, C. and Cheng, B. (2022). The impact of forest certification on the ternary margins of China's forest product export. *Forests, 13*(8). doi:10.3390/f13081313

Zubizarreta, M., Arana-Landín, G., Wolff, S. and Egiluz, Z. (2023). Assessing the economic impacts of forest certification in Spain: a longitudinal study. *Ecological Economics, 204*. doi:10.1016/j.ecolecon.2022.107630

Part 3

Measuring compliance and improving performance

Chapter 13

The role of technology in monitoring and verifying safety, quality and sustainability in agri-food supply chains

Aleksandra Kowalska, Maria Curie-Skłodowska University, Poland; Laura Clapham, University of Cambridge, UK; and Louise Manning, University of Lincoln, UK

1 Introduction

Over the last three decades, the design and implementation of regulatory and private governance approaches has supported a high degree of confidence in the food safety and quality management systems employed in food supply chains (Manning et al., 2019). The focus on embedding effective food safety management systems within food businesses and ensuring better regulatory control of complex food supply chains was driven by a number of high-profile public health scares in the 1980–1990s. These included the occurrence of bovine spongiform encephalopathy (BSE) with cattle, predominantly reported in the United Kingdom (UK) with 4 million cattle slaughtered, and its links with the 178 deaths from new variant Creutzfeldt-Jakob disease (vCJD)[1]. This incident followed on from antifreeze being found in Austrian and Italian wines

1 https://www.cjd.ed.ac.uk/sites/default/files/figs.pdf.

http://dx.doi.org/10.19103/AS.2024.0133.06

in 1985–1986 which was associated with multiple deaths (Manning & Luning, 2018), the 1988 UK Salmonella in eggs scare and the 1999 European dioxin incident centred on Belgium.

The concept of safe food, which is reinforced by the Global Food Safety Initiative (GFSI), has focused on the prevention of both unintentional (contamination) and intentional adulteration of food products (GFSI, 2018). Despite the existence of both public regulatory and private governance systems, food safety incidents, poor quality products and disingenuous, even false, declarations about the environmental and social sustainability of food products continue to occur. These problems have highlighted the importance of material and product traceability. Effective food traceability systems (FTSs) are becoming more and more important for stakeholders, particularly those stakeholders who seek to operate sustainably and communicate this to their customers.

Product analysis and failure detection is one element within the complex validation, monitoring and verification programmes that are adopted to assure food safety and consistent compliance with quality criteria defined within product specifications (including natural, sustainable or organic product claims). However, final product testing as a food safety, quality and food integrity control in itself is never effective enough and has to be combined with preventative, systems-based approaches to minimise the likelihood of product incidents occurring in the first place (Henson & Caswell, 1999; Zwietering et al., 2016). This is why the Hazard Analysis and Critical Control Point (HACCP) system has become mandatory along the post-harvest food supply chain in the European Union (EU) and the United States of America (USA). This has been the case since the 1990s when HACCP principles were adopted by the Codex Alimentarius Commission (CAC). In 2020, CAC adopted the revised Code of Practice (General Principles of Food Hygiene (CXC 1-1969) and its HACCP annex), which included additional guidance on allergen management, management commitment to food safety (including the maintenance of a positive food safety culture) and promoting continual improvement as science and technology evolves (FAO, 2023).

The outbreak of the COVID-19 pandemic and the implementation of pandemic-related restrictions affected the operation of food safety procedures along the supply chain, and consequently, exposed the need to reinforce good hygiene practices and systematically improve the traceability/product tracing systems in place, especially to enable quick and effective recall of noncompliant products. A convergence of crises (e.g. Ukraine/Russian conflict), COVID-19 and the climate crisis, makes the role of technology in ensuring effective traceability and transparency essential. Industry 4.0 technologies, including blockchain technology, cloud computing and the Internet of Things

(IoT) can have a specific role in making the agri-food supply chain 'smart enough' to address these contemporary challenges (Yadav et al., 2022). The use of these technologies could make the agri-food supply chain more data-driven, intelligent and agile, crucial both in routine supply operations, and in the event of an emergency situation, such as a product recall or a pandemic (Kowalska et al., 2023). In a dynamic and sometimes challenging market environment, business intelligence on potential geopolitical and supply shocks and their impact is important. Indeed, the use of technology to derive business intelligence and then use it effectively in decision-making is becoming more and more relevant, even essential.

This chapter will consider the developing role of technology in monitoring and verifying safety, quality and sustainability and delivering transparency in agri-food supply chains. The chapter is structured as follows: this section provides an introduction to the chapter. The following section (Section 2) considers the value of FTS. The next sections (Sections 3–6) then give an overview of smart technology-based solutions for effective traceability and greater transparency. These technologies include blockchain technology, the IoT, artificial intelligence (AI) and digital twins (DTs). Section 7 considers the issue of transparency in food supply chains and Section 8 concludes the chapter.

2 The value of food traceability systems

Traceability shows how materials, products, packaging and processing aids have moved through an individual business and across supply chains (Keogh et al., 2023) and is a critical aspect of food supply chain management (Sufiyan et al., 2019). Indeed, tracking and tracing of these items is both a legal and market requirement (Islam et al., 2021a). Regulation (EC) No 178/2002 of the European Parliament and of the Council of 28 January 2002 laying down the general principles and requirements of food law, establishing the European Food Safety Authority and laying down procedures in matters of food safety defines traceability as the ability to trace and follow a food, feed, food-producing animal or substance intended to be or expected to be incorporated into a food or feed through all stages of production, processing and distribution.

Traceability involves the transactional processes of *tracing* materials and ingredients forward to the final product and alternatively *tracking* food products back to initial ingredients via all production steps. Demonstrating traceability underpins a range of aspects such as consumer confidence, openness and transparency that in turn create trust relationships across food supply chains and ultimately to the consumer (Manning et al., 2022).

Regulation (EC) No 178/2002 introduces, in Section 2, the principles of transparency. These principles are as follows:

- **public consultation** (there shall be open and transparent public consultation, directly or through representative bodies, during the preparation, evaluation and revision of food law, except where the urgency of the matter does not allow it); and
- **public information** (public authorities shall take the appropriate steps to inform the general public where there are reasonable grounds to suspect that a food may present a risk to health).

Whilst transparency in this context focuses on public consultation and public information from a regulatory stance, transparency within the food supply chain can have additional perceived attributes. To disclose information not only about the intrinsic nature of the product but also the activities that have occurred at each production step extends far beyond the transactional nature of batch traceability. Transparency focuses on being able to map the whole supply chain and the tasks within the activities that occur at each step. Thus, traceability is an antecedent of supply chain transparency. Traceability information too is only one element of the information that may need to be publicly disclosed to stakeholders and consumers to demonstrate transparency, especially in the event of a food recall.

FTSs, their design, implementation and evolution, are influenced by a combination of different drivers including regulations and the need to capture information on food safety, quality and integrity, worker welfare, animal welfare, environmental protection and sustainability requirements (Islam & Cullen, 2021). Islam et al. (2022) differentiate between internal FTSs that operate within an organisation and external FTSs that operate outside of the organisation. These internal FTSs need to link together (interoperability) to create a chain of custody for a product. The benefits and barriers associated with the adoption and implementation of FTSs have been collated in Table 1. Barriers to the adoption of FTSs include a lack of resources, funding and capacity to develop systems awareness; a lack of governance and agreement on common standards; a lack of skills and knowledge about traceability and traceability technologies; inefficient data and knowledge management; a lack of clear implementation strategies and pathways and technology-scaling issues (Badia-Melis et al., 2015; Islam & Cullen, 2021).

Traceability has been achieved for many years using Enterprise Resource Planning (ERP) platforms that connect core business processes with strategic management of product and supply chain data (Martindale et al., 2018). However, these internal FTSs are not always interlinked between all actors within the supply chain so information loss can occur in either internal or external FTS

Table 1 Benefits and barriers associated with FTS

Benefits	Sources
Brand protection	Charan & Panghal (2018)
Collation of data	Alfian et al. (2017)
Compliance with legislation and certification requirements	Islam & Cullen (2021)
Effective functioning of supply chains	Islam & Cullen (2021)
Effective operational management, logistics and distribution management	Bourlakis & Bourlakis (2006); Jedermann et al. (2014); Mishra et al. (2020); Islam & Cullen (2021)
Faster more accurate collection, retention and reuse of data	Bertolini et al. (2006); Óskarsdóttir & Oddsson (2019); Feng et al. (2020)
Inventory management and resource efficiency	Lin et al. (2020); Islam & Cullen (2021)
Reduced data loss, data security	Karlsen et al. (2011); Wang et al. (2022)
Barriers	
Inefficient data and knowledge management	Badia-Melis et al. (2015); Islam & Cullen, (2021)
Information loss and lack of interoperability	Karlsen & Olsen (2016)
Lack of governance and agreement on common standards	Badia-Melis et al. (2015); Islam & Cullen (2021)
Lack of clear implementation strategies and pathways and technology scaling issues	Badia-Melis et al. (2015); Islam & Cullen (2021)
Lack of resources, funding and capacity to develop systems awareness	Badia-Melis et al. (2015); Islam & Cullen (2021)
Lack of skills and knowledge about traceability and traceability technologies	Badia-Melis et al. (2015); Islam & Cullen (2021)

(Karlsen & Olsen, 2016). FTSs range from being solely paper-based (analogue) to hybrid systems that are both paper-based and digitalised and then to fully digital systems that use information and communication technologies (ICTs) within what is described as Industry 4.0 or the fourth industrial revolution marked by the automation and digitalisation of production systems (Islam et al. 2021b). Most FTSs have been implemented through extensive use of ICTs (Olsen & Borit, 2018) that have contributed to the efficient collection, storage, processing, analysis and use of data.

However, ICTs are not able to avoid the potential for bias in the use of training datasets and the supply chain data itself. How can bias arise in training datasets for applications such as machine learning in food supply chain applications? The data used for training machine learning models will come from existing datasets, e.g. the data may be derived from both cognitive and non-cognitive influences on the food system itself, and there can be bias in terms of how the original data was collected (Manning et al., 2023). Examples of bias in the training datasets include human conscious and unconscious bias within their work environments that then introduces bias in the derived data, say e.g., in recruitment data. This may introduce 'fairness bias' in the training dataset through either behavioural bias or content production bias (Mehrabi et al., 2021). Bias in the sampling processes that have created the dataset mean that the group of farmers who have shared data may, or may not, be representative of all farmers in the supply chain. This is described as representation bias by Mehrabi et al. (2021).

FTSs require interoperability both within individual organisations and across supply chains to go beyond tracking individual traceable resource units (TRUs) to developing a full transactional history for the lifecycle of the product (Newsome et al., 2013). This is especially important if aspects of sustainability, such as greenhouse gas emissions (GHGEs), are being considered. The process of determining the environmental footprints of products 'from farm to fork' requires that data that is collected is usable and can be exchanged and collated to provide a meaningful dataset about the product. Interoperability has been defined as the 'ability of two or more systems or applications to exchange information and to mutually use the information that has been exchanged' (see ISO/IEC 22123-1:2023).

Information used in FTSs can come from a variety of sources and can be applied to different TRU within a supply chain. Granularity, in terms of traceability, describes specific aspects of the TRU including the *time element*, e.g., is the TRU one hour's or one day's production, and the *distance* in terms of the boundaries of the FTS from simply 'one step forward, one step back' through to full 'field to fork' traceability (Keogh et al., 2023). Granularity can also relate to the *depth of product identification* that is recorded by a given technology and is influenced by the memory capacity of the technologies used (Karlsen et al., 2011; Islam et al., 2021b).

Chain of custody extends beyond traceability and encompasses the controls and responsibilities at each step of the supply chain in terms of all inputs and outputs that occur together with the information that is transferred, and the impacts they all can have on food safety, quality and sustainability (Chopra, 2020). Indeed, chain of custody certification is becoming increasingly important in carbon certification schemes and tracing certified sustainability benefits associated with food ingredients and packaging materials through supply chains (Vidal et al., 2005; Santoso et al., 2019); aquaculture and worker welfare (Funge-Smith et al., 2007; Seo et al., 2015; Sparks et al., 2022); and with agri-food certification standards such as LEAFMarque, which promote climate positive farming and biodiversity (Pérez Perales et al., 2019). FTSs also have the potential to facilitate sustainability reporting which is becoming mandatory for certain food business operators from 2024 under the Directive (EU) 2022/2464 regarding corporate sustainability reporting. It can be reasonably expected that the adoption of blockchain technology-based FTSs should make it easier for enterprises to gather sufficient information from actors throughout their value chain, especially from small- or medium-sized suppliers and from suppliers in emerging markets and economies. Collection of such data has been identified as one of the difficulties that organisations may encounter in their sustainable reporting, as referred to in the preamble to the Directive (EU) 2022/2464 and then in Article 29b sustainability reporting standards. The next section of the chapter considers how, by making FTSs smart, through the application and integration of technology, this can more effectively deliver a product-related chain of custody, internally within a business and externally across a supply chain.

3 Making food traceability systems smart: blockchain and other distributed ledger technology solutions for food traceability

Smart systems can digitally connect supply chain actors from product development through production planning process to the consistent production of food products that are safe and comply with product specifications (Schmidt et al., 2022). These product specifications will include food safety, quality and sustainability requirements. Rabobank (2015) describes smart food systems as being 'more productive, more (globally) integrated, less wasteful and more profitable, more efficient in using resources to produce and deliver the food consumers' need, where and when they need and want it, making it [the system] more sustainable.' Adoption of ICTs has delivered business benefits across a range of business sectors through fast and accurate communication that supports 'improved decision making, product development, quality of product, and service often measured through increased customer satisfaction' (Schmidt et al., 2022, p. 276).

There are several smart technology solutions that facilitate smart food traceability, e.g. blockchain technology, cloud computing, smart contracts, IoT, AI (particularly machine learning), portable detection devices, smart indicators and sensors integrated on food packaging (Yu et al., 2022). Some of these are summarised in Table 2. This section looks at blockchain technology with following sections looking at other technology solutions.

Blockchain technology was invented by Satoshi Nakamoto in 2008 for use in the crypto currency bitcoin (Al Sherbaz et al., 2018; Baralla et al., 2019). Blockchain, which has gained significant research attention in the context of food supply chains, is one of the distributed ledger technologies (DLTs) solutions which can provide a cryptographically secure and immutable record of transactions and associated metadata (origin, contracts, process steps, environmental variations, microbial records etc.) linked across whole

Table 2 Advantages of technology use for traceability

Technology	Advantages	Source
Barcodes and tags e.g. 1D, 2D barcodes, QR codes, RFID tags	Technology that can travel with the batch and store the history of the batch through the supply chain	Alfian et al. (2020)
Blockchain	Immutable, tamperproof, records of transactions across a supply chain	Pearson et al. (2019)
Cloud computing	Remotely stored data which is accessible and shareable on demand	ISO/IEC 22123-1:2023
Digital information platforms	Permissioned access to traceability data	Tian (2018); Machado et al. (2020)
Digital twins	Modelling and mimicking of the physical environment supporting information flow and traceability practices	Autiosalo et al. (2020)
IoT (including RFID, barcodes, sensors and software)	Real-time collation of data for a particular batch(es)	IBM (2023)
Machine learning, algorithms and AI	Combining technologies can support traceability systems and decision tools	De Nadai Fernandes et al. (2020)
Smartphone applications	Opportunity for in-situ monitoring and verification activities to support food safety assessments and traceability systems	Lopez-Ruiz et al. (2014)

supply chains (Pearson et al., 2019). Using blockchain together with an IoT data collation smart system in a food supply chain means that a decentralised information platform can be built for all supply chain members to have permissioned access to the data (including enterprises, government bodies, third-party regulators, certification bodies and food safety auditors) (Tian, 2018; Machado et al., 2020). In the last few years, some major international standards regarding blockchain technology and other DLTs have been developed in order to facilitate the application of these new technologies, i.e. ISO 22739:2020 blockchain and DLTs – Vocabulary, ISO 23257:2022 Blockchain and DLTs – reference architecture among others.

DLTs such as blockchain, together with cloud computing and smart contract technologies, are opening up opportunities for more efficient FTSs, transparency and tamperproof record keeping (protected from unauthorised access, copying or exfiltration) (Pearson et al., 2019; Wang et al., 2022). DLT systems maintain a ledger in a distributed network and allow a wide range of parties to have a shared view of the ledger and to make their own changes to that ledger (ISO 23257:2022). A smart contract is 'a computer programme stored in a DLT system wherein the outcome of any execution of the programme is recorded on the distributed ledger'

(ISO/TR 23455:2019), so it is an example of an application of blockchain technology and other DLT systems.

Cloud computing is 'a paradigm for enabling network access to a scalable and elastic pool of shareable physical or virtual resources with self-service provisioning and administration on-demand' (ISO/IEC 22123-1:2023). IoT configurations support a range of processes at steps in the food supply chain including the management of environmental conditions (e.g. temperature monitoring), food safety and quality monitoring and verification, and assuring product traceability (Iftekhar & Cui, 2021). IoT and cloud computing, with the aid of wireless technologies, enable the collection and analysis of traceability data in real-time, thereby easing mitigation and investigation of foodborne outbreaks, food incidents and so on (Yu at al., 2022). However, often there is neither the digital infrastructure nor the hardware to facilitate automated data collection via blockchain technology at meso (business) and/or micro (individual) level where economic access to IoT is limited. This means that human data entry within these organisations is still necessary, and the threat of human error or deliberate fraud persists. Furthermore, at a macro supply chain level, application of blockchain technology may lack adequate data and input standards, creating governance issues and preventing learning from one use case being communicated elsewhere (Rogerson & Parry, 2020). Blockchain technology can support mitigation of food safety incidents regarding the protection of public health, corporate image, the potential loss of revenue etc. and fosters trust between supplier and consumer and other supply chain actors. The application of blockchain technology for FTSs should improve the performance of recall processes, thereby reducing food loss and waste, and resulting in a decrease in food supply chain management costs (Bosona & Gebresenbet, 2023).

DLTs including blockchain have the potential to reduce information asymmetry which is a major barrier to effective crisis management and food product recalls (Manning & Kowalska, 2021). Using blockchain technology enables data management to be divided among a number of supply chain members who can act as nodes in a peer-to-peer (P2P) blockchain network (Bosona & Gebresenbet, 2023). Inequality of access to data, and as a result inequality of respective bargaining power, can occur between the supply actor(s) that contribute data and information e.g. the farmer and the other actors that aggregate, control and share the data (Andrejevic, 2014; van de Burg et al., 2020). Thus, due to the increased ability to access multi-stakeholder data and information, blockchain technology-based traceability systems can support the transition towards more sustainable food systems.

The challenges in developing blockchain technology and applying it throughout the food industry across the world are, inter alia, labour and skills shortages, particularly in developing economies (Kshetri, 2019). Furthermore, there are scalability problems that might emerge over time. Thakur et al. (2020) suggest using offline channels to improve scalability by reducing the number of transactions that need to be recorded by blockchain technology. Further challenges in the use of blockchain technology in FTSs include high investment costs, limitations on data storage capacity and the possible leakage of private sensitive data (Rejeb at al., 2020; Bosona & Gebresenbet, 2023).

As conventional laboratory-based methods of detection of food adulterants and contaminants are time-consuming, labour and cost intensive, portable detection technologies have been emerging. These include array sensors, portable spectrometers, microfluidic lab-on-a-chip (LOC) and smartphone-based analysis (Yu et al., 2022). The growing prevalence of smartphones across the world gives a strong rationale for developing a smartphone-based platform to detect both food safety-related and fraudulent types of noncompliance, e.g. nitrite in meat products (Lopez-Ruiz et al., 2014), *Salmonella* spp. (Wang et al., 2019) and so on. Ma et al. (2022) suggested that there are three types of applications of smartphones related to food quality, safety and sustainability: firstly, food verification e.g. details on composition and date coding, secondly FTS and then personalised diet guidance. All food supply chain actors need cost-effective rapid methods of detection, and such technology could provide opportunity for food businesses and consumers to trace the origin, production date etc. of food 'from farm to fork'. The use of smartphones in food inspection could also allow food producers to meet the varying needs of their customers and clients, official inspectors, customs service etc. In this regard, the QR code is a leading option for communication between various stakeholders within and outside of food supply chains. It is important that the smartphone-based FTSs have the function of recording each traceability link and traceability chain in the network in order to implement product recall processes effectively (Ma et al., 2022).

4 Making food traceability system smart: Internet of Things

IoT refers to a network of 'physical devices [smart objects], vehicles, appliances and other physical objects that are embedded with sensors, software and network connectivity that allows them to collect and share data' (IBM, 2023). These smart objects can include identification and traceability technologies such as radio-frequency identification (RFID) tags, barcodes, smart indicators on packaging and sensors for temperature, environment and product assessment. The 'things' in IoT can include a wide variety of technologies from a smart-watch to a control system that contains sensors (e.g. temperature, location, light and motion) to links to machine learning applications and algorithms within models that can use the data collected. It is essential to have

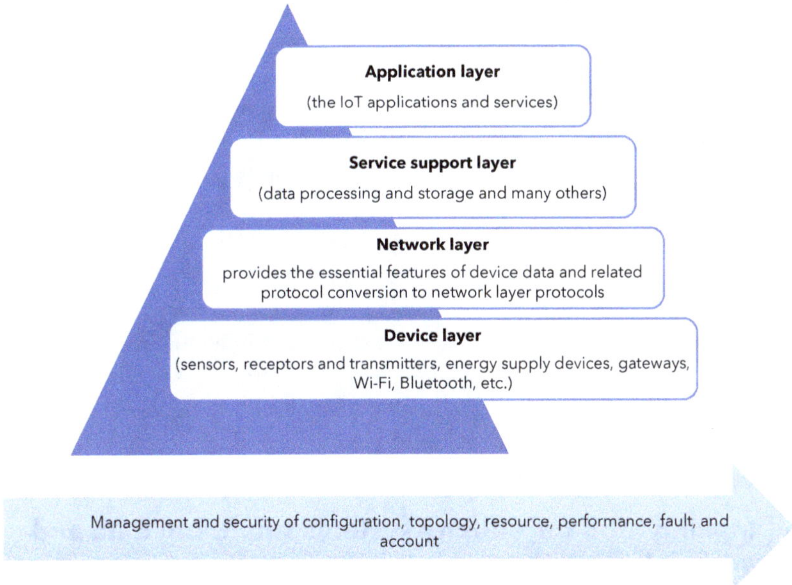

Figure 1 IoT architecture. Source: Adapted from: Bouzembrak et al. (2019).

a communication medium (Bluetooth, Radio-Frequency Identification (RFID), 4G and so on), which facilitates communication with other people or machines and the computing resources.

There are several layers in the architecture within an IoT system (Figure 1) including the device layer, network layer, service support layer, application layer, and management and security layer (Bouzembrak et al., 2019). IoT increases the regularity, speed and accuracy with which the information is collected and recorded. It provides for effective data communication and supports the benefits of using blockchain technology in food supply chains (Rejeb et al., 2020).

Alfian et al. (2020) proposed a traceability system that uses RFID to track and trace perishable food, and IoT sensors to measure temperature and humidity during storage and transportation. Shahbazi and Byun (2021) have proposed the development of blockchain machine learning-based FTSs in the perishable food e-commerce business. However, the adoption of IoT in food supply chains to promote real-time monitoring and field-to-fork traceability also creates greater vulnerability to cyberattacks (Latino & Menegoli, 2022).

A cyberattack at one processing step in the food supply chain can have an amplifying effect causing disruption of the whole supply chain (Cartwright & Cartwright, 2023), which can take weeks, even months to put right if digital systems fail. The cyber threats that can occur in food supply chains include malware (ransomware and worms) which account for 38% of attacks, IoT-based attacks (25%) where IoT devices are an entry point for attack, distributed denial of service or DDoS (19%) and code injection attacks (SQL Injection, XSS), identity-based attacks (man in the middle) and phishing attacks all being 6% each (Rani et al., 2023; Manning & Kowalska, 2023). Combined attacks can occur involving ransomware combined with DDoS (Paloaltonetworks 2023). More generally, as digitalisation increases the attack surface extends (FBI, 2021). It is crucial to implement managerial, procedural or other security (counter)measures designed to prevent and mitigate cyberattacks, and/or reduce the likelihood of food defence threats being realised. These include verification protocols, cybersecurity training programmes, password protection, virus software, updates and patches (Manning & Kowalska, 2023; Mitenius et al., 2014; van Ruth et al., 2017; Voyiatzis & Serpanos 2022), among others. An effective food defence strategy at organisational and industry level is a necessity in order to reduce vulnerability to cyberattacks.

5 Making food traceability systems smart: big data and artificial intelligence

Smart FTSs consist of three main elements: IoT, big data and AI which have a role in shaping the future of agricultural and food systems. Increasingly, interconnected sensors and other devices are used for the monitoring of food safety, quality and sustainability and for collecting a wide range of data for reasons of traceability and logistics. The records relating to primary production, food processing, transport and storage of agri-food products, and food retail are increasingly switching to digital records that are more accessible for data processing. AI and big data are particularly useful in this regard (Misra et al., 2022). IoT is a crucial tool for automated data capture and transfer. Verified and standardised smart contract data can be entered into the blockchain system and stored in the chronological order in which the transactions were completed (Bosona & Gebresenbet, 2023).

De Mauro et al. (2015, p. 103) propose the following definition of big data: 'Big Data represents the information assets characterised by such a High Volume, Velocity and Variety to require specific Technology and Analytical Methods for its transformation into Value,' where velocity is the speed at which data is generated, processed and analysed. To meet the challenge of analysing the extremely large and fast-growing data, AI and its subfield, machine learning, were proposed as viable options by the U.S. Food and Drug Administration (FDA) in a 2019 statement (Sharpless & Yiannas, 2019). AI has been described as the science and engineering that is combined to develop 'intelligence' through either using software on its own, or software built into hardware, that can sense, interact, learn and provide feedback or make predictions (Friedlander & Zoellner, 2020). Deng et al. (2021, p. 515) explain that AI 'simulates behaviours of intelligent beings, such as learning, problem-solving, and decision-making'. Machine learning utilises data and algorithms to mimic the way that humans learn with accuracy improving over time as the application is used. Machine learning algorithms, including Bayesian networks, Neural networks and so on, learn from big data and support making predictions and/or decisions on the basis of the data (Marvin et al., 2017). Machine learning does not attempt to programme an exhaustive list of explicit instructions or rules, instead it learns from examples and generalises to new cases based on their closeness to learned examples or alternatively, trains a model with data to learn its parameters through optimisation and then makes predictions using new data (Deng et al. 2021). The application of machine learning algorithms in Brazilian beef traceability has been considered by De Nadai Fernandes et al. (2020) as an opportunity to develop and maintain reliable and internationally accepted FTS.

6 Making food traceability systems smart: digital twins

Agriculture 4.0 is concerned with making use of automation, digital technologies and AI to improve efficiency, reduce waste and increase yields (Valle & Kienzle, 2020). DTs promise to deliver significant advantages in digital assessment, management and improvement of food systems. Often said to have been first introduced by the United States (US) National Aeronautics and Space Administration (NASA), the DT concept has been utilised for exploring industry 4.0 through the combination of physical operations and increasing automation with smart digital technology to provide all the tools needed to deliver a smart manufacturing environment. It is believed that the DT concept also offers great opportunities for the agricultural sector in delivering smart agriculture, especially climate smart agriculture.

DTs are a virtual representation of a physical object developed through collecting and manipulating data from sensors in real time in order to aid decisionmaking and reduce the workload of the user. IoT is very important in the world of DTs where devices can communicate via a network, allowing the flow of data from the physical twin (animal, plant, field, store, machine, and farm) to the DT. The definition of a DT changes depending on the context in which it is applied. Reviewing existing literature suggests that a DT can be a construct (Grieves & Vickers, 2017), a virtual model of a process, product or service (Marr, 2017) or reflect a real-world object (Qi et al., 2021), which it is connected to. A DT is also accurate (Autiosalo et al., 2020), updated in real-time (Defraeye et al., 2021), is more than simply a model (Harper et al., 2019; Autiosalo et al., 2020; Wright & Davidson, 2020) and exists simultaneously alongside its physical twin whilst containing all necessary information about it (Autiosalo et al., 2020; Qi et al., 2021).

DTs are technologically advanced and intelligent in some way, perhaps using machine learning or mechanistic algorithms (Autiosalo et al., 2020). DTs require a user interface where the user can access all elements of the DT such as computer-assisted design (CAD) models, tables, plots or live video recordings (Palit & Datta, 2017). Some DTs focus more on modelling and aim to mimic the exact physical environment with regard to a set of scenarios, whereas other DTs focus more on the information management aspects of smart systems and information flow (Autiosalo et al., 2020). In terms of lifecycles, some DTs support recollection, i.e. they keep a history about the physical object which can be interrogated (Verdouw et al., 2021), essential to provide traceability and transparency.

A DT can be made of digital building blocks (DTBs) and digital twin features (DTFs), which are the different technical functionalities of the DT, and a network of DTs can be created where each DT acts as a node within the network, where functionalities include security, data storage, user interface, simulation models, data analysis, AI functions and computation (Autiosalo et al., 2020). Developing a complex, highly accurate DT is not an easy task, and many challenges exist (Tao & Qi, 2019). Moreover, there is a high upfront cost, and extensive computational resources are required (Autiosalo et al., 2020). These factors may mean that DTs are only accessible to a small number of large companies with the resources required unless open-access repositories can be made available (Palit & Datta, 2017). Further challenges include the type of data and the challenges of merging different types of data, the potential for capturing the wrong types of data that cannot be combined to form metadata or is at the wrong frequency, is discrete rather than continuous, or lacks the level of accuracy required (Tao & Qi, 2019). There may be privacy issues associated with the data which cannot be overcome reducing the ability for it to be openly shared (Tao & Qi, 2019), or consent may be given to share data and then that permission is withdrawn at a later date (Tigard, 2021).

In order to deliver traceability and transparency, different IoT devices and technologies must be able to communicate and transport data to allow the DT to exist alongside its physical counterpart in the real world. In larger scale DTs, interoperability becomes even more fundamental as systems and data owned by different parties may have to come together. The use of application programming interfaces (APIs) can ensure that different organisations (e.g. different companies along the supply chain) can access or input the data into a DT or can connect their own DTs together in a network. Blockchain as an example of DLT may be of value when organisations along the supply chain need to collate information about the physical object during its lifecycle and make this available to others either in an open access system or within a private data consortium. Digital platforms combine software and technology and allow the exchange of data and information from a variety of sources, organisations and if agreed with a wider range of stakeholders. However, with digital agriculture platforms, concerns do arise with regard to data quality, data privacy, intellectual property and data ownership and how to optimise interoperability (Runck et al., 2022).

When considering combining the resources from different stakeholders, it can be technically difficult to merge the data. Moshrefzadeh et al. (2020) explored the problems associated with interoperability when creating a DT of the agricultural landscape. They highlight a lack of transparency that can be an obstacle as different stakeholders may hold different information about the same physical object but not know what information each has or how it can be combined or interconnected. Further, they may define different objects in a dissimilar way or may record the same object or event in different units, frequencies or in different ways complicating the process of pooling data together. Moshrefzadeh et al. (2020) suggest that a standard infrastructure is developed to store data that is used by all stakeholders, where objects have stable identifiers and data are always labelled in the same way.

Digital future farms have been proposed for arable and dairy (van Evert et al., 2021), and DTs for livestock farming (Jo et al., 2018), aquaponics (Ahmed et al., 2019; Ghandar et al., 2021) and hydroponics (Sreedevi & Kumar, 2020). DTs have also been proposed for smart applications in food ingredients manufacture (Koulouris et al., 2021). Examples of scenarios where DTs can be applied in the agri-food context include management of nitrogen application (Cesco et al., 2023); smart irrigation on farm (Alves et al., 2023), assessing crop yields (Angin et al., 2020); real-time monitoring of energy consumption, system failure analysis and prediction, optimisation and as a hub or technology integration tool (Pylianidis et al., 2021); and to reduce food waste and losses (Tagliavini et al., 2019). For example, a DT can be developed using data from multiple sensors (temperature, humidity, soil moisture, soil nutrient levels, pH), using wireless sensor networks and multispectral imaging to detect crop diseases and nutrient deficiencies. Data can be obtained from smart irrigation equipment and multispectral imaging for signs of disease, combined with drone imagery and satellite data (see Angin et al., 2020 for a wider explanation).

DTs have been proposed for traceability post-harvest (Dyck et al., 2023) and for traceability and dynamic process control within processing (Liu et al., 2021) and real-time monitoring and traceability in aircraft assembly. A DT, in this context, must also be able to integrate different IoT devices in multiple stages of the supply chain. Power imbalances within supply chain relationships, often driven by information asymmetry, can cause unease during data sharing. For example, if large agribusinesses or retailers have more digital expertise and resources than farmers this can mean they are more powerful in the supply relationship when the farmers share data (van de Burg et al., 2020). It is essential for the future of DTs in food supply chains that farming enterprises have access to the data required, in the format they can use, to improve efficiency, resource management and sustainability of farm practices. A trust framework may be necessary to ensure that all actors conduct themselves in an ethical way (Temoshok & Abruzzi, 2018; Brewer et al., 2021), when sharing such data.

7 Transparency in food supply chains

Existing systems of communication have a crucial role in ensuring greater traceability and transparency of information sharing between manufacturers, distributors, consumers, food control institutions and other stakeholders. These include:

- the Rapid Alert System for Food and Feed (RASFF);
- the Trade Control and Expert System (TRACES); and
- the European Union Notification System for Plant Health Interceptions (EUROPHYT).

These systems have been integrated into Information Management System for Official Controls (IMSOC) in order to make information exchange timely and efficient. Both the EUROPHYT and TRACES databases run by the Directorate General for Health and Food Safety (DG SANTE) of the European Commission facilitate the exchange of data and documents between trading parties and control authorities, contributing to increased transparency.

TRACES has been compulsory since 2015. The electronic certification capability of TRACES NT (New Technology) has revolutionised traditional sanitary and phytosanitary certification practices. It has enabled both EU and non-EU authorities to stamp digitally official documents and certificates required for the importation of animals, animal products, food, feed and plants into the EU, and the intra-EU trade and EU exports of animals and certain animal products (ESCAP, 2023). The electronic transmission of documents makes international food trade more transparent and sustainable regarding saving paper and energy, lower emissions etc. Digitalisation means reduced administrative burden for food business operators and it may contribute to the development of entrepreneurship and promote fair practices along FSC. Thus, the future of paper-based FTS is neither reliable nor sustainable. Furthermore, responsible use of the newest technology (DLTs, IoT, AI and so on) should serve to monitor and verify food quality, safety and sustainability at a regulatory level.

Transparency in a supply chain is the extent to which all its stakeholders have a shared understanding of, and access to, the product-related information that they request without loss, noise, delay and distortion (Hofstede, 2003; Deimel et al., 2008). Transparency in agri-food supply chains can be demonstrated via the application of organisational arrangements for information flow and exchange within organisations and between actors (Pant et al., 2015). Supply chain transparency is both an important and a difficult goal to achieve (Abeyratne & Monfared, 2016: Astlll et al., 2019). Furthermore, organisations are faced with the need to choose between retaining some information as confidential and thus not shared, e.g. the company secrets that are the source of their competitiveness and, in other contexts, sharing data with others and its meaning and acting with full transparency. Supply chain transparency is delivered through enhancing supply chain visibility and creating a chain of custody through effective end-to-end supply chain information sharing (Smart, 2008). Multiple factors contribute to the drivers for greater transparency in supply chains such as the need to have a robust chain of custody so sustainability claims can be verified; for effective management and optimisation of resource use and in so doing reducing food loss and food waste; for efficient management of risks and recalls; and prevention of foodborne illness outbreaks and satisfying consumers' changing demand (Astill et al., 2019). It has been argued that blockchain technology-based FTS can increase transparency (Gupta & Shankar, 2023). Transparency in essence is providing greater visibility of FTSs, the data associated with them, and the implications that can be drawn from the data (Sunny et al., 2020).

Transparency is therefore associated with the degree of accessibility there is individually, and collectively, for supply chain actors to information and data. Transparency in this context increases productivity, improves services to supply chain partners and consumers and can reduce transaction and operational costs especially through the use of blockchain technology (Centobelli et al., 2021). Centobelli et al. (2021) differentiate between trust, traceability and transparency in the context of blockchain technology, where trust is promoted through features such as smart contracts, immutable recording of transactions and authentication processes; traceability through the capabilities of tracking and tracing; and transparency in terms of accessibility to information and data (Centobelli et al., 2021). Whilst transparency can be defined through governance structures and trust frameworks (Brewer et al., 2021), transparency is also framed by normative standards such as responsibility, rights, justice and accountability (Mol, 2015). Indeed, Jakku et al. (2019) argue that a lack of trust and transparency, especially with regard to data ownership, and who benefits from integration of smart farming technologies could limit the implementation of smart farming systems.

8 Conclusion

Effective FTSs are crucial for maintaining market position and achieving competitive advantage (Islam et al., 2021a). A significant enhancement of FTSs requires the use of the new technologies, such as blockchain technology and other DLTs, IoT, AI, DTs, fifth-generation (5G) mobile communication systems and so on. AI and the use of big data contribute to decision-making especially in situations of great uncertainty (Jahani et al., 2023). Automated data capture and transfer and digitalisation of the food system should help the conservation of supply chain resources and improve organisational and supply chain energy efficiency. Agriculture 4.0 has the technological capacity to deliver climate smart agriculture. Developing smart FTS, e.g. blockchain machine learning-based and IoT enabled FTS, should create deeper trust relationships within food supply chains although further research is needed to explore this.

There are challenges in developing smart FTSs, which include insufficient financial and/or human resources, limited data storage capacity and increased vulnerability to cyberattacks that need to be regularly monitored, evaluated and overcome. There are also other problems associated with the data collated across supply chains, i.e. scalability, different types of data that needs to be merged, privacy issues etc. Ethical conduct of farmers and food business operators is crucial to prevent unauthorised access to private data and to ensure the ethical dissemination or use of any shared data. Efficient FTSs provide more transparency across food supply chains which is increasingly demanded by business partners and consumers. Transparency creates essential trust relationships across food supply chains from farm to fork, and it is these trust relationships that underpin sustainable development of business enterprises and continued delivery of business customer and consumer requirements.

Improving transparency about activities and operations that occur in food supply chains through building chains of custody can build value (product value, reputational value and social value). Developing digital chains of custody could deliver more equitable sharing of data and information, efficient assessment and mitigation of complex risk scenarios and, if required, effective management of product recalls. Digital chains of custody can also optimise the use of resources (and reduce emissions) and improve the effectiveness of validation and verification of product claims regarding safety, quality and other environmental and social criteria.

9 References

Abeyratne, S. A. and Monfared, R. P. (2016). Blockchain ready manufacturing supply chain using distributed ledger. *International Journal of Research in Engineering and Technology*, 5(9), 1–10.

Ahmed, A., Zulfiqar, S., Ghandar, A., Chen, Y., Hanai, M. and Theodoropoulos, G. (2019). Digital twin technology for aquaponics: Towards optimizing food production with dynamic data driven application systems. In Tan, G., Lehmann, A., Teo, Y., Cai, W. (eds), *Asian Simulation Conference* (pp. 3–14). Singapore: Springer.

Alfian, G., Rhee, J., Ahn, H., Lee, J., Farooq, U., Ijaz, M. F. and Syaekhoni, M. A. (2017). Integration of RFID, wireless sensor networks, and data mining in an e-pedigree food traceability system. *Journal of Food Engineering, 212*, 65–75.

Alfian, G., Syafrudin, M., Farooq, U., Ma'arif, M. -R., Syaekhoni, M. A., Fitriyani, N. L, Lee, J. and Rhee, J. (2020). Improving efficiency of RFID-based traceability system for perishable food by utilizing IoT sensors and machine learning model. *Food Control, 110*, 107016. https://doi.org/10.1016/j.foodcont.2019.107016.

Al-Sherbaz, A., Turner, S. J., Dohan, M. and Clarke, E. (2018, March). Blockchain: no thunder no wonder. In *British Computing Society (BCS) Northampton Talk*. https://doi.org/10.6084/m9.figshare.5979625.v2

Alves, R. G., Maia, R. F. and Lima, F. (2023). Development of a digital twin for smart farming: irrigation management system for water saving. *Journal of Cleaner Production, 388*, 135920.

Andrejevic, M. (2014). Big data, big questions| the big data divide. *International Journal of Communication, 8*, 17.

Angin, P., Anisi, M. H., Göksel, F., Gürsoy, C. and Büyükgülcü, A. (2020). AgriLoRa: a digital twin framework for smart agriculture. *Journal of Wireless Mobile Networks, Ubiquitous Computing, and Dependable Applications, 11*(4), 77–96.

Astill, J., Dara, R. A., Campbell, M., Farber, J. M., Fraser, E. D., Sharif, S. and Yada, R. Y. (2019). Transparency in food supply chains: ā review of enabling technology solutions. *Trends in Food Science & Technology, 91*, 240–247.

Autiosalo, J., Vepsäläinen, J., Viitala, R. and Tammi, K. (2020). A feature-based framework for structuring industrial digital twins. *IEEE Access, 8*, 1193–1208.

Badia-Melis, R., Mishra, P. and Ruiz-García, L. (2015). Food traceability: new trends and recent advances. A review. *Food Control, 57*, 393–401.

Baralla, G., Ibba, S., Marchesi, M., Tonelli, R. and Missineo, S. (2019). A blockchain based system to ensure transparency and reliability in food supply chain. In *Euro-Par 2018: Parallel Processing Workshops: Euro-Par 2018 International Workshops, Turin, Italy, August 27-28, 2018, Revised Selected Papers24* (pp. 379–391). Springer International Publishing.

Bertolini, M., Bevilacqua, M. and Massini, R. (2006). FMECA approach to product traceability in the food industry. *Food Control, 17*(2), 137–145.

Bosona, T. and Gebresenbet, G. (2023). The role of blockchain technology in promoting traceability systems in agri-food production and supply chains. *Sensors, 23*, 5342. https://doi.org/10.3390/s23115342.

Bourlakis, M. and Bourlakis, C. (2006). Integrating logistics and information technology strategies for sustainable competitive advantage. *Journal of Enterprise Information Management, 19*(4), 389–402.

Bouzembrak, Y., Klüche, M., Anand Gavai, A., Hans J. P. and Marvin, H. J. P. (2019). Internet of things in food safety: literature review and a bibliometric analysis. *Trends in Food Science & Technology, 94*, 54–64. https://doi.org/10.1016/j.tifs.2019.11.002.

Brewer, S., Pearson, S., Maull, R., Godsiff, P., Frey, J.G, Zisman, A., Parr, G., McMillan, A., Cameron, S., Blackmore, H., Manning, L. and Bidaut, L. (2021). A data trust and trust framework for digital food systems. *Nature Food*. https://doi.org/10.1038/s43016-021-00346-1.

Cartwright, A., & Cartwright, E. (2023). The economics of ransomware attacks on integrated supply chain networks. *Digital Threats: Research and Practice*, *4*(4), 1–14.

Centobelli, P., Cerchione, R., Del Vecchio, P., Oropallo, E. and Secundo, G. (2021). Blockchain technology for bridging trust, traceability and transparency in circular supply chain. *Information & Management*, 103508. https://doi.org/10.1016/j.im.2021.103508.

Cesco, S., Sambo, P., Borin, M., Basso, B., Orzes, G. and Mazzetto, F. (2023). Smart agriculture and digital twins: applications and challenges in a vision of sustainability. *European Journal of Agronomy*, *146*, 126809.

Charan, S. and Panghal, A. N. I. L. (2018). Importance of traceability in food supply chain for brand protection and food safety systems implementation. *Ann Biol*, *34*(2), 111–118.

Chopra, A. (2020, June). Blockchain technology in food industry ecosystem. In *IOP Conference Series: Materials Science and Engineering* (Vol. 872, No. 1, p. 012005). IOP Publishing. https://10.1088/1757-899X/872/1/012005.

De Mauro, A., Greco, M. and Grimaldi, M. (2015). What is big data? A consensual definition and a review of key research topics. *AIP Conference Proceedings,1644*, 97–104. https://doi.org/10.1063/1.4907823

De Nadai Fernandes, E. A., Sarriés, G. A., Bacchi, M. A., Mazola, Y. T., Gonzaga, C. L. and Sarriés, S. R. V. (2020). Trace elements and machine learning for Brazilian beef traceability. *Food Chemistry, 333*, 127462. https://doi.org/10.1016/j.foodchem.2020.127462.

Defraeye, T., Shrivastava, C., Berry, T., Verboven, P., Onwude, D., Schudel, S., Buhlmann, A., Cronje, P. and Rossi, R. M. (2021). Digital twins are coming: will we need them in supply chains of fresh horticultural produce? *Trends in Food Science and Technology, 109*, 245–258. https://doi.org/10.1016/j.tifs.2021.01.025.

Deimel, M., Frentrup, M. and Theuvsen, L. (2008). Transparency in food supply chains: empirical results from German pig and dairy production. *Journal on Chain and Network Science*, *8*(1), 21–32.

Deng, X., Cao, S. and Horn, A. L. (2021). Emerging applications of machine learning in food safety. *Annual Review of Food Science and Technology*, *12*(1), 513–538.

Directive (EU) 2022/2464 of the European Parliament and of the Council of 14 December 2022 amending Regulation (EU) No 537/2014, Directive 2004/109/EC, Directive 2006/43/EC and Directive 2013/34/EU, as regards corporate sustainability reporting (OJ L 322, 16.12.2022, p. 15–80).

Dyck, G., Hawley, E., Hildebrand, K. and Paliwal, J. (2023). Digital twins: a novel traceability concept for post-harvest handling. *Smart Agricultural Technology*, *3*, 100079.

ESCAP. (2023). TRACES – Trade Control and Expert System. Available at: https://www.digitalizetrade.org/projects/traces-trade-control-and-expert-system (June 16, 2023).

FAO. (2023). *Codex Alimentarius. International Food Standards. General Principles of Food Hygiene CXC 1-1969*. Available at: https://www.fao.org/fao-who-codexalimentarius/sh-proxy/en/?lnk=1&url=https%253A%252F%252Fworkspace.fao.org%252Fsites%252Fcodex%252FStandards%252FCXC%2B1-1969%252FCXC_001e.pdf (June 13, 2023).

Federal Bureau of Investigation (FBI). (2021). *Private Industry Notification: Cyber Criminal Actors targeting the Food and Agriculture Sector with Ransomware Attacks. 1 September 2022.* Available at: https://s3.documentcloud.org/documents/21053957/fbi-tlp-white-pin-cyber-criminal-actors-targeting-food-agriculture-sector-ansomware-attacks-9-1-21.pdf (August 8, 2022).

Feng, H., Wang, X., Duan, Y., Zhang, J. and Zhang, X. (2020). Applying blockchain technology to improve agri-food traceability: a review of development methods, benefits and challenges. *Journal of Cleaner Production*, *260*, 121031.

Food & Agribusiness Research and Advisory. (2015) *Building a Smarter Food System*. Rabobank. Available at: http://agfundernews.com/wp-content/uploads/2015/10/Rabobank-Building-Smarter-Food-Systems-report.pdf.

Friedlander, A. and Zoellner, C. (2020). Artificial intelligence opportunities to improve food safety at retail. *Food Protection Trends*, *40*(4), 272–278.

Funge-Smith, S., Corsin, F. and Clausen, J. (2007, March). Overview of aquaculture certification. In *Expert Workshop on Guidelines for Aquaculture Certification* (pp. 27–30).

GFSI. (2018). *Tackling Food Fraud through Food Safety Management Systems*. Available at: https://mygfsi.com/wp-content/uploads/2019/09/Food-Fraud-GFSI-Technical-Document.pdf (June 19, 2023).

Ghandar, A., Ahmed, A., Zulfiqar, S., Hua, Z., Hanai, M. and Theodoropoulos, G. (2021). A decision support system for urban agriculture using digital twin: a case study with aquaponics. *IEEE Access*, *9*, 35691–35708. https://doi.org/10.1109/ACCESS.2021.3061722.

Grieves, M. and Vickers, J. (2017). Digital twin: Mitigating unpredictable, undesirable emergent behavior in complex systems. In Kahlen, J., Flumerfelt, S., and Alves, A. (eds), *Transdisciplinary Perspectives on Complex Systems* (pp. 85-113). Cham: Springer.

Gupta, R. and Shankar, R. (2023). Managing food security using blockchain-enabled traceability system. *Benchmarking: An International Journal, Vol. ahead-of-print No. Ahead-of-print.* https://doi.org/10.1108/BIJ-01-2022-0029

Harper, K., Ganz, C. and Malakuti, S. (2019). Digital twin architecture and standards. *IIC Journal of Innovation*, *12*, 72–83. https://www.iiconsortium.org/news/joi-articles/2019-November-JoI-Digital-Twin-Architecture-and-Standards.pdf

Henson, S. and Caswell, J. (1999). Food safety regulation: an overview of contemporary issues. *Food Policy*, *24*(6), 589–603.

Hofstede, G. J. (2003). Transparency in netchains. *Information Technology for a Better Agri-Food Sector, Environment and Rural Living* (pp. 17–29). Debrecen, Hungary: Debrecen University.

IBM. (2023). What is Internet of Things? Available at: https://www.ibm.com/topics/internet-of-things (November 12, 2023).

Iftekhar, A. and Cui, X. (2021). Blockchain-based traceability system that ensures food safety measures to protect consumer safety and COVID-19 free supply chains. *Foods*, *10*(6), 1289.

Islam, S. and Cullen, J. M. (2021). Food traceability: a generic theoretical framework. *Food Control*, *123*, 107848. https://doi.org/10.1016/j.foodcont.2020.107848

Islam, S., Cullen, J. M. and Manning, L. (2021a). Visualising food traceability systems: a novel system architecture for mapping material and information flow. *Trends in Food Science & Technology*, *112*, 708–719.

Islam, S., Manning, L. and Cullen, J. M. (2021b). A hybrid traceability technology selection approach for sustainable food supply chains. *Sustainability*, *13*(16), 9385.

Islam, S., Manning, L. and Cullen, J. M. (2022). Systematic assessment of food traceability information loss: a case study of the Bangladesh export shrimp supply chain. *Food Control*, *142*, 109257.

ISO 22739:2020 Blockchain and distributed ledger technologies – Vocabulary

ISO 23257:2022 Blockchain and distributed ledger technologies – Reference architecture

ISO/IEC 22123-1:2023(en) Information technology – Cloud computing – Part 1: Vocabulary

ISO/TR 23455:2019 Blockchain and distributed ledger technologies – Overview of and interactions between smart contracts in blockchain and distributed ledger technology systems

Jahani, H., Jain, R. & Ivanov, D. (2023). Data science and big data analytics: a systematic review of methodologies used in the supply chain and logistics research. *Annals of Operations Research*. https://doi.org/10.1007/s10479-023-05390-7

Jakku, E., Taylor, B., Fleming, A., Mason, C., Fielke, S., Sounness, C. and Thorburn, P. (2019). "If they don't tell us what they do with it, why would we trust them?" Trust, transparency and benefit-sharing in Smart Farming. *NJAS-Wageningen Journal of Life Sciences*, *90*, 100285.

Jedermann, R., Nicometo, M., Uysal, I. and Lang, W. (2014). Reducing food losses by intelligent food logistics. *Philosophical Transactions of the Royal Society A: Mathematical, Physical and Engineering Sciences*, *372*(2017), 20130302.

Jo, S. K., Park, D. H., Park, H. and Kim, S. H. (2018). Smart livestock farms using digital twin: feasibility study. In *2018 International Conference on Information and Communication Technology Convergence (ICTC)* (pp. 1461–1463). IEEE. https://doi.org/10.1109/ICTC.2018.8539516.

Karlsen, K. M., Donnelly, K. M. and Olsen, P. (2011). Granularity and its importance for traceability in a farmed salmon supply chain. *Journal of Food Engineering*, *102*(1), 1–8.

Keogh, J., Simske, S. and Manning, L. (2023). Global food chain traceability – reflections on the past, present and insights into future directions. *Food Safety Magazine*. https://digitaledition.food-safety.com/april-may-2023/column-supply-chain/.

Koulouris, A., Misailidis, N. and Petrides, D. (2021). Applications of process and digital twin models for production simulation and scheduling in the manufacturing of food ingredients and products. *Food and Bioproducts Processing*, *126*, 317–333. https://doi.org/10.1016/j.fbp.2021.01.016.

Kshetri, N. (2019). Blockchain and the economics of food safety. *IEEE IT Professional*, *21*(3), 63–66. https://doi.org/10.1109/MITP.2019.2906761.

Latino, M. E. and, Menegoli, M. (2022). Cybersecurity in the food and beverage industry: a reference framework. *Computers in Industry*, *141*, 103702. https://doi.org/10.1016/j.compind.2022.103702.

Li, Y., Liu, W., Zhang, Y., Zhang, W., Gao, C., Chen, Q. and Ji, Y. (2023). Interactive real-time monitoring and information traceability for complex aircraft assembly field based on digital twin. *IEEE Transactions on Industrial Informatics*. https://ieeexplore.ieee.org/document/10008048.

Lin, K., Chavalarias, D., Panahi, M., Yeh, T., Takimoto, K. and Mizoguchi, M. (2020). Mobile-based traceability system for sustainable food supply networks. *Nature Food*, *1*(11), 673–679.

Liu, J., Cao, X., Zhou, H., Li, L., Liu, X., Zhao, P. and Dong, J. (2021). A digital twin-driven approach towards traceability and dynamic control for processing quality. *Advanced Engineering Informatics*, *50*, 101395.

Lopez-Ruiz, N., Curto, V.F., Erenas, M. M., Benito-Lopez, F., Diamond, D., Palma, A. J. and Capitan-Vallvey, L. F. (2014). Smartphone-based simultaneous pH and nitrite colorimetric determination for paper microfluidic devices. *Analytical Chemistry, 86*(19), 9554–9562. https://doi.org/10.1021/ac5019205.

Karlsen, K. M. and Olsen, P. (2016). Problems and implementation hurdles in food traceability. In Espiñeira, M. and Santaclara, F. J. (eds), *Advances in Food Traceability Techniques and Technologies* (pp. 35–46). Cambridge: Woodhead Publishing.

Kowalska, A., Lingham, S., Maye, D. and Manning, L. (2023). Food insecurity: is leagility a potential remedy? *Foods*, *12*(16), 3138. https://doi.org/10.3390/foods 12163138.

Ma, T., Wang, H., Wei, M., Lan, T., Wang, J., Bao, S., Ge, Q., Fang, Y. and Sun, X., (2022). Application of smart-phone use in rapid food detection, food traceability systems, and personalized diet guidance, making our diet more health. *Food Research International*, *152*, 110918.

Machado, T. B., Ricciardi, L. and Oliveira, M. B. P. (2020). Blockchain technology for the management of food sciences researches. *Trends in Food Science & Technology, 102*, 261–270.

Manning, L., Brewer, S., Craigon, P.J., Frey, J., Gutierrez, A., Jacobs, N., Kanza, S., Munday, S., Sacks, J. and Pearson, S. (2023). Reflexive governance architectures: considering the ethical implications of autonomous technology adoption in food supply chains. *Trends in Food Science & Technology*, *133*, 114–126.

Manning, L. and Kowalska, A. (2023). The threat of ransomware in the food supply chain: a challenge for food defence. *Trends in Organized Crime*, 1–29. https://doi.org/10 .1007/s12117-023-09516-y.

Manning, L., Brewer, S., Craigon, P.J., Frey, J., Gutierrez, A., Jacobs, N., Kanza, S., Munday, S., Sacks, J. and Pearson, S. (2022). Artificial intelligence and ethics within the food sector: Developing a common language for technology adoption across the supply chain. *Trends in Food Science & Technology*, *125*, 33–42.

Manning, L. and Kowalska, A. (2021). The role of technology in crisis management/recall in food supply chains. In Manning L. (Ed.), *Developing Smart Agri-food Supply Chains: Using Technology to Improve Safety and Quality* (pp. 239–259). Cambridge, UK: Burleigh Dodds Science Publishing.

Manning, L., Luning, P. A. and Wallace, C. A. (2019). The evolution and cultural framing of food safety management systems–Where from and where next?. *Comprehensive Reviews in Food Science and Food Safety*, *18*(6), 1770–1792.

Manning, L. and Luning, P. (2018). Chapter 16: Determining farm derived food safety risk Food safety for the 21st century: managing HACCP and food safety through the global chain. In Wallace, C. A., Sperber, W. H., and Mortimore, S. E. (eds), *Food safety for the 21st century: Managing HACCP and food safety throughout the global supply chain*, 2nd edition. Chichester West Sussex: John Wiley & Sons.

Marr, B. (2017). What is digital twin technology - and why is it so important? *Forbes*. Available at: https://www.forbes.com/sites/bernardmarr/2017/03/06/what-is-digital -twin-techn ology-and-why-is-it-so-important/

Martindale, W., Hollands, T., Swainson, M. and Keogh, J. G. (2018). Blockchain or bust for the food industry? *Food Science and Technology*, *33*(4), 40–45.

Marvin, H. J., Janssen, E. M., Bouzembrak, Y., Hendriksen, P.J. and Staats, M. (2017). Big data in food safety: an overview. *Critical Reviews in Food Science and Nutrition*, *57*, 2286–2295.

Mehrabi, N., Morstatter, F., Saxena, N., Lerman, K. and Galstyan, A. (2021). A survey on bias and fairness in machine learning. *ACM Computing Surveys (CSUR)*, *54*(6), 1–35.

Mishra, N., Mistry, S., Choudhary, S., Kudu, S. and Mishra, R. (2020). Food traceability system using blockchain and QR code. In Patel, D. et al. (eds), *IC-BCT 2019: Proceedings of the International Conference on Blockchain Technology* (pp. 33–43). Singapore: Springer.

Misra, N.N., Dixit, Y., Al-Mallahi, A., Bhullar, M.S., Upadhyay, R., and Martynenko, A. (2022). IoT, big data, and artificial intelligence in agriculture and food industry. *IEEE Internet of Things Journal*, *9*(9), 6305–6324. https://doi.org/10.1109/JIOT.2020.2998584.

Mitenius, N., Kennedy, S. P. and Busta, F. F. (2014). Food defense. In Y. Motarjemi, H. Lelieveld H (Eds.), *Food Safety Management*(pp. 937–958) San Diego: Academic Press.

Mol, A. P. (2015). Transparency and value chain sustainability. *Journal of Cleaner Production*, *107*, 154–161. https://doi.org/10.1016/j.jclepro.2013.11.012.

Moshrefzadeh, M., Machl, T., Donaubauer, A., Gackstetter, D. and Kolbe, T. H. (2020). Towards a distributed digital twin of the agriculture landscape. *Journal of Digital Landscape Architecture*, 5, 173–186. https://doi.org/10.14627/537690019.

Newsome, R. L., Bhatt, T. and McEntire, J. C. (2013). Proceedings of the July 2011 traceability research summit. *Journal of Food Science*, *78*(s2), B1–B8.

Olsen, P. and Borit, M. (2018). The components of a food traceability system. *Trends in Food Science & Technology*, *77*, 143–149.

Óskarsdóttir, K. and Oddsson, G. V. (2019). Towards a decision support framework for technologies used in cold supply chain traceability. *Journal of Food Engineering*, *240*, 153–159.

Palit, S. and Datta, A. (2017). Digital twins. *Journal of Innovation Management*, *5*, 14–34.

Paloaltonetworks. (2023). What Is Multi-Extortion Ransomware? Available at: https://www.paloaltonetworks.com/cyberpedia/what-is-multi-extortion-ransomware (November 12, 2023).

Pant, R. R., Prakash, G. and Farooquie, J. A. (2015). A framework for traceability and transparency in the dairy supply chain networks. *Procedia-Social and Behavioral Sciences*, *189*, 385–394.

Pearson, S., May, D., Leontidis, G., Swainson, M., Brewer, S., Bidaut, L., Frey, J. G., Parr, G., Maull, R. and Zisman, A. (2019). Are distributed ledger technologies the panacea for food traceability? *Global Food Security*, *20*, 145–149.

Pérez Perales, D., Alarcón Valero, F., Drummond, C. and Ortiz, Á. (2019). Towards a sustainable agri-food supply chain model. the case of LEAF. In Ortiz, Á., Andrés Romano, C., Poler, R., and García-Sabater, J.P. (eds), *Engineering Digital Transformation: Proceedings of the 11th International Conference on Industrial Engineering and Industrial Management* (pp. 333–341). Cham: Springer International Publishing.

Pylianidis, C., Osinga, S. and Athanasiadis, I. N. (2021). Introducing digital twins to agriculture. *Computers and Electronics in Agriculture*, *184*, 105942. https://doi.org/10.1016/j.compag.2020.105942.

Qi, Q., Tao, F., Hu, T., Anwer, N., Liu, A., Wei, Y., Wang, L. and Nee, A. (2021). Enabling technologies and tools for digital twin [Digital Twin towards Smart Manufacturing and Industry 4.0]. *Journal of Manufacturing Systems, 58*, 3–21. https://doi.org/10.1016/j.jmsy.2019.10.001.

Rani, I. B., Ewards, S. V., Palmer, G. M. and Kathrine, G. J. W. (2023, June). Detection of cyber-attacks in food industry using multi-layer perceptron and convolution neural network. In *2023 International Conference on Sustainable Computing and Smart Systems (ICSCSS)* (pp. 1405–1409). IEEE. https://10.1109/ICSCSS57650.2023.10169805.

Regulation EC/178/2002 laying down the general principles and requirements of food safety law, establishing the European Food Standards Agency and laying down procedures in matters of food safety OJ L/31 1.2.2002 pp. 001–024.

Rejeb, A., Keogh, J.G., Zailani, S., Treiblmaier, H. and Rejeb, K. (2020). Blockchain technology in the food industry: a review of potentials, challenges and future research directions. *Logistics, 4*, 27. https://doi.org/10.3390/logistics4040027.

Rogerson, M. and Parry, G. C. (2020). Blockchain: case studies in food supply chain visibility. *Supply Chain Management, 25*(5), 601–614. https://doi.org/10.1108/SCM-08-2019-0300.

Runck, B. C., Joglekar, A., Silverstein, K. A., Chan-Kang, C., Pardey, P. G. and Wilgenbusch, J. C. (2022). Digital agriculture platforms: driving data-enabled agricultural innovation in a world fraught with privacy and security concerns. *Agronomy Journal, 114*(5), 2635–2643.

Santoso, P., Purwanto, A. and Asbari, M. (2019). Influence of implementation chain of custody forest management system FSC-STD-40-004 V3-0 to business performance of paper industriesia in Banten Indonesia. *International Journal of Management and Humanities (IJMH), 4*(4), 32–36.

Schmidt, I., Morris, W., Thomas, A. and Manning, L. (2022). Smart -systems: the role of advanced technologies in improving business quality, performance and supply chain integration. *Standards, 2*(3), 276–293.

Seo, J. S., Seo, Y. H., Yoon, D. H., Seo, W. C. and Ock, Y. S. (2015). A study on food safety approach for seafood Eco-label chain of custody: focused on requirement analysis by AHP method. *The Journal of Fisheries Business Administration, 46*(3), 51–61.

Shahbazi, Z. and Byun, Y. C. (2021). A procedure for tracing supply chains for perishable food based on blockchain, machine learning and fuzzy logic. *Electronics, 10*, 41. https://doi.org/10.3390/electronics10010041.

Sharpless, N. and Yiannas, F. (2019). Statement from acting FDA Commissioner Ned Sharpless, M.D., and Deputy Commissioner Frank Yiannas on steps to usher the U.S. into a new era of smarter food safety. Press Rel., April 30. Available at: https://www.fda.gov/news-events/press-announcements/statement-acting-fda-commissionerned-sharpless-md-and-deputy-commissioner-frank-yiannas-steps-usher (July 5, 2023).

Smart, A. (2008). eBusiness and supply chain integration. *Journal of Enterprise Information Management, 21*(3), 227–246. https://doi.org/10.1108/17410390810866619.

Sparks, J. L. D., Matthews, L., Cárdenas, D. and Williams, C. (2022). Worker-less social responsibility: how the proliferation of voluntary labour governance tools in seafood marginalise the workers they claim to protect. *Marine Policy, 139*, 105044.

Sreedevi, T. R. and Kumar, M. S. (2020, July). Digital twin in smart farming: a categorical literature review and exploring possibilities in hydroponics. In *2020 Advanced Computing and Communication Technologies for High Performance Applications (ACCTHPA)* (pp. 120–124). IEEE. https://doi.org/10.1109/ACCTHPA49271.2020.9213235.

Sufiyan, M., Haleem, A., Khan, S. and Khan, M. I. (2019). Analysing attributes of food supply chain management: a comparative study. In K. Shanker, R. Shankar, and R. Sindhwani (Eds.), *Advances in Industrial and Production Engineering. Lecture Notes in Mechanical Engineering* (pp. 515–523). Singapore: Springer. https://doi.org/10.1007/978-981-13-6412-9_50.

Sunny, J., Undralla, N. and Pillai, V. M. (2020). Supply chain transparency through blockchain-based traceability: an overview with demonstration. *Computers & Industrial Engineering*, *150*, 106895.

Tagliavini, G., Defraeye, T. and Carmeliet, J. (2019). Multiphysics modeling of convective cooling of non-spherical, multi-material fruit to unveil its quality evolution throughout the cold chain. *Food and Bioproducts Processing*, *117*, 310–320. https://doi.org/10.1016/j.fbp.2019.07.013.

Tao, F. and Qi, Q. (2019). Make more digital twins. *Nature*, *573*, 490–491. https://doi.org/10.1038/d41586-019-02849-1.

Temoshok, D. and Abruzzi, C. (2018). Developing trust frameworks to support identity federations. US Department of Commerce, National Institute of Standards and Technology. Available at: https://csrc.nist.rip/library/NIST%20IR%208149.pdf.

Thakur, S., John G. and Breslin, J.G. (2020). Scalable and secure product serialization for multi-party perishable good supply chains using blockchain. *Internet of Things, 11*, 100253. https://doi.org/10.1016/j.iot.2020.100253.

Tian, F. (2018). An information system for food safety monitoring in supply chains based on HACCP, blockchain and internet of things. Doctoral thesis, WU Vienna University of Economics and Business. Available at: https://epub.wu.ac.at/6090/.

Tigard, D. W. (2021). Digital twins running amok? open questions for the ethics of an emerging medical technology. *Journal of Medical Ethics*, *47*(6), 407–408. https://doi.org/10.1136/ medethics-2021-107526.

Valle, S. S. and Kienzle, J. (2020). Agriculture 4.0 – agricultural robotics and automated equipment for sustainable crop production. *Integrated Crop Management, 24*. Available at http://www.fao.org/3/ cb2186en/CB2186EN.pdf.

van de Burg, S., Wiseman, L. and Krkeljas, J. (2020). Trust in farm data sharing: reflections on the EU code of conduct for agricultural data sharing. *Ethics and Information Technology, 23*, 185–198. https://doi.org/10.1007/s10676-020-09543-1.

van Evert, F. K., Berghuijs, H. N. C., Hoving, I. E., de Wit, A. J. W. and Been, T. H. (2021). A digital twin for arable and dairy farming. In Stafford, J.V. (ed), *Precision Agriculture'21* (pp. 364–377). Wageningen Academic Publishers. Wageningen, Netherlands.

Van Ruth, S. M., Huisman, W. and Luning, P. A. (2017). Food fraud vulnerability and its key factors. *Trends in Food Science & Technology, 67*, 70–75. https://doi.org/10.1016/j.tifs.2017.06.017.

Verdouw, C., Tekinerdogan, B., Beulens, A. and Wolfert, S. (2021). Digital twins in smart farming. *Agricultural Systems*, *189*, 103046. https://doi.org/https://doi.org/10.1016/j.agsy.2020.103046.

Vidal, N., Kozak, R. and Cohen, D. (2005). Chain of custody certification: an assessment of the North American solid wood sector. *Forest Policy and Economics*, *7*(3), 345–355.

Voyiatzis, A. G. and Serpanos, D. N. (2022). Active hardware attacks and proactive countermeasures. In *Proceedings ISCC 2002 Seventh International Symposium on Computers and Communications* (pp. 361–366). https://doi.org/10.1109/ISCC.2002.1021702.

Wang, L., He, Y. and Wu, Z. (2022). Design of a blockchain-enabled traceability system framework for food supply chains. *Foods, 11*, 744. https://doi.org/10.3390/foods11050744.

Wang, S., L. Zheng, L., G. Cai, G., N. Liu, N., M. Liao, M., Y. Li, Y., X. Zhang, X. and Lin, J. (2019). A microfluidic biosensor for online and sensitive detection of Salmonella typhimurium using fluorescence labelling and smartphone video processing. *Biosensors & Bioelectronics, 140*, 111333. https://doi.org/10.1016/j.bios.2019.111333.

Wright, L. and Davidson, S. (2020). How to tell the difference between a model and a digital twin. *Advanced Modelling and Simulation in Engineering Sciences, 7*(13). https://doi.org/10.1186/s40323-020-00147-4.

Yadav, V. S., Singh, A. R., Raut, R. D., Mangla, S. K., Luthra, S. and Kumar, A. (2022). Exploring the application of Industry 4.0 technologies in the agricultural food supply chain: a systematic literature review. *Computers & Industrial Engineering, 169*, 108304. https://doi.org/10.1016/j.cie.2022.108304.

Yu, Z., Jung, D., Park, S., Hu, Y., Huang, K., Rasco, B.A., Wang, S., Ronholm, J., Lu, X. and Chen, J. (2022). Smart traceability for food safety. *Critical Reviews In Food Science And Nutrition, 62*(4), 905–916. https://doi.org/10.1080/10408398.2020.1830262.

Zwietering, M. H., Jacxsens, L., Membré, J. M., Nauta, M. and Peterz, M. (2016). Relevance of microbial finished product testing in food safety management. *Food Control, 60*, 31–43.

Chapter 14

Measuring and improving good agricultural practices (GAP) related to safety of fresh produce: the case of controlled environment agriculture

Louise Manning, Lincoln University, UK; and Jim Monaghan, Harper Adams University, UK

1 Introduction

The United Nations Food and Agriculture Organisation (FAO, 2016, p. 1) defines good agricultural practices (GAP) as a 'collection of principles to apply for on-farm production and post-production processes, resulting in safe and healthy food and non-food agriculture products, while taking into account economic, social and environmental sustainability.' The scope of GAP, described by Bamunuarachchi et al. (2019) as the four 'pillars' of GAP, are as follows:

- food safety and quality,
- economic viability,
- environmental sustainability and
- social acceptability.

http://dx.doi.org/10.19103/AS.2024.0133.31

Food security is also an aspect of implementing GAP and, together with delivering food safety, is foundational to meeting the UN's Sustainable Development Goals (SDGs) (Ranjbar et al., 2022). Whilst GAP has been instrumentalised and codified into international, public and private (third-party) standards; effective implementation of GAP needs a more holistic approach as farmers' skill levels, attitudes and perceptions all influence the adoption of GAP at the farm level (Ranjbar et al., 2022).

Over the last 20 years, commodity-specific GAP guidelines have emerged, for example, the International Fresh Produce Association Romain Lettuce Standard.[1] The introduction In the United States (US) in 1998 of the Guide to Minimise Microbial Food Safety Hazards for Fresh Fruits and Vegetables (FDA, 1998) had a clear focus on food safety. In more recent times, GAP guidelines have been realigned to also focus on sustainability, where food safety plays a part, rather than a central focus solely on the safety of the fresh produce itself. Indeed, López-Gálvez et al. (2021, p. 1) argue that whilst 'priority must be given to [food] safety aspects, food safety policies should be calibrated in order to avoid unnecessary deleterious effects on the environment.' Some studies focus on the use of refrigeration as a food safety control measure whilst also impacting on the greenhouse gas emissions (GHGEs) associated with a product. Thus, there is a contested space between, on the one hand, implementing procedures to ensure that food is safe and, on the other addressing pressing issues such as reducing food waste and optimising resource use especially to reduce the associated greenhouse gas emissions (GHGEs) (Guillier et al., 2016a, 2016b; Duret et al., 2019).

Food safety risk assessment and risk analysis, and the resultant development of a food safety management system (FSMS), should therefore not be undertaken in isolation of consideration of other sustainability factors. There is an increasing realisation that food safety aspects are not the sole consideration within prerequisite programmes such as GAP, good hygienic practices (GHP) or good manufacturing practices (GMP). Instead, integrated risk analysis approaches, such as multiple criteria decision analysis, can be used to integrate qualitative and quantitative measures to develop a transparent decision framework covering multiple objectives related to economic, environmental and social outcomes (Guillier et al. 2016a, 2016b).

This transition to a broader concept of what GAP should cover can be seen in the evolution of the GLOBALG.A.P. suite of farm assurance standards supporting global trade of fresh produce such as fruits, vegetables and flowers.[2] GLOBALG.A.P. standards have now been adopted by over 200 000 certified producers in more than 135 countries farming over 4 million hectares.

1 https://www.freshproduce.com/resources/food-safety/commodity-specific-resources/
2 https://www.globalgap.org/

GLOBALG.A.P.'s lead standard is the Integrated Farm Assurance (IFA) standard (latest version IFA V6).[3] The IFA standard has evolved into a holistic standard with core criteria which include not just food safety and traceability, but also issues such as worker health, safety and welfare; responsible use of fertilisers and biostimulants; water management; responsible use of plant protection products and integrated pest management; energy efficiency; waste management; and protection of biodiversity and habitat. GLOBALG.A.P. has also developed a number of add-on or 'bolt on' standards that are focused particularly on environmental sustainability and social responsibility. This includes the GLOBALG.A.P. Farm Sustainability Assessment (GGFSA) which allows producers to combine achieving IFA certification with compliance with the Farm Sustainability Assessment (FSA) programme of the Sustainable Agriculture Initiative (SAI).[4]

As the evolution of GLOBALG.A.P shows, there is a shift from food safety as the sole objective and this creates new questions as to how GAP is conceptualised and defined, now and going forward including as new types of food production system emerge. This chapter considers how previous approaches to food safety-related hazard analysis and risk assessment in fresh produce GAP frameworks are being revisited and repositioned, and the contested issues that can arise. The next section considers how food safety objectives, performance objectives and food safety performance criteria are developed for food supply chains as an element of GAP. This is followed by consideration of the challenges and opportunities of positioning GAP in production systems and the controls that could be adopted to minimise food safety risks. The chapter then concludes and considers future opportunities for research.

2 Food safety objectives (FSO), performance objectives (PO) and other measures

Safe food is produced by 'adhering to good hygienic practices (GHP), good manufacturing practices (GMP), good agricultural practices (GAP), etc. and implementation of food safety risk management systems such as hazard analysis critical control point (HACCP), but the level of safety that these food safety systems are expected to deliver has seldom been defined in quantitative terms.' (Van Schothorst et al. 2009, p. 967).

In order to seek to quantify food safety objectives (FSO) and performance criteria (PC), the term 'appropriate level of sanitary or phytosanitary protection" or ALOP (see Table 1) was introduced in the World Trade Organisation Sanitary

3 https://www.globalgap.org/what-we-offer/solutions/
4 https://www.globalgap.org/what-we-offer/solutions/ggfsa/

Table 1 Definitions

Term	Definition	Sources
Appropriate level of performance (ALOP)	Level of protection deemed appropriate by the member (country) establishing a sanitary or phytosanitary measure to protect human, animal or plant life or health within its territory.	(WHO, 1997)
Control measure (CM)	Any action and activity that can be used to prevent or eliminate a food safety hazard or to reduce it to an acceptable level. The acceptable level can be set microbiological specifications, guidelines on pathogen control, hygiene codes, microbiological criteria. The CM can relate to procedures or processes such as pasteurisation, cooking, metal detection or filtering. A control measure can operate at one or more process step and if the activity or step is determined as a critical control point (CCP) then critical limits can be set, validated, monitored and verified.	(ICMSF, 2002)
Food safety objective (FSO)	The maximum frequency and/or concentration of a hazard in a food at the time of consumption that provides or contributes to the ALOP.	(ICMSF, 2002; CAC, 2007, 2019)
	The stringency that government food safety controls deem necessary for operational food safety management to deliver.	(Gorris, 2005)
Microbiological criteria (MC)	Set of parameters used to determine whether a specific batch of food is acceptable or not.	(Whiting et al., 2006)
Performance criterion (PC)	The effect in the frequency and/or concentration of a hazard in a food that must be achieved by the application of one or more CM to provide or contribute to a PO or an FSO.	(ICMSF, 2002; CAC, 2007, 2019)
Performance objective (PO)	The maximum frequency and/or concentration of a hazard in a food *at a specified step in the food supply chain before the time of consumption* that provides or contributes to an FSO or ALOP.	(ICMSF, 2002; CAC, 2007, 2019)

and/or Phytosanitary (SPS) Measures, i.e. the WTO/SPS agreement in 1997 (WHO, 1997). Developing an ALOP as a food safety or public health objective at the population level shows what level of risk the competent authorities, on behalf of society, either consider to be achievable or acceptable (Gorris, 2005) and establishes a sanitary measure to protect human, animal or plant life or health within its borders (WHO, 1997; Manning and Soon, 2014). An ALOP can be set as the frequency of incidence of reported illness in a population for a given pathogen in a given year and, as a result, can drive public health strategies, but it is not a performance indicator that specifically informs either food safety management practices or GAP at the organisational level (Gorris, 2005).

In the setting of an FSO within a particular country or a region such as the European Union (EU), competent authorities can determine a risk-based limit that should be achieved operationally within the food chain, whilst providing flexibility in how that limit is achieved through specific production, preparation, manufacturing, distribution, and marketing measures (CAC, 2007), e.g. in applying specific elements within GAP. The FSO is a target or the degree of stringency, *'for operational food safety management, leaving flexibility in the way equivalent food safety levels are achieved by different* [approaches in different] *food chains'* (Gorris, 2005, p. 801). The FSO is set at the point of human consumption as it is the last stage where a food safety performance standard can be set (ICMSF, 2002). A food safety margin (FSM) is the standard set which provides assurance that the FSO will be met and will be determined based on the degree of uncertainty in the level of knowledge that informs a risk assessment and also the potential for deviation from the norm in the food production process based on expected performance and behaviour (Doménech and Martorell, 2016).

A performance objective (PO) is proposed by either the competent authority or the operational food safety managers. The PO is developed for a specified step in the supply chain before consumption, e.g. pasteurisation, cooking or chlorination of fresh produce, where achieving the PO at that step contributes to delivering an FSO or ALOP (Gorris, 2005; CAC, 2007, 2019). Control measures (CM) are interventions implemented at a given step or steps in the supply chain to deliver the PO. Operationally when implementing a hazard analysis critical control point (HACCP), the PO can relate to a critical limit (CL) or a target level/tolerance which is set to give a greater degree of food safety control to ensure a CL will not be breached, where this is deemed appropriate. However, the FSO and PO can only be set if the food safety hazard or contaminant is 'known' and there has been a scientific risk-based determination of the scientifically or socially acceptable level of the hazard in a given context (Manning et al., 2020).

A food organisation must validate its FSMS. The validation process needs to ensure that the FSMS has the capability to, where possible, eliminate or otherwise control food safety hazards to an acceptable level (Van Schothorst et al. 2009; Manning et al. 2019). Monitoring and verification activities ensure the CM are consistently implemented and are effective so that food safety standards are always met (Manning, 2018). The PC describes the effect or result in terms of the frequency and/or concentration of a hazard in a specific food that must be achieved through the application of one or more CM that will contribute to delivering a PO or an FSO (ICMSF, 2002; CAC, 2007, 2019). The definitions used in this section are drawn together in Table 1.

In summary, the FSO, e.g. non-detection of foreign bodies in the finished product can be achieved through multiple means depending on the nature of the food and the processes that occur in the supply chain. Thus, whilst the FSO may be consistent across a range of foods, the POs and the PC will be different in given food product and food production contexts, e.g. production in the field versus controlled environment agriculture or CEA. Thus, what is considered an effective GAP framework with one fresh produce type or production system, such as field-grown potatoes that will always be cooked before consumption, will not be considered appropriate for products that are eaten raw and sold as 'ready-to-eat', e.g. leafy salads. The FSO is an 'anchor-point' but how the FSO is achieved may vary and the combined development of FSO, PO and PC will inform appropriate guidance on GAP and GMP with an expectation of compliance with that guidance, albeit that it is described as voluntary (Gorris, 2005).

In order for food business operators (FBOs) to determine the level of compliance of materials and products at the farm level and during food processing with POs and PC they can use a variety of monitoring and verification methods. These include microbiological, chemical and physical assessment of ingredients and product samples and testing of performance parameters such as temperature testing. Determining hygiene standards through the swabbing of facilities, especially food and hand contact surfaces is also important. Alternatively, monitoring and verification can be undertaken by assessment of process data and/or supply chain documentation. Documentation can be assessed in person or through remote assessment of manufacturing facilities, where aspects such as the review of the FSMS and HACCP-related records can determine the adequacy of CMs adopted (Van Schothorst et al., 2009).

Audits of farms, food manufacturers, processors and so on are only a 'snapshot' of actual performance (Powell et al., 2011). Audits can be internal, where the organisation undertakes audits of its own performance, second-party where organisations are audited by their customers, or third-party where an independent organisation undertakes audits. Attaining third-party certification can provide supporting evidence of the degree of formalisation of the FSMS, assuring customers (retailers, manufacturers or food service) or competent authorities that POs and PC can be achieved (Manning, 2018). The integrity of, and trust in, the efficacy of audits, relies on the integrity of the auditors and the auditee organisation, especially since the latter reflects their daily practices and behaviours during the actual audit (Manning et al., 2019). However, one of the challenges that third-party certification faces is that standards are designed to deliver systems and processes that are auditable and observable rather than a bespoke FSMS developed through effective identification, assessment and analysis of food safety risks and the means for their elimination, or where this is not possible acceptable levels of control.

3 Risk analysis and risk assessment in fresh produce production

HACCP-derived operational prerequisite programmes such as GAP have traditionally been adopted to minimise, but not eliminate, food safety risk (Monaghan et al., 2016; Manning and Luning, 2018). Risk analysis is a process consisting of three components:

- risk assessment
- risk management and communication
- the risk profile

The risk profile (i.e. the description of the food safety problem (CAC, 2019) and its context) needs to be clearly defined for effective risk analysis to occur. Context factors influence the extent of the food safety risk associated with a product and as a result the associated POs, PC and GAP frameworks that need to be implemented (Manning and Luning, 2018). These context factors will be both production system generic and also specific to the given organisation and the environment in which it operates. Context factors can be internal, e.g. product type, procedures undertaken or external to the organisation, e.g. customer requirements, weather conditions, etc. Luning et al. (2011) defined the context factors as follows:

- **Product characteristics:** the intrinsic properties of initial materials and final products;
- **Production characteristics:** the extrinsic conditions utilised during primary production, processing, or handling);
- **Organisational characteristics**: namely **individual** (people) characteristics, **group** characteristics associated with culture; **organisational** characteristics (tasks, responsibilities, rules, procedures) and **information systems**, which decision-making;
- **Control characteristics**: the elements of the food safety control system, e.g. protocols, capabilities and control systems; and
- **Chain characteristics:** the conditions and relationships in supply chains (Luning and Marcelis, 2007; Luning et al., 2011; Kirezieva et al., 2015a; Manning and Luning, 2018).

One viewpoint would be that as the degree of FSMS formalisation increases this should lower the risk of a food safety incident occurring (Sawe et al., 2014; Kirezieva et al., 2015a, 2015b). However, FSMSs are developed based on known food safety issues and so may not lower the risk of a food safety incident associated with a novel, emergent food safety issue. The role of objective evidence is therefore important in the development of FSMSs for fresh produce and also in understanding where there may be gaps in food safety knowledge, e.g. with regard to new growing and production techniques, or where there is a level of uncertainty in the reliability of the evidence base that is available.

The starting point for managing the risk of microbiological hazards in fresh produce is an understanding that complete elimination of contamination with microbial hazards in fresh produce may be impossible, particularly where these products are grown in a field environment (De Keuckelaere et al., 2015). As described, the FSMSs developed by growers are built upon risk assessment procedures that encompass the whole crop production cycle, i.e. field history, water sources, animal manures and worker hygiene. These assessments are then used to define CM to reduce the risk of biological, chemical or physical contamination of the crop; these CM can be both internally and/or independently audited (Monaghan et al., 2012), often forming elements of GAP frameworks. The risk assessment approaches used have been refined over the last 30 years and there is a generally accepted approach that relies mainly on qualitative rather than quantitative assessment of risk factors.

The term 'risk assessment' in fresh produce FSMSs development and in third-party standards, generally refers to a subjective, qualitative approach based more on expert opinion and experience, e.g. as required by GlobalG.A.P. (Monaghan et al. 2016), in contrast to a scientific process consisting of formal components and quantification of levels of risk as outlined by the Codex Alimentarius Commission (CAC, 2019). The key aspects that differ between these two risk assessment approaches are summarised in Table 2.

In essence, growers are commonly completing a generic hazard identification and exposure assessment through the application of HACCP principles (Manning et al., 2019). HACCP principles are then used to frame the definition of CMs and preventive actions to reduce the risk of biological, chemical or physical contamination of the crop. Whilst fresh produce growers employ HACCP principles in the development of their food safety plans and the associated FSMSs, specific process steps at the farm level constituting CCPs (where activities at the process step can effectively eliminate the hazard or reduce it to an acceptable level) are very limited (CAC, 2020).

Instead, the growers will implement CM, procedures and protocols which may reduce the risk of microbial contamination, often termed as 'hurdles' (Mogren et al., 2018). The *hurdle effect* is based on the supposition that for most foods several factors (hurdles) will contribute to the overall stability, potential spoilage and safety of food (Leistner, 1992).

Key hurdles that can be used in food preservation or in ensuring food safety include temperature control (Leistner, 2000; Wadamori et al., 2017); or through 'true' CCPs such as the use of sanitisation in the packhouse (Wadamori et al., 2017), irrigation water disinfection by chemicals or ultraviolet treatment

Table 2 Risk assessment for fresh produce and how it compares to CAC defined risk assessment approaches (adapted from Monaghan et al. 2016)

CAC risk assessment components (CAC, 2019)	Common approach for risk assessment for fresh produce
Hazard identification	Undertaken at a grower level. All microbial hazards are considered together in a generic approach at farm level, i.e. relative hazards may be considered as a group rather than different species, e.g. verotoxigenic *E. coli* versus *Salmonella*. Chemical and physical hazards are considered and a screening process is undertaken to see if they are likely to occur at the farm level (as applicable).
Hazard characterisation	Addressed by food safety enforcement and government agencies and developed by academics and researchers.
Exposure assessment	Undertaken at a grower level. Consumer exposure to microbial hazards is considered qualitatively as the probability or possibility that any microbial contamination on the product could lead to illness in a consumer and the severity should it occur. The grower may, or may not, have access to information on hazard characterisation. Growers would tend to undertake hazard analysis as part of a HACCP process.
Risk characterisation	Addressed by food safety enforcement and government agencies and developed by academics and researchers.

(UV) and irradiation with leafy green vegetables (Mogren et al. 2018). At the farm level, hurdles range from prerequisites, such as hygiene training for field workers and hygiene protocols for equipment, or fencing to ensure wild animals are excluded, which are designed to minimise the microbiological burden on fresh produce (Mogren et al., 2018). With regard to chemical contamination or physical contamination, adoption of practices as part of GAP can minimise the likelihood of contamination in field, storage or transport.

4 Good agricultural practices (GAP) in the context of controlled environment agriculture (CEA)

The production system through which these aspects of GAP are explored in this chapter is controlled environment agriculture (CEA). CEA encompasses both covered production in fully enclosed transparent growing structures, including glasshouses and multi-tunnel structures, and vertical farming in fully enclosed buildings, typically lit with LED lights. CEA provides the opportunity for year-around production, with increased production control, potentially greater yields and the ability to produce food in urban locations much closer to the consumer (Gómez et al., 2019). The range of crops that can be grown in CEA is wide, but in practice, the crops being grown commercially are mainly high-risk category fresh produce that is usually eaten uncooked, particularly leafy salads and microgreens (Beacham et al., 2019).

In CEA, seeds may be germinated in the production unit in soilless substrates and grown hydroponically or aeroponically in shallow trays under LED lighting at a controlled temperature and humidity. The optimised growing conditions hasten crop growth and shorten production cycles. For example, microgreens are harvested when they have developed 1–2 true leaves, 7–21 days from germinating, with their cotyledons fully expanded and still turgid (Kyriacou et al., 2016). Larger crops such as rocket/rucola or spinach have longer crop cycles but are still being harvested in 2–4 weeks.

As noted earlier, complete elimination of contamination with microbial hazards in fresh produce may be impossible, particularly where these products are grown in a field environment (De Keuckelaere et al., 2015). One of the drivers for growers of high-risk leafy crops to move production away from field environments into CEA is the ability to eliminate or reduce microbial contamination from soil, water, livestock and wildlife, and to a limited extent, equipment and workers.

In CEA systems, crops are grown in the absence of soil (Mogren et al. 2018). Open-field lettuce production in the United States has been shown to be more prone to faecal contamination (measured as *E. coli*) than lettuce grown in greenhouses (Holvoet et al., 2015). This reduction in risk of contamination has commercial importance especially in the United States where product recalls have had a significant impact on the field-grown leafy green sector (Spalding et al., 2022). The choice of hurdle, or combination of hurdles, depends on the pathogen(s) of interest, the expected pathogen burden on harvested leafy green vegetables or other fresh produce crops, and the FSO established for the point of consumption (Mogren et al., 2018).

5 Managing contamination risks in CEA systems

CEA production reduces the opportunities for microbial contamination from the external environment. Conversely, the production conditions that allow intensive, year-round production are supportive of the survival, growth and spread of any zoonotic agents that do enter the production environment. The potential sources of microbial contamination in CEA include substrates, seeds, water, equipment and workers.

The persistence of zoonotic agents in CEA systems is an area of growing scientific and practical interest. The short production duration of leafy crops can pose a risk to the microbial status of the crop. Zoonotic agents such as *E. coli* O157 and *Salmonella* spp. decline over time and, in field crops, persist on the leaves of spinach and lettuce for 1–3 weeks (Hutchison et al., 2008). Exposure to solar UV and wind-mediated drying are key factors in bacterial decline on leaf surfaces in the field. In CEA, levels of solar UV are either low, if sunlight is coming through glass or plastic roofs, or is absent in a vertical farm, unless added with UV-LED. This can, potentially, allow zoonotic agents to survive in CEA conditions.

The persistence of zoonotic agents in the soil varies greatly with environmental conditions. In cool damp conditions, *E coli* O157 persists in soil for >5 months after application of contaminated compost or irrigation water (Islam et al., 2004). In warm summer conditions with strong sunlight and low rainfall (e.g. in the United Kingdom), the decline of the zoonotic agents is rapid and a moderate level of contamination falls below the levels of detection within approximately 1 week (Monaghan and Hutchison, 2012). Pathogenic and generic *E. coli* have been observed to persist in contaminated nutrient solutions for up to 4 weeks and to contaminate the edible parts of radish and lettuce microgreens (Işık et al., 2020).

5.1 Substrates

Many hydroponic systems use small volumes of soil-less substrate to supply anchorage and stability for plant roots, with the roots either being contained in the substrate or growing out into water (hydroponic) or a misted environment (aeroponics) (Niu and Masabni, 2022). A variety of substrates are used in CEA including inorganic sterile substrates such as rockwool or fibre mats as well as organic-derived substrates including peat and coir.

Soil-less substrate (e.g. peat or coir) can be colonised by zoonotic agents such as *E. coli* O157:H7 (Wright and Holden, 2018), *Listeria monocytogenes* and *Salmonella enterica* (Misra and Gibson, 2020) and contaminated substrates can be transferred to the edible parts of leafy crops (Xiao et al., 2015). Unsterilised substrates such as peat are a potential source of contamination (Di Gioia et al., 2017). Dankwa et al. (2020) demonstrated that peat used in hydroponic lettuce production had detectable levels of coliform bacteria which were transferred from the substrate to the harvested lettuce leaves. The sterilisation of growing substrates, through heat or chemical disinfection (Dankwa et al., 2020) is, therefore, an important step in reducing the risk of introducing microbial contamination to a CEA production environment.

5.2 Seeds

A particular challenge in CEA production of rapidly growing leafy crops such as microgreens is that contaminated seeds can lead to microbial bacteria and/or viruses becoming internalised from the beginning of the growth cycle which is then very difficult to eliminate (Wang and Kniel, 2016). Contaminated seeds have been implicated in a number of large-scale foodborne illness outbreaks linked to sprouted seeds (EFSA, 2011). There is, therefore, a focus on sterilising seeds used for sprouting.

There are some similarities between sprouted seed and microgreen production. Both are grown in CEA, have relatively short production cycles and are consumed when physiologically immature (Turner et al., 2020), often raw or without a heat treatment step. Seed contamination can also pose a risk to longer-growing leafy crops and can persist for weeks, longer than the CEA production cycle of many leafy crops. For example, hydroponically grown lettuce from lettuce seed inoculated with *Salmonella* led to contamination of the inoculated crop as well as cross-contamination of uninoculated plants through the circulation of the hydroponic solution. The *Salmonella* persisted in the plants and growing environment for the 6 weeks of lettuce growth (Li et al., 2022).

Particular care is needed in the production and storage of seed destined for CEA production to prevent contamination. Microbial seed decontamination is required for the production of sprouted seed and its efficacy has been demonstrated but is not routinely required for CEA leafy crops (Misra and Gibson, 2021). The most common chemical approach as a GAP hurdle for decontamination is the use of chlorine washes (Wadamori et al., 2017), but alternative approaches are developing, including organic acids washes, ozone (both gaseous and water-based), electrolysed water, and atmospheric cold plasma discharges or plasma-activated water (Mogren et al., 2018; Machado-Moreira et al., 2019). Alternative physical approaches are also being developed such as thermal inactivation of microorganisms, UV irradiation and high hydrostatic pressure. The use of bacteriophages, bacteriocins, or protective cultures is at an early stage of development and may be effective (Machado-Moreira et al., 2019).

5.3 Water

CEA systems rely on hydroponics and aeroponics to provide water for plant growth. In small-scale CEA units, the volume of water required may be adequately supplied using municipal treated water, i.e. tap water. Larger-scale units including commercial greenhouses will require large volumes of water and may utilise ground water from a private borehole, for example. If the water is contaminated, there is a risk that this contamination will be transferred to the harvested crop. This risk is increased in closed hydroponic systems which continuously recycle the nutrient solution (Riggio et al., 2019) and in short duration crops, particularly micro-greens that are harvested ~7–14 days after sowing. A recent survey of U.S. CEA producers of microgreens showed that municipal water and well/bore hole water was utilised for the majority of production (Misra and Gibson, 2021), both of which are lower-risk water sources than open water sources such as river or reservoir water that are commonly used in field production (Allende and Monaghan, 2015).

The risk of microbial contamination of hydroponic systems is not as well studied as the potential for field crop contamination. It is clear, however, that once a hydroponic system is contaminated, then the risk of cross-contamination via the system is high. In an experimental study, a nutrient flow technique system was inoculated with *Salmonella* Typhimurium and *Listeria monocytogenes*. Both pathogens persisted in the system throughout the growth cycle of lettuce (28 days) following both a simulated sporadic (~10^4 CFU/ml) and extreme (~10^7 CFU/ml) contamination event and pathogens accumulated in the rockwool medium and on lettuce roots and were transferred to the leaves at quantifiable levels (Ilic et al., 2022). If these results were replicated in a commercial unit, they would pose a potential risk to both consumers and workers in the unit.

In order to minimise the risk of introducing pathogens, growers can select water sources with lower levels of contamination and combine this with one or several types of water treatment. A number of options are open to CEA producers (see Riggio et al., 2019). These include approaches such as slow sand filtration usually followed by membrane filtration and/or treatments such as reverse osmosis, UV light, chlorination, ozonation or the use of hydrogen peroxide. The decision on the methods used should should be based on risk assessment of incoming water quality and determination of the hurdle effect required.

It is also standard practice to validate the functioning of water UV or ozone treatment units by quantifying the reduction in indicator species. This will inform the performance criterion considered important in consistently delivering safe food. The validation samples are still relatively expensive to analyse and time-consuming to collect. Regular monitoring of water quality before and after treatment is also an important component in FSMSs within wider GAP that should be applied to CEA production. The UK Red Tractor GAP farm assurance scheme states that primary water treatments (e.g. UV, chlorine, chlorine dioxide, ozonation) can be applied to high-risk crops as long as the treatment process has been validated, i.e. has been proven to be effective in reducing microbial loading to an acceptable level (Red Tractor, 2022).

5.4 Equipment hygiene

In field production, equipment can be contaminated by the environment, such as soil, water etc. posing a high risk of cross-contamination of the crop. This is a particular risk where equipment is moved between field locations and where it may move through contaminated areas (e.g. where there is manure run-off) to get to the field for harvest, or be exposed to contamination by rodents or birds. Equipment used in CEA can be protected in terms of reducing opportunities for contamination much more than in the field. Nevertheless, microbial pathogens have been sporadically isolated from greenhouse production units

(Orozco et al., 2008). Routine cleaning and disinfection in both field and CEA environments is necessary to maintain high levels of hygiene. A challenge for CEA is that bespoke equipment may be designed for a novel production system without a consideration for hygiene and access for cleaning. This requires validation of the efficacy of proposed cleaning protocols for new machinery. These lessons have been learned in food factories, with *Listeria* spp. contamination persisting in hard to clean locations (Carpentier and Cerf, 2011) leading to equipment which is designed to reduce potential surface contamination and to optimise cleaning/disinfection. It is therefore important that new equipment designed and built specifically for CEA considers microbial food safety as part of the design brief to avoid potential harbourage of microbial contaminants, especially *Listeria monocytogenes*. Validation and verification of hygiene measures are essential to ensure that PCs are met.

5.5 Worker hygiene

Whilst there is a move towards automating many steps in the production and harvest of leafy crops in CEA units, manual labour is still required. In particular, harvesting and packing is still predominantly a manual process. In contrast to field production, workers can be more easily managed in CEA facilities with controlled access to the growing area requiring high levels of hygiene, particularly in vertical farming units. The use of control measures such as health screening, personal hygiene protocols, and protective clothing requirements all provide a hurdle effect to minimise risk. Hand washing is the key CM and washing using use of soap and water has been shown to be effective at removing norovirus (Sobolik et al., 2021) and bacteria (Prince-Guerra et al., 2020) in farm settings.

6 Hygiene indicators for CEA systems

A particular challenge of more qualitative approaches to risk assessments for growing produce in fields or CEAs is understanding or ranking the effectiveness of individual and combined CM, i.e. the hurdle effect, in order to both prioritise actions as well as demonstrate due diligence. With regard to collecting evidence, quantifying the reduction in microbial contamination of a crop as a consequence of field worker hygiene training at field level, e.g. is impractical, since there are too many variables involved. It is equally difficult to quantify the 'real-time' reduction in microbial contamination through actions such as fencing of livestock. In research/experimental systems, the efficacy of CM in order to inform guidance on best practice in terms of GAP is often quantified through changes in either microbial pathogens or microbial indicator species such as *E. coli* (Monaghan and Hutchison, 2016). This data can be utilised in quantitative microbial risk assessments (QMRA) to inform the development of system-specific risk assessments.

Hygiene indicators are routinely used in food factories to validate cleaning protocols and check on the efficacy of a cleaning process. These approaches have crossed over into the production and harvest of high-risk fresh produce crops such as watercress with guidance on critical values for harvest equipment for both pathogenic *Listeria monocytogenes* and the general hygiene indicator *Enterobacteriaceae* (AHDB, 2017). The routine monitoring of commercial crop production by frequent measurement of *E. coli* has been proposed as a hygiene indicator and PC for primary production (EFSA, 2014), but more data and standardisation of sampling procedures would be needed before critical values could be identified to validate the effectiveness of GAP approaches (Allende et al., 2018). Currently, the cost of monitoring hygiene indicators is high. However, as automated monitoring systems are developed within CEA environments, the cost of sampling and analysis will decline, and the possibility of using continuous monitoring systems (see Sidhu et al., 2020) would enable a much clearer understanding of the level of risk within different stages of primary production. There are also real opportunities in the development of internet of things (IoT) and artificial intelligence (AI) driven monitoring which combine large data sets with environmental data to both model risk, monitor microbial risk in 'real-time' (Allende et al., 2022) and/or feed into digital twins that can predict periods of risk within the growing cycle (Qian et al., 2022). GAP recommends multiple CM/hurdles as standard approaches to manage food safety, but few scientific studies have quantified the effect of these actions.

Recently published research has focused on the hurdle effect of CM through the use of the Swiss Cheese Model (SCM), e.g. da Cuhna et al. (2022) and Wiśniewska (2023). As a consequence of existing approaches, risk assessments are being built on assumptions rather than evidence which become hard to justify beyond industry best practice (Monaghan et al., 2017). As sustainability aspects such as seeking to extend shelf-life to reduce food waste, more sustainable packaging options, reducing energy and water use are considered with CEA production, the impact on the risk assessment processes as well as the outputs from the risk analyses must be addressed.

7 Conclusion

This chapter has considered how previous approaches to food safety-related hazard analysis and risk assessment in fresh produce GAP frameworks are being revisited and repositioned, as we consider CEA production and the potential issues that can arise. The development of food safety objectives, performance objectives and food safety performance criteria have been explored. This requires reimagining of GAP in the context of CEAs rather than just translating thinking about field-based food safety aspects to controlled environment conditions. The challenges and opportunities of positioning GAP in production systems such as CEA, and the controls that could be adopted to minimise food safety risks, have also been considered.

Approaches to risk analysis may also have to be updated to recognise the need for a food safety hurdle effect via CM with clear, transparent associated PC that integrates into bespoke, effective FSMSs for CEA considering the processes involved, the crops are grown and the inherent and associated risks that can arise. The use of IoT and AI, even the development of digital twins of the CEA environment to promote and deliver GAP has also been explored. In terms of future challenges and opportunities for ensuring food safety in CEA environments, further work needs to be done to build the data on the food safety issues associated with CEA (Ilic et al. 2022), refocus GAP standards on CEA rather than field environments (see Ilic et al., 2017), and undertake work to embed new IoT and AI-enabled technology into both QMRA (Allende et al, 2022) and agile digital FSMSs (Qian et al., 2022) for CEA production. The use of multiple CM hurdles within the CEA environment to minimise food safety risks is also worthy of future study.

8 Where to look for further information

Detailed guidance on managing food safety on CEA is not yet available. Broader approaches are to be found iin guidance documents for GLOBALG.A.P (https://www.globalgap.org/) and the specific standards for fruit and vegetables (https://www.globalgap.org/what-we-offer/solutions/ifa-fruit-and-vegetables/).

The UK Red Tractor Assurance also has guidance encompassing protected production systems, which may have some relevance. (https://redtractorassurance.org.uk/standard-categories/fresh-produce/)

9 References

Allende, A., Bover-Cid, S. & Fernández, P.S. (2022). Challenges and opportunities related to the use of innovative modelling approaches and tools for microbiological food safety management. *Current Opinion in Food Science*, *45*, 100839.

Allende, A., Datta, A.R., Smith, W.A., Adonis, R., MacKay, A. & Adell, A.D. (2018). Implications of new legislation (US FSMA) and guidelines (EC) on the establishment of management systems for agricultural water. *Food Microbiology*, *75*, 119–125.

Allende, A. & Monaghan, J. (2015). Irrigation water quality for leafy crops: a perspective of risks and potential solutions. *International Journal of Environmental Research and Public Health*, *12*(7), 7457–7477.

AHDB. (2017) Industry Guide to Good Hygiene Practice: Watercress. FSA/FSS. pp43. Available at: https://archive.ahdb.org.uk/knowledge-library/industry-guide-to-good-hygiene-practice-watercress [Accessed 12 August 2022]

Bamunuarachchi, B.A.D.S., Hitihamu, S., & Lurdu, M.S. (2019). Good Agricultural Practices (GAP) in Sri Lanka: Status, Challenges and Policy. Research Report No. 227. December 2019. Available at: http://www.harti.gov.lk/images/download/reasearch_report/new1/report_no_227.pdf [Accessed 20 August 2022]

Beacham, A.M., Vickers, L.H. & Monaghan, J.M. (2019). Vertical farming: a summary of approaches to growing skywards. *The Journal of Horticultural Science and Biotechnology*, *94*(3), 277–283.

CAC (Codex Alimentarius Commission). (2007). Joint FAO/WHO Food Standards Programme. Procedural Manual. Seventeenth edition. ISSN 1020-8070.

CAC (Codex Alimentarius Commission). (2019). The 27th Edition of the Procedural Manual of Codex Alimentarius Commission. Available at: https://www.fao.org/3/ca2329en/CA2329EN.pdf [Accessed 20 August 2022]

CAC (Codex Alimentarius Commission). (2020). General Principles of Food Hygiene. CXC 1-1969. Adopted in 1969. Amended in 1999. Revised in 1997, 2003, 2020. Editorial corrections in 2011. Available at: https://www.fao.org/fao-who-codexalimentarius/sh-proxy/en/?lnk=1&url=https%253A%252F%252Fworkspace.fao.org%252Fsites%252Fcodex%252FStandards%252FCXC%2B1-1969%252FCXC_001e.pdf [Accessed 20 September 2022]

Carpentier, B. & Cerf, O. (2011). Persistence of *Listeria monocytogenes* in food industry equipment and premises. *International Journal of Food Microbiology*, *145*(1), 1–8.

Dankwa, A.S., Machado, R.M. & Perry, J.J. (2020). Sources of food contamination in a closed hydroponic system. *Letters in Applied Microbiology*, *70*(1), 55–62.

da Cunha, D.T., Hakim, M.P., Soon, J.M., & Stedefeldt, E. (2022). Swiss -cheese model of food safety incidents: preventing foodborne illness through multiple layers of defence. *Food Control*, *139*, 109053.

De Keuckelaere, A., Jacxsens, L., Amoah, P., Medema, G., McClure, P., Jaykus, L.A., & Uyttendaele, M. (2015). Zero risk does not exist: lessons learned from microbial risk assessment related to use of water and safety of fresh produce. *Comprehensive Reviews in Food Science and Food Safety*, *14*(4), 387–410.

Di Gioia, F., De Bellis, P., Mininni, C., Santamaria, P. & Serio, F. (2017). Physicochemical, agronomical and microbiological evaluation of alternative growing media for the production of rapini (*Brassica rapa* L.) microgreens. *Journal of the Science of Food and Agriculture*, *97*(4), 1212–1219.

Doménech, E. & Martorell, S. (2016). Definition and usage of food safety margins for verifying compliance of food safety objectives. *Food Control*, *59*, 669–674.

Duret, S., Hoang, H.M., Derens-Bertheau, E., Delahaye, A., Laguerre, O. & Guillier, L. (2019). Combining quantitative risk assessment of human health, food waste, and energy consumption: The next step in the development of the food cold chain? *Risk Analysis*, *39*(4), 906–925.

EFSA (European Food Safety Authority). (2014). Scientific opinion on the risk posed by pathogens in food of non-animal origin, part 2. *Salmonella* and norovirus in leafy greens eaten raw as salads. *EFSA Journal*, 12(3), 3600.

EFSA Panel on Biological Hazards (BIOHAZ). (2011). Scientific Opinion on the risk posed by Shiga toxin-producing Escherichia coli (STEC) and other pathogenic bacteria in seeds and sprouted seeds. *EFSA Journal*, *9*(11), 2424.

FDA. (1998). Guide to Minimise Microbial Food Safety Hazards for Fresh Fruits and Vegetables. Available at: https://www.fda.gov/regulatory-information/search-fda

-guidance-documents/guidance-industry-guide-minimize-microbial-food-safety -hazards-fresh-fruits-and-vegetables [Accessed 20 August 2022].

Gómez, C., Currey, C.J., Dickson, R.W., Kim, H.J., Hernández, R., Sabeh, N.C., Raudales, R.E., Brumfield, R.G., Laury-Shaw, A., Wilke, A.K. & Lopez, R.G. (2019). Controlled environment food production for urban agriculture. *HortScience*, *54*(9), 1448–1458.

Gorris, L.G. (2005). Food safety objective: An integral part of food chain management. *Food Control*, *16*(9), 801–809. https://doi.org/10.1016/j.foodcont.2004.10.020

Guillier, L., Duret, S., Hoang, H.M., Flick, D., & Laguerre, O. (2016a). Is food safety compatible with food waste prevention and sustainability of the food chain? *Procedia Food Science*, *7*, 125–128.

Guillier, L., Duret, S., Hoang, H.M., Flick, D., Nguyen-Thé, C., & Laguerre, O. (2016b). Linking food waste prevention, energy consumption and microbial food safety: the next challenge of food policy? *Current Opinion in Food Science*, *12*, 30–35.

Holvoet, K., Sampers, I., Seynnaeve, M., Jacxsens, L. & Uyttendaele, M. (2015). Agricultural and management practices and bacterial contamination in greenhouse versus open field lettuce production. *International Journal of Environmental Research and Public Health*, *12*(1), 32–63.

Hutchison, M.L., Avery, S.M. & Monaghan, J.M. (2008). The air-borne distribution of zoonotic agents from livestock waste spreading and microbiological risk to fresh produce from contaminated irrigation sources. *Journal of Applied Microbiology*, *105*(3), 848–857.

ICMSF (International Commission on Microbiological Specifications for Foods). (2002). *Microorganisms in Food 7. Microbiological Testing in Food Safety Management*. Kluwer Academic/Plenum, NY.

Ilic, S., Moodispaw, M.R., Madden, L.V. & Lewis Ivey, M.L. (2022). Lettuce Contamination and Survival of *Salmonella* Typhimurium and *Listeria monocytogenes* in Hydroponic Nutrient Film Technique Systems. *Foods*, *11*(21), 3508.

Ilic, S., LeJeune, J., Ivey, M.L.L. & Miller, S. (2017). Delphi expert elicitation to prioritize food safety management practices in greenhouse production of tomatoes in the United States. *Food Control*, *78*, 108–115.

Işık, H., Topalcengiz, Z., Güner, S. & Aksoy, A. (2020). Generic and Shiga toxin-producing Escherichia coli (O157: H7) contamination of lettuce and radish microgreens grown in peat moss and perlite. *Food Control*, *111*, 107079.

Islam, M., Doyle, M.P., Phatak, S.C., Millner, P. & Jiang, X. (2004). Persistence of enterohemorrhagic Escherichia coli O157: H7 in soil and on leaf lettuce and parsley grown in fields treated with contaminated manure composts or irrigation water. *Journal of Food Protection*, *67*(7), 1365–1370.

Kirezieva, K., Luning, P.A., Jacxsens, L., Allende, A., Johannessen, G.S., Tondo, E.C., Rajkovic, A., Uyttendaele, M. & van Boekel, M.A. (2015a). Factors affecting the status of food safety management systems in the global fresh produce chain. *Food Control*, *52*, 85–97.

Kirezieva, K., Jacxsens, L., Hagelaar, G.J., van Boekel, M.A., Uyttendaele, M. & Luning, P.A. (2015b). Exploring the influence of context on food safety management: Case studies of leafy greens production in Europe. *Food Policy*, *51*, 158–170.

Kyriacou, M.C., Rouphael, Y., Di Gioia, F., Kyratzis, A., Serio, F., Renna, M., De Pascale, S. & Santamaria, P. (2016). Micro-scale vegetable production and the rise of microgreens. *Trends in Food Science & Technology*, *57*, 103–115.

Leistner, L. (1992). Food preservation by combined methods. *Food Research International*, *25*(2), 151–158.

Leistner, L. (2000). Basic aspects of food preservation by hurdle technology. *International Journal of Food Microbiology*, *55*(1–3), 181–186.

Li, Y., Zwe, Y.H., Tham, C.A.T., Zou, Y., Li, W. & Li, D. (2022). Fate and mitigation of Salmonella contaminated in lettuce (Lactuca sativa) seeds grown in a hydroponic system. *Journal of Applied Microbiology*, *132*(2), 1449–1456.

López-Gálvez, F., Gómez, P.A., Artés, F., Artés-Hernández, F. & Aguayo, E. (2021). Interactions between microbial food safety and environmental sustainability in the fresh produce supply chain. *Foods*, *10*(7), 1655.

Luning, P.A. & Marcelis, W.J. (2007). A conceptual model of food quality management functions based on a techno-managerial approach. *Trends in Food Science & Technology*, *18*(3), 159–166.

Luning, P.A., Marcelis, W.J., Rovira, J., Van Boekel, M.A.J.S., Uyttendaele, M. & Jacxsens, L. (2011). A tool to diagnose context riskiness in view of food safety activities and microbiological safety output. *Trends in Food Science & Technology*, *22*, S67–S79.

Machado-Moreira, B., Richards, K., Brennan, F., Abram, F. & Burgess, C.M. (2019). Microbial contamination of fresh produce: What, where, and how? *Comprehensive Reviews in Food Science & Food Safety*, *18*(6), 1727–1750.

Manning, L. (2018). Triangulation: Effective verification of food safety and quality management systems and associated organisational culture. *Worldwide Hospitality and Tourism Themes*, *10*(3), 297–312.

Manning, L. & Luning, P.A. (2018). *Food Safety in Agriculture: Determining Farm-Derived Food Safety Risk. Food safety for the 21st Century: Managing HACCP and Food Safety through the Global Chain, Wallace et al., Wiley*. Blackwell, Oxford.

Manning, L., Birchmore, I. & Morris, W. (2020). Swans and elephants: a typology to capture the challenges of food supply chain risk assessment. *Trends in Food Science & Technology*, *106*, 288–297.

Manning, L., Luning, P.A. & Wallace, C.A. (2019). The evolution and cultural framing of food safety management systems–Where from and where next? *Comprehensive Reviews in Food Science & Food Safety*, *18*(6), 1770–1792. https://doi.org/10.1111/1541-4337.12484

Manning, L. & Soon, J.M. (2014). Developing systems to control food adulteration. *Food Policy*, *49*, 23–32.

Misra, G. & Gibson, K.E. (2020). Survival of *Salmonella enterica* subsp. enterica serovar Javiana and *Listeria monocytogenes* is dependent on type of soil-free microgreen cultivation matrix. *Journal of Applied Microbiology*, *129*(6), 1720–1732.

Misra, G. & Gibson, K.E. (2021). Characterization of microgreen growing operations and associated food safety practices. *Food Protection Trends*, *41*(1), 56–69.

Monaghan, J.M., Augustin, J.C., Bassett, J., Betts, R., Pourkomailian, B. & Zwietering, M.H. (2016). Risk assessment or assessment of risk? Developing an evidence-based approach for primary producers of leafy vegetables to assess and manage microbial risks. *Journal of Food Protection*, *80*(5), 725–733.

Monaghan, J.M. & Hutchison, M.L. (2012). Distribution and decline of human pathogenic bacteria in soil after application in irrigation water and the potential for soil-splash-mediated dispersal onto fresh produce. *Journal of Applied Microbiology*, *112*(5), 1007–1019.

Monaghan, J.M., Thomas, D.J.I., Hutchison, M.L. & Goodburn, K. (2012). Overview of current UK fresh produce farming practices that minimise the risk of foodborne illness outbreaks associated with ready to eat fresh produce. *Acta Hortic*, *19*, 26.

Mogren, L., Windstam, S., Boqvist, S., Vågsholm, I., Söderqvist, K., Rosberg, A.K., Lindén, J., Mulaosmanovic, E., Karlsson, M., Uhlig, E. & Håkansson, Å. (2018). The hurdle approach–A holistic concept for controlling food safety risks associated with pathogenic bacterial contamination of leafy green vegetables. A review. *Frontiers in Microbiology*, *9*, 1965.

Monaghan, J.M. & Hutchison, M.L. (2016). Ineffective hand washing and the contamination of carrots after using a field latrine. *Letters in Applied Microbiology*, *62*(4), 299–303.

Niu, G. & Masabni, J. (2022). Hydroponics. In *Plant Factory Basics, Applications and Advances* (pp. 153–166). Academic Press. https://doi.org/10.1016/B978-0-323-85152-7.00023-9

Orozco, L., Rico-Romero, L. & Escartin, E.F. (2008). Microbiological profile of greenhouses in a farm producing hydroponic tomatoes. *Journal of Food Protection*, *71*(1), 60–65.

Powell, D.A., Jacob, C.J. & Chapman, B.J. (2011). Enhancing food safety culture to reduce rates of foodborne illness, *Food Control*, *22*, 817–822.

Prince-Guerra, J.L., Nace, M.E., Lyles, R.H., Fabiszewski de Aceituno, A.M., Bartz, F.E., Arbogast, J.W., Gentry-Shields, J., Jaykus, L.A., Heredia, N., García, S. & Leon, J.S. (2020). Both handwashing and an alcohol-based hand sanitizer intervention reduce soil and microbial contamination on farmworker hands during harvest but produce type matters. *Applied and Environmental Microbiology*, *86*(18), e00780-20.

Qian, C., Liu, Y., Barnett-Neefs, C., Salgia, S., Serbetci, O., Adalja, A., Acharya, J., Zhao, Q., Ivanek, R. & Wiedmann, M. (2022). A perspective on data sharing in digital food safety systems. *Critical Reviews in Food Science and Nutrition*, 1–17. https://doi.org/10.1080/10408398.2022.2103086

Ranjbar, B., Naeimi, A., & Badsar, M. (2022). Designing an integrated model for strawberry growers' behavior toward implementation of good agricultural practices in Iran. *Environment, Development and Sustainability*, *24*(9), 10924–10944.

Red Tractor. (2022). Fresh Produce Standards. Available at: https://redtractorassurance.org.uk/wp-content/uploads/2022/08/FreshProduce-V5-Standards-Aug-22-FINAL.pdf [Accessed 25 October 2022]

Riggio, G.M., Wang, Q., Kniel, K.E. & Gibson, K.E., (2019). Microgreens–A review of food safety considerations along the farm to fork continuum. *International journal of food microbiology*, *290*, 76–85.

Sawe, C.T., Onyango, C.M. & Njage, P.M.K. (2014). Current food safety management systems in fresh produce exporting industry are associated with lower performance due to context riskiness: case study. *Food Control*, *40*, 335–343.

Sidhu, R.K., Cavallaro, N.D., Pola, C.C., Danyluk, M.D., McLamore, E.S. & Gomes, C.L. (2020). Planar interdigitated aptasensor for flow-through detection of Listeria spp. in hydroponic lettuce growth media. *Sensors*, *20*(20), 5773.

Sobolik, J.S., Newman, K.L., Jaykus, L.A., Bihn, E.A. & Leon, J.S. (2021). Norovirus transmission mitigation strategies during simulated produce harvest and packing. *International Journal of Food Microbiology*, *357*, 109365.

Spalding, A., Goodhue, R.E., Kiesel, K. & Sexton, R.J. (2022). Economic impacts of food safety incidents in a modern supply chain: E. coli in the romaine lettuce industry. *American Journal of Agricultural Economics*, 1–27. https://doi.org/10.1111/ajae.12341

Turner, E.R., Luo, Y. & Buchanan, R.L. (2020). Microgreen nutrition, food safety, and shelf life: A review. *Journal of Food Science*, *85*(4), 870–882.

UN FAO (Food and Agriculture Organisation). (2016). A Scheme and Training Manual on Good Agricultural Practices (GAP) for Fruits and Vegetables. Volume 1. The scheme – standard and implementation structure. Available at: https://www.fao.org/3/i6677e/i6677e.pdf [Accessed 20 August 2022]

Van Schothorst, M., Zwietering, M.H., Ross, T., Buchanan, R.L. & Cole, M.B. (2009). Relating microbiological criteria to food safety objectives and performance objectives. *Food Control*, *20*(11), 967–979.

Wang, Q. & Kniel, K.E. (2016). Survival and transfer of murine norovirus within a hydroponic system during kale and mustard microgreen harvesting. *Applied Environmental Microbiology*, *82*, 705–713.

Wadamori, Y., Gooneratne, R. & Hussain, M.A. (2017). Outbreaks and factors influencing microbiological contamination of fresh produce. *Journal of the Science of Food and Agriculture*, *97*(5), 1396–1403.

Whiting, R.C., Rainosek, A., Buchanan, R.L., Miliotis, M., LaBarre, D., Long, W., Ruple, A. & Schaub, S. (2006). Determining the microbiological criteria for lot rejection from the performance objective or food safety objective. *International Journal of Food Microbiology*, *110*(3), 263–267.

WHO. (1997). *Food Safety and Globalization of Trade in Food, a challenge to the public health sector*. WHO/FSF/FOS/97.8 Rev. 1, WHO, Geneva.

Wiśniewska, M. (2023). Just culture and the reporting of food safety incidents. *British Food Journal*, *125*(1), 302–317.

Wright, K.M. & Holden, N.J. (2018). Quantification and colonisation dynamics of *Escherichia coli* O157:H7 inoculation of microgreens species and plant growth substrates. *International Journal of Food Microbiology*, *273*, 1–10.

Xiao, Z., Bauchan, G., Nichols-Russell, L., Luo, Y., Wang, Q. & Nou, X. (2015). Proliferation of *Escherichia coli* O157: H7 in soil-substitute and hydroponic microgreen production systems. *Journal of Food Protection*, *78*(10), 1785–1790.

Chapter 15

Measuring and modelling on-farm greenhouse gas emissions : the basis for achieving climate-neutral farming

Matthias Kuhnert, Mohamed Abdalla, and Sylvia H. Vetter, University of Aberdeen, UK; and Durba Kashyap, INRA, France

1 Introduction

The agricultural sector, forestry and other land use contribute around 24% of the global greenhouse gas emissions (GHGEs) (IPCC, 2014), with 10–14% directly coming from agriculture (Paustian et al., 2016). The major gases contributing to GHGEs from agriculture are carbon dioxide (CO_2), methane (CH_4) and nitrous oxides (N_2O). While CO_2 shows the largest emissions, the other two gases show a stronger impact on climate change due to their larger global warming potential (GWP). The GWP is a value that indicates the impact on the GHGE effect, with CH_4 showing about 28 times, and N_2O 273 times the impacts compared to CO_2 over a hundred-year time period (IPCC, 2021).

http://dx.doi.org/10.19103/AS.2024.0133.29

All three GHGEs are emitted from livestock, grasslands and croplands, but their relevance for different land uses varies. While CH_4 is more relevant for the livestock production, CO_2 and N_2O are of higher relevance for crop and grassland systems. There are exceptions to this as, for example, CH_4 is more relevant in rice cultivation and wetlands than the other GHGEs.

The agricultural sector is of special interest when considering climate change mitigation strategies. One aspect is the relevant contribution of the sector to the global GHGEs, the other is the potential through soil carbon sequestration to achieve negative emissions that act as a carbon sink. Another crucial aspect is that the sector produces food, with increasing demand due to the increasing world population. As a consequence, food production must increase, while the emissions have to decrease, which requires a balancing strategy (Frank et al., 2017).

There are different suggestions about how to achieve reduced or negative emissions in agriculture (Paustian et al., 2016, 2017; Vetter et al., 2022), but political or economic legislation or incentives are not yet established. Carbon certificate trading for carbon emissions from the agricultural sector has started already, but clear regulations are not yet in place, especially for the negative emissions in the form of carbon capture from the atmosphere and storage in the soil (Kuhnert et al., 2022). A key bottleneck is the methodologies for quantification of GHGEs and soil organic carbon for individual farms or farm communities. The certification of carbon savings requires a functional Measurement, Reporting and Verification (MRV) system to monitor emission changes. These systems require on-farm measurements, which are costly, labour intensive and often require expert knowledge. Modelling is an obvious alternative but modelling also requires specific input data from measurements for location specific parameterisation, calibration and evaluation (Kuhnert et al., 2022). Additionally, modelling might not be able to provide the same level of accuracy for GHGEs as with on-farm measurements.

The objective of this chapter is to provide an overview of the available approaches to quantify GHGEs at the farm level, including measurement-based and modelling approaches. Different approaches are separated by having a livestock focus and/or cropland or grassland focus. A clear separation is not possible for all presented options, as GHGEs from both farming systems (e.g. from grazing animals on grassland) are partly measured or modelled together. Measurement options from croplands and grasslands are mainly the same, but the way they are used might differ due to differences in the management of the two systems. The chapter includes a short discussion about the contribution of these methods to achieve net zero GHGEs at the farm scale.

2 Greenhouse gas emissions on the farm

Farms can be distinguished into three categories:

- Crop production farms;
- Livestock and/or poultry farms; and
- Farms that produce both crops and livestock.

For the rest of this chapter, poultry is not considered individually but is still included in the category if livestock is mentioned. Depending on the farm type, GHGEs originate from animals (housed or grazing animals), from the soil (grassland or cropland), from machinery or from other sources (e.g. vehicles, manure storage and plants). All these carbon sources and sinks need to be compared and summarised to quantify the carbon balance for the farm. The farm is a net carbon source if the overall balance shows more emissions to the atmosphere than carbon fixation. This will be calculated in CO_2-equivalent, which means for non-CO_2 gases the climate impacts will be calculated based on the impact of CO_2, using the GWP of each of the GHGEs. A farm is acknowledged as a carbon sink if the emission balance shows more fixation than production of emissions. The major process of fixing atmospheric carbon is through photosynthesis by plants and biomass production and carbon storage in the soil, in the form of plant residues, animal excreta or manure.

Negative emissions (reduction or sequestration) are the major target of recent initiatives to achieve climate-neutral farming or so-called net zero-emission targets. Obviously, not all GHGEs can be avoided but more carbon can be captured or emissions reduced via specific land and livestock management changes. A further discussion of the different available options to achieve negative emissions is beyond the scope of this chapter, but this theme is explored by Paustian et al. (2016) and Vetter et al. (2022). Significant management changes can make the farm a carbon sink allowing farms to enter into carbon certificate trading.

The carbon balance is the net carbon balance (all gains and losses included) on the farm and its calculation requires a clear definition of the farm and its activities. While the operational area of the farm is well-defined, products that are imported and exported from the farm need to be considered as well. All these goods (e.g. imported fertiliser or seedlings) have a specific carbon footprint. Imported goods, e.g. fertiliser or seedlings, have a carbon footprint from their production, and potentially exported goods like manure will have either a positive or negative impact on the farm carbon balance. On-field application can increase the soil organic carbon content, but long storage will result in emissions, and sale of manure and resultant crops if it is used on will reduce the carbon balance. Exported products will be also transported, stored

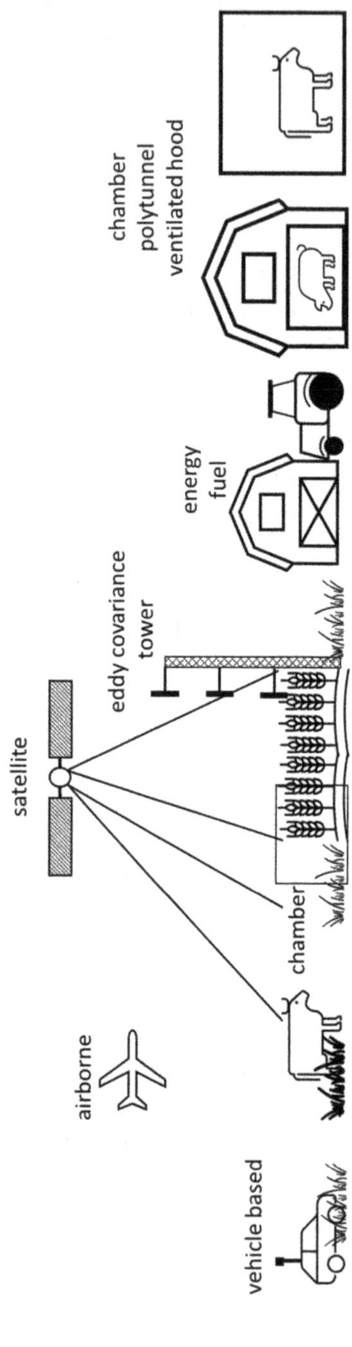

Figure 1 Overview about measurement options of GHG emissions on the farm. Small-scale measurement systems like handhold system, masks for animals and other measurements in animals or on the microscale are not included in this figure.

and packed, which is difficult to predict at the farm stage. Therefore, farm-specific emissions are often calculated as emissions up to the farmgate, which includes all emissions until the goods leave the farm, but no more.

This chapter focuses on on-farm GHGEs which can be separated into different categories:

- Imported products;
- Emissions produced on the farm; and
- Products that capture carbon from the atmosphere (e.g. harvested crops) which are then exported from the farm.

Measurement approaches target different scales, with different levels of accuracy. There is a wide range of different measurement options (Fig. 1) that consider single or several parts of the farm system. In some approaches, specific distinct measurement calculations are combined to achieve the overall GHGE balance for the farm. The different methodologies are separated into measurements based on cropland and grassland and then livestock-related measurements.

3 Methods to measure greenhouse gas emissions from cropland and grasslands: chamber measurements

Farms often include many fields, which creates challenges in measuring everything at that level of detail. There are therefore different options to measure GHGEs on different scales and in varying resolutions. The target scales include:

- the plot scale (chamber measurement);
- the field scale (eddy covariance (EC) measurements); and
- the landscape scale (remote sensing).

This means the methods consider a couple of square metres, a couple of hectares or a couple of square kilometres, respectively. The landscape scale can exceed the individual farm scale, depending on the size of the farm, but remote sensing can also be applied on a large mono-crop farm area like large crop or grassland areas (e.g. farms with >100 ha fields). In general, the fluxes of the three most prominent GHGEs, i.e. CO_2, N_2O and CH_4, are all measured with similar approaches, but the technical details can differ. The different methods used for measuring GHGEs from cropland and grassland soils can have a significant effect on the volume and quality of data for subsequent analysis. It is important to note that no standard approaches for GHGE measurements have yet been agreed. Different methods are summarised in this and the following sections. This section focuses on chamber measurements.

Two well-known types of flux chambers have been used for measuring GHGEs from different ecosystems (Maier et al., 2022):

- Manual flux chambers; and
- Automatic flux chambers.

Globally, there are many different flux chamber techniques, approaches and designs available depending on the research objectives. Flux chambers are easy to apply and inexpensive and, therefore, widely used for estimating the exchange of GHGs between soils, vegetation and the atmosphere. Chambers for measuring CO_2 are usually connected to a gas analyser in the field (Rochette, 2011). However, measurements for N_2O and most of CH_4 require samples to be collected and sent to a laboratory for analysis by gas chromatography (GC). This can result in a significant error in the data due to errors arising from gas sampling and from gas storage; these potential sources of error should be carefully considered and uncertainties reduced.

Flux chambers have different requirements, limitations, advantages and disadvantages, which have relevant effects on upscaling of data, calculation of emission factors and comparability to data of other studies (Hutchinson and Livingston, 2001). To avoid having poor quality data which are also less suitable for modelling or meta-analyses, good decisions about study design, chamber design and collection of ancillary data are crucial. To improve the quality of individual flux measurements (and thus the GHG analysis on which they are based), sources of uncertainty and errors in these chambers should be reduced. This could be done by handling them carefully, carrying out quality assurance on gas measurements, increasing the number of measurement replicates and improving data analyses.

The recommended method for measuring gas fluxes at long-term sites is the non-steady-state method. Most flux chambers therefore adopt the non-steady-state technique, where the soil surface is temporarily covered with a chamber and the gas flux across the surface is calculated from the chamber headspace concentration change over time (Livingston et al., 2006; Maier et al., 2022).

Handling of the flux chamber by an experienced expert is very important. When placing the flux chambers in the field, fluctuations of initial pressure should be avoided and chamber deployment time should be optimised. Sampling from automated chambers is continuous but, for manual chambers, regular samples should be collected weekly or fortnightly throughout the experimental period and more intensively when management practices are being carried out (e.g. fertiliser application). Syringes or vials can be used for manual sampling and at least four replicates should be collected. GC measurement quality and calibrations should be optimised and checked regularly. If using an online gas analyser, which increases efficiency and quality of chamber measurements, the potential for leaks, as well as the data being collected, should be checked directly in the field (Maier et al., 2022).

There is controversy in the literature about using a chamber vent. Using a vent can result in an equal pressure in and outside the chamber. However, use of a vent can also result in depressurising the chamber when the wind blows over it (Maier et al., 2022; Rochette, 2011). Some studies found that vent chambers can overestimate the N_2O gas flux due to the upward mass flow of soil air in response to depressurisation. Gas leaks could also result from chamber walls if they are not properly inserted in soils. A good seal between the chamber and its surrounding is required. Although most of the evidence on its use supports the use of a vent tube in non-steady-state chambers, the vent should have optimised size, be mounted near the ground and with a horizontal outlet pointed downwind to avoid adverse chamber depressurisation (Hutchinson and Livingston, 2001; Livingston et al., 2006). The vent can be shielded by the open end of the tube when the wind is blowing strong.

4 Methods to measure greenhouse gas emissions from cropland and grasslands: eddy covariance measurements

The EC technique allows long-term, direct and continuous measurements of turbulent fluxes over an ecosystem, like a grassland or a field, to quantify the net CO_2 budget among other variables (Aubinet et al., 2000; Baldocchi et al., 2001). Turbulent flux refers to the complex atmospheric processes occurring just above land surfaces. Through capturing fluctuations in temperature, trace gases and particles, the turbulent flux is determined between the land surface and the atmosphere.

The measurement system consists typically of a 3-D sonic anemometer/thermometer and one or more additional gas analysers (e.g. CO_2, CH_4) to capture high-frequency fluctuations. These instruments are placed on a meteorological mast above the field to sample the turbulence in the field assuming turbulent air eddies are carried along in the wind (Fig. 2). Other requirements for a measurement site are horizontal homogeneity, steady-state conditions and well-developed turbulence (Aubinet et al., 2000; Mauder et al., 2021). The equipment should be placed high enough above the surface to ensure a large enough footprint, capturing the fluxes from the ecosystem of interest (e.g. the field) but not too high that the footprint during stable night-time conditions extends beyond the boundary layer above the ecosystem of interest (Foken et al., 2012). Besides direct measurements, the method requires

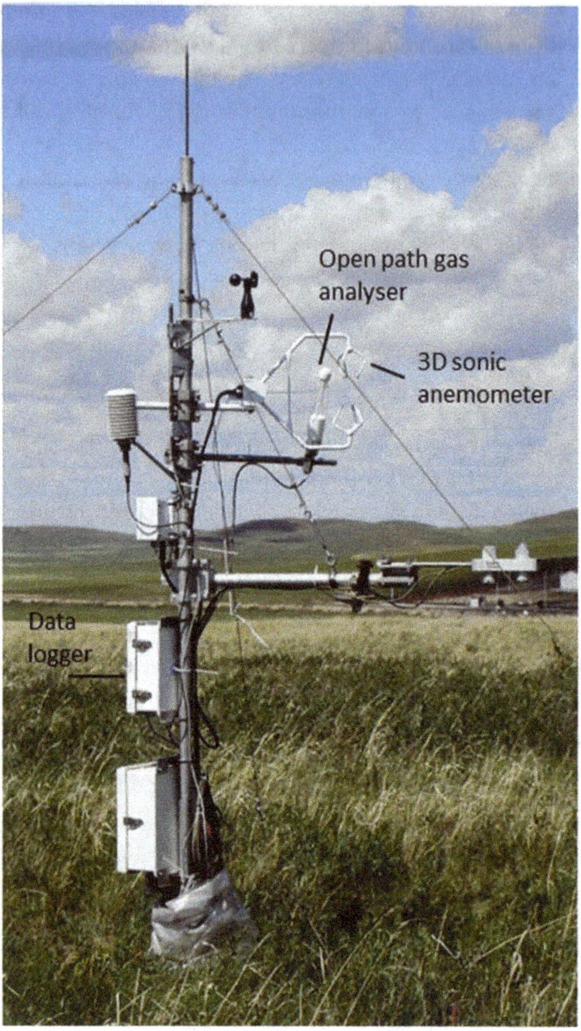

Figure 2 Eddy covariance measurement tower, open path gas analyser, 3D sonic anemometer; meteorological measurements. Source: Adapted from: picture by S. Vetter.

comprehensive post-processing analysis including a series of quality tests and flux corrections (Aubinet et al., 2000). Meteorological measurements like radiation, precipitation, temperature and soil temperature are needed for flux correction, gap filling and data analysis.

Based on the raw measurements, the vertical flux is then calculated within an averaging interval of approximately 10–30 min by

$$F_x = \overline{wx} = \overline{w}\overline{x} + \overline{w'x'}$$

where w is the vertical wind velocity component and x is either the horizontal wind velocity u in the direction of the mean flow or a scalar in the case of scalar fluxes, such as the sensible heat flux, the latent heat flux or trace gas fluxes (Aubinet et al., 2000; Mauder et al., 2021); $\overline{w'x'}$ represents the covariance.

The net ecosystem exchange of CO_2 (NEE) derives from the calculated flux and the storage component. These typically half-hour datapoints are used for subsequent processing and gap filling (Falge et al., 2001; Reichstein et al., 2005) to calculate daily, monthly and annual balances. Over the last two decades, software packages have been developed to help with processing including handbooks on good practice (Aubinet et al., 2000, 2012).

Continuous EC measurements became possible with the development of a newer generation of sonic anemometers and infrared gas analysers for water vapour and CO_2 in the 1990s (Foken et al., 2012). The first EC measurement towers were installed in 1990s (Baldocchi et al., 2001) and, with appropriate guidance, began to be used by non-micrometeorologists (Aubinet et al., 2000). Further developments such as tunable diode lasers and quantum cascade laser spectrometers allowed measurements of methane (CH_4) and nitrous oxide (N_2O) (Foken et al., 2012).

The infrared gas analyser can be used in an open- or closed-path configuration. The open-path system analyses the reduced intensity in the light path as light is absorbed by CO_2 or H_2O (for example) in the open air. A closed-path analyser has an internal sample cell that is flushed by sampled air, transported from the top of the tower (next to the sonic anemometer) and a second cell purged by a small flow of air with a known (can be zero) CO_2 and H_2O molar concentration (Munger et al., 2012).

Each system has its own advantages and disadvantages (Mauder et al., 2021; Munger et al., 2012) and the choice of a system depends on the measuring site, accessibility and the power supply. An open-path system has the advantage that it can measure air eddies simultaneously next to the wind anemometer. This system has a low-energy requirement and reduces the need for extensive calibration. The main disadvantage is the loss of flux data due to effects of rain, fog, mist and snow, which can impair open path measurements.

The closed path system can be used under all weather conditions but, due to the time differences in measurements between the sonic anemometer and the infrared gas analyser, extensive calibration and corrections are needed to correct the fluxes based on tube length and pump rate. The system needs to be maintained regularly (e.g. each week) and a continuous power supply need to be guaranteed.

EC measurements over short vegetation like crops and grasslands are made closer to the ground as compared to forest ecosystems. This corresponds to a smaller storage component for shorter vegetation and minimises errors when analysing the NEE of carbon fluxes (Wohlfahrt et al., 2012). However, croplands and some grasslands are managed ecosystems and present challenges for continuous long-term measurements due to different conditions over the course of a year. The soil e.g. might lay bare before planting or have full growth crop vegetative cover before harvest. Consequently, net CO_2 fluxes evolve during the cropping period and are subject to large fluctuations (Moureaux et al., 2012). The main challenges for EC measurements over croplands derive from the potential interference from management practices and crop development over a growing season (Moureaux et al., 2012). For a correct analysis of the data, additional information and data on crop rotations, and management dates are required. Knowledge about sowing, harvest and tillage dates, plant density, leaf area index (LAI) and biomass distribution are important to fully understand the measured fluxes.

In order to reduce disruption, the measurement tower can be installed on the field border to reduce obstacles to the farmer but also close enough to measure the field footprint. If the tower is installed in the middle of the field, the farmer will have to adapt smart driving strategies. Another option is mobile measurement systems where the mast is not fixed permanently at one site.

Vehicle-based measurements have also been developed using spectroscopy to quantify the CH_4 fluxes. One example is based on a three Aerodyne single-laser quantum cascade laser spectrometer to measure CH_4 and ethane fluxes (Arndt et al., 2018). The system was combined with a nondispersive infrared LI-6262 gas analyser to measure CO_2 and H_2O fluxes. The system requires additional measurement of wind speed and calibration to account for hot spots. This system does not allow continuous measurements but is able to measure spatial distribution of GHGEs.

5 Methods to measure greenhouse gas emissions from cropland and grasslands: remote sensing

Remote sensing is a general term for non-invasive measurements made from a distance. This includes approaches that measure near the ground, e.g. using unmanned aerial vehicles (e.g. drones) or from a further distance above the ground e.g. by airplanes or satellites. EC closed path CH_4 measurement systems e.g. are used by planes to measure emissions from the ground. These can be used to measure emissions from livestock (Arndt et al., 2018) or for crop- and grasslands (Shen et al., 2022). Remote sensing is most suitable for larger areas of several 100 ha.

As noted, drones can potentially be used to measure GHGEs (Tedeschi et al., 2022). However, few examples for the use of this measurement approach in crop- and grasslands have been reported in the scientific literature to date (Tedeschi et al., 2022). Drones are currently used primarily for monitoring crops (Roth et al., 2020) and livestock (Behjati et al., 2021). There are some examples of CH_4 measurements by drones over larger areas (Gålfalk et al., 2021) or for detecting gas leaks in pipelines (Barchyn et al., 2019), which indicate the potential for the use to measure GHGEs on farms. Drones will be an attractive option to quantify farm-derived GHGEs over larger areas in the future.

Measurements using instrumentation mounted on airplanes are already used on large-scale farms as in the USA. Arndt et al. (2018) used a closed path EC system for CH_4 measurements with an airplane over a large herd of cattle on two farms. The measurement approach used was found to be comparable in accuracy with an open-path EC measurement system on the ground and with vehicle-based measurements using the tracer ratio method. Pressure and temperature information are used to convert the measured CH_4 mixing ratio to CH_4 density and, by considering the wind speed, it is possible to determine the CH_4 flux vector.

Remote sensing approaches can potentially use also precision imagery to derive different ground-based variables or fluxes. While the technology is already established for biomass estimates using the normalised difference vegetation index (NDVI), the estimation of GHG fluxes is still under development and requires further validation before it can be used.

6 Methods to measure greenhouse gas emissions from livestock

Measurement options for GHGEs in livestock can be broadly categorised into:

- animal-based;
- facility-based; and
- large-scale measurements.

These methods differ in complexity, accuracy, labour requirements, cost, types of animal management for which they are relevant, duration of measurement and repeatability over time (Jonker and Waghorn, 2020). Each technique has its own specific requirements, advantages and limitations which, when used outside of the intended purpose or required conditions, affects the validity of the final result.

6.1 *Animal-based greenhouse gas emission measurement techniques*

Animal-based measurements involve techniques that measure GHGEs emitted by individual animals. These can be further classified into three categories:

- Direct gas measurement techniques;
- Spot sampling techniques; and
- Indirect measurement techniques.

Direct gas measurement techniques estimate CH_4 in ambient air by calculating the difference in CH_4 concentrations with and without the animal. Among these, respiration chambers are the most common technique that has been used for decades and is regarded as the 'gold standard' for measuring individual livestock emissions (Goopy et al., 2011). Other methods are explained in Table 1.

Spot sampling techniques are short-term measurement techniques that measure CH_4 concentration in the breath of individual animals for short periods of time. Over the last decade, several new methods such as Greenfeed®, sniffers, hand-held lasers and portable chambers (Table 2) have been developed. These methods are simpler to use, less invasive, less labour and cost-intensive compared to more established methods like respiration chambers and the (sulphur hexafluoride) SF_6 tracer technique (Jonker and Waghorn, 2020).

Indirect measurement techniques like *in vitro* incubation and diet estimations provide an alternative to the high complexity and cost associated with the direct techniques, and can be used for initial screening of feed-related solutions/treatments (Table 3) (Tedeschi et al., 2022).

Apart from these techniques, new approaches are listed in Table 4. Similar to croplands and grasslands, these measurements address different scales and provide more specific or aggregated results for calculating emissions.

6.2 Facility-based greenhouse gas emission measurement techniques

In addition to the measurements at the individual animal and herd scale, there are facility-based techniques that cover measurements of emissions from animal housing and manure storage facilities. In the case of animal housing, GHGE measurement of barns housing livestock relies on understanding and measuring the ventilation rate of the barn. There are different ways to identify air exchange or ventilation rates:

- Naturally ventilated barns: Indirect gas balancing methods with internal (e.g. CO_2) and external tracer gas (e.g. gas not produced in barn like SF_6) can be used (Ogink et al., 2013). Air exchange rates are derived by measuring the dilution of tracer gas with a known release rate.
- Mechanically ventilated barns: Continuous monitoring through sensors is used to estimate the ventilation rate. Emission rate of a target gas inside a livestock facility is determined by multiplying air exchange rate and target gas concentration.

Table 1 Methods for direct gas measurements on livestock

Technique	Level	Description	Advantage	Limitation
Respiration chambers (open and closed circuit)	A	Sealed and environmentally controlled chambers where test animals are kept while gas composition is analysed using indirect calorimetry. Open circuit chambers use air inlet and exhaust to maintain one-way stream of air, while accumulated air over time is analysed in closed circuit chambers (Johnson and Johnson, 1995).	Accurate and continuous data; can be used to measure effects of treatment diets	High cost; does not reflect normal behaviour, technically demanding, monitoring limited to a few animals at a given time
Polytunnels	A	Large (tent type or inflatable) polyethylene tunnels where animals are housed in controlled conditions; an alternative to respiration chambers with simpler operation.	Portable, allows normal grazing, can be used to measure effects of treatment diets	Requires frequent calibration, difficult to maintain temperature and humidity inside the tunnel
Ventilated hood and/or headbox systems	A	Simplified version of respiration chambers which measures gas exchange from head only (placed as airtight box around animal's head)	Portable, less expensive than respiration chambers	Hindgut emissions not measured; adaptation required
SF_6 tracer technique	A	SF_6 permeation tube is inserted into rumen of the animal (orally administered). CH_4 is estimated using known release rate of the tracer gas and the ratio of expired CH_4 and tracer gas concentrations near mouth and nostrils (Johnson et al., 1994).	Accurate, relatively inexpensive, allows: (1) large no. of animals to be monitored, (2) free movement	SF_6 is a highly potent GHG with GWP 22800; can disrupt normal animal behaviour, skill and labour intensive, less precise, high equipment failure risk, hindgut emissions not measured
Open-path laser	H	Uses infrared diode laser and micrometeorological dispersion approach to send beams of light to retro reflector which reflects beam back to the detector. Intensity of received light indicates CH_4 concentrations.	Measures herd scale emissions; facilitates whole farm measurements	High cost; high data loss, not very accurate; restricted by environmental factors

The levels include animal (A) and herd (H). Source: Adapted from: Tedeschi et al., 2022; Kumari et al., 2020; Goopy et al., 2016; Bekele et al., 2022.

Table 2 Spot sampling measurements on livestock

Technique	Level	Description	Advantage	Limitation
Inline sniffers	A	Captures and measures breath CH_4 and CO_2 concentrations, while animal visits a feed bin or dish (equipped with sensor) of an automatic milking system, milking parlour, feed kiosk or automated feed bin.	No interference with animal normal habits/routine; allows for repeated measurements over long periods	High variation (within- and between-animal CH_4 concentration), less precise, flux not measured
Portable accumulation chambers	A	Similar working principal as closed-circuit respiration chambers. Animal is place in a plexiglass/polycarbonate bottom-less box sealed on the animal. Measures short-term (1–2 h) CH_4 from grazing sheep (Goopy et al., 2011).	Portable, applicable for large number of animals for genetic screening (of methane production)	Only sheep, strict and consistent measurement protocol has to be followed
Face mask	A	Placed over animal muzzle for short time periods (e.g. 30 min), simpler.	Repeatable over time	Discomfort for animal (restricts eating/drinking) and manual (labour intensive).
Hand-held laser methane detectors	A	Uses infrared absorption spectroscopy to measure CH_4 plume concentrations in ppm from animal's breathe. It is pointed towards nostrils at fixed distance.	Quick, accurate, portable, easy to use	challenging to get good-quality and consistent samples (e.g. due to environmental conditions and operator variability), as well as to analyse and interpret data. Only concentration (and not flux) is measured.
Application of CH_4: CO_2	A	Ratio of CH_4: CO_2 in exhaled breathe is used to estimate CH_4 production. Relies on feed intake (energy content) and incremental heat generated from it (Madsen et al., 2010).	Potential to measure large no. of animals	Unsuitable for precision measurements
Greenfeed ®	A	Patented gas-flux quantification system consisting of head chamber with an auto-feeder (overhead hopper for bait) that measures CO_2, CH_4 and airflow. CH_4 concentrations recorded in short term 3–6 min) for every 24 h for each animal.	Ease of use, suitable for large number of animals in both research and commercial settings, suitable for grazing conditions, indoors and outdoors group-housing and individually penned animals (e.g. tie-stalls).	Animal training required, bait may interfere with the dietary treatments, hindgut emissions not considered, measurement every 24 h.

The levels include animal (A) and herd (H). Source: Adapted from: Tedeschi et al., 2022; Kumari et al., 2020; Goopy et al., 2016; Jonker and Waghorn (2020).

Table 3 Methods for indirect gas measurements on livestock

Technique	Level	Description	Advantage	Limitation
In vitro incubation	*In vitro*	Ruminal fermentation of feedstuff is evaluated by incubation of rumen inoculum with feed substrates under anaerobic conditions in air-tight bottles leading to gas accumulation and subsequent analysis.	High reproducibility allows evaluation of different rumen environments	Outcome can be different from actual measurements, relies on rumen of donor animal, not always representative of live animal emissions, standardisation can be difficult
Diet estimation		Uses models to estimate CH_4 from feed intake and quality	Allows estimation when measurement is not possible	Requires multiple measurements to account for seasonal differences; may require validation with respiration chamber measurements

Source: Adapted from: Tedeschi et al., 2022; Kumari et al., 2020.

Techniques like GC, spectroscopic techniques and lasers are used to measure CH_4 inside and outside the barn for emission rate measurements. Apart from these measurements, other methods like inverse modelling are also used to estimate CH_4 in animal housing (Tedeschi et al., 2022).

Manure storage can contribute significantly to the overall GHGE balance. Livestock produces manure, that has to be stored for further use, either on the farm or for export. Some livestock production generates slurry that is transferred out of the housing area to storage in tanks or lagoons. N_2O, CH_4 and CO_2 from these storage areas are measured using various techniques, with flux chambers being the most commonly used method (Kupper et al., 2020). Other measurement methods include dispersion modelling based on a backward Lagrangian stochastic (bLS) or UK-ADMS atmospheric dispersion model, flux gradient method, micrometeorological mass balance method (e.g. integrated horizontal flux, IHF; vertical radial plume mapping, VRPM), sampling at exhaust chimney, and tracer gas (rarely applied) (Kupper et al., 2020).

6.3 Large-scale greenhouse gas emission measurement techniques

Measurement techniques addressing GHGEs at a large scale use remote sensing approaches e.g. to quantify GHGEs from large herds grazing pasture.

Table 4 Recent and new methods for gas measurements on livestock

Technique	Level	Description	Advantage	Limitation
Infrared thermography	A	Thermal imager is used to detect radiation (explores relationship between heat and methane production) (Montanholi et al. (2008).	Simple, inexpensive, non-invasive	Temperature and emission relationship not yet well established
Intraruminal telemetry	A	Uses intra-rumen device including gas sensors and wireless networks to measure real time concentrations of CH_4, CO_2 and H_2 inside rumen.	Accurate, small changes (<50 µmol/L) could be detected (Gibbs et al., 2008), large no. of free grazing animals can be monitored	Still in exploratory stages; rumen environment could corrode electric circuits (https://www.wolf-hound.com/)
Eddy covariance	E	Micrometeorological method directly recording gas, energy and momentum exchange between ecosystem and atmosphere.		High cost, subject to atmospheric/weather conditions
Quantitative molecular biology	A	Uses correlation between methanogen activity and CH_4 production in short time intervals		
Blood methane concentration	A	Using GC, enteric CH_4 is estimated as it absorbs into rumen wall and gets transported to blood stream and respired by lungs (Ramírez-Restrepo et al., 2010).	Potential to measure large no. of animals	Could be logistically challenging and labour intensive

The levels include animal (A) and ecosytem (E). Source: Adapted from: Kumari et al., 2020; Goopy et al., 2016; Bekele et al., 2022.

Aircraft, drones and satellites have been used in association with the tracer flux method, inverse modelling, imagery and LiDAR (light detection and ranging) to monitor emissions at herd level over large areas like feedlots and pastures. However, research is a still lagging in validating these techniques (Tedeschi et al., 2022). While these methods provide aggregate estimates, they have has limited use for segregating source-specific emissions (Kumari et al., 2020).

7 Models to estimate greenhouse gas emissions on the farm

Taking actual measurements is costly as well as labour and time intensive. Models provide a cheaper and faster alternative to estimate GHGEs instead of detailed measurements. Models exist for different scales of assessment, different complexities and for different target variables. Typical scales are field, farm, landscape or national scale, depending on different stakeholder different objectives. The complexity of tools and models can be categorised by the tier system, introduced by the Intergovernmental Panel on Climate Change (IPCC). The system includes three categories:

- Tier 1: Indicates the lowest level of complexity (a generic factor, which provides an estimate with a high uncertainty);
- Tier 2: Indicates medium complexity (national specific factor or simple equation, that allows a more specific estimate); and
- Tier 3: Indicates complex process-based models.

Broadly speaking, increased complexity is associated with potentially more accuracy but usually also with an increased data requirement. Tier 3 models are mainly used in research by experts, while policymakers and other stakeholders mainly use Tier 1 or Tier 2 approaches.

Based on the simple categorisation of a model's complexity, the use of the Tier 3 models suggest potentially the best results will be obtained but this is not necessarily the case. Potential accuracy depends on a high volume and quality of data to parameterise the model. More data is associated with increased potential for errors and uncertainty (Wallach and Thorburn, 2017). Additionally, the models should be calibrated, which requires data derived from measurements. Not only is this associated with additional work but also the way the model is calibrated might affect the results (Albanito et al., 2022).

Models are developed for different scales. Biogeochemical and crop models are usually developed for point scale (one-dimensional) but can be applied to individual fields (assuming constant representation of the considered parameters and variables). Tools target mainly the farm scale. Livestock models often use simulations of a herd based on up-scaled measurements of individual animals or measurements on the entire herd that are then used in developing regressions describing the processes. Several models include all-farm emissions, from fuel and energy consumption to footprints from imported and exported goods. Their structure is usually based on smaller tools or models that describe the individual processes and combine the results for the overall net carbon balance. Due to the number of processes that are included, most processes are represented as simple functions (often generic factors or regression functions). However, some models include more complex process-based physical equations and there are developments to include more complex functions in the farm models.

It should be noted that a best model does not exist and that qualitative comparison of different models is sometimes difficult. Each model is developed and used for different purposes. The suitability of a model depends on the objective, the target variables, data availability, accuracy requirements, time steps, skill set and experience of the user and other factors that might restrict the use of specific models. Depending on the environmental conditions, model performance can vary and different models might perform better in specific conditions. Research has shown that ensemble runs usually provide a very good result (Fronzek et al., 2018; Rodríguez et al., 2019).

Models are often categorised based on their main objective e.g. simulating livestock emission, cropland emissions or both. All systems produce GHGEs but, for livestock, CH_4 emissions are of special interest, especially from enteric fermentation and N_2O emissions from manure. On croplands, N_2O from the fertiliser use and the CO_2 from the soil respiration are the most relevant emissions.

7.1 Livestock models

Models for GHGEs from livestock show different complexities and structure. There are simpler empirical models that quantify the emissions based on feed characteristics. Processes are described using empirical models based on and calibrated using sample measurements (e.g. regression functions). These models are restricted in application as they can only be used in conditions in which the original measurements on which they are based were taken (Jose et al., 2016).

Process-based models are built on mathematical description of the processes. These models often include climate and other environmental factors as well as factors that affect processes (e.g. microbial communities that affects the methane emissions). Developing process-based models requires a complete or at least a sufficient understanding of the processes involved, which needs to build on measurements and scientific knowledge. For processes that are not sufficiently understood to be described using mathematical equations, empirical models can be used to fill these gaps. Sanjo Jose et al. (2016) identifies late pregnancy and early lactation as crucial stages involving rapid metabolic changes as critical processes which models need to address.

While emission sources, information about feeding and manure production are easier to account for in indoor settings, the dynamics in grasslands is more complex. Animals move around an extensive pasture, so feeding rates and manure inputs on the soil are often unclear. Several models consider the variability as input data (e.g. amount of grazing or manure production), while some models simulate the availability of pasture based on area, weather conditions and nitrogen (N) availability. These processes are simplified to avoid an unnecessary increase in model complexity and to reduce the number of input variables. Dynamics (movement, grazing and defecation of animals) are difficult to predict in extensive systems.

7.2 Cropland and grassland models

In the case of croplands and grasslands, biogeochemical and soil vegetation atmosphere transfer (SVAT) models simulate carbon, and some also include nitrogen dynamics (Berardi et al., 2020; Petropoulos et al., 2009). There are also other models, which are not focused on GHGE simulation, but still include these processes, which can be used, e.g. soil models. Different models might focus more on the pedosphere (soil), the atmosphere or the biosphere (plant), respectively.

As processes involve dynamics on different time scales, time steps vary, with SVAT models using often sub-daily (hourly or half-hourly) times steps, while other models simulate on daily, monthly or annually time steps. The most suitable time step depends on what processes are included. While carbon dynamics show good representation in monthly time steps, nitrogen emissions require daily or sub-daily time steps, even though some models include the N simulation in models using monthly time steps.

As farms usually contain several fields, each field needs to be simulated individually, though some fields can be grouped together where conditions are similar (e.g. same soil, crop, rotation, management). Most models simulate dynamics for both croplands and grasslands. The major differences are the different management options of these systems, which includes duration of surface cover, sowing and harvest, crop rotation, tillage and fertiliser and manure applications. Some management options (e.g. intercropping and legumes as break crops) are still challenging to incorporate into models.

Arguably, biogeochemical models are most suitable for the simulation of GHGEs at the farm level. These models work on the point scale, which means they simulate the fluxes only one-dimensionally. This requires aggregating the input data accordingly. Usually, a field is assumed as a homogeneous area, with all relevant input parameters averaged over the field. If there is a need to consider differences within the field, the input parameters need to be measured accordingly. However, these models are complex and require input data that are not necessarily available for all farms (e.g. precise soil information, daily weather data). Less complex tools that are designed for farm applications might be more suitable for the application by non-expert modellers, including farmers or policymakers.

8 Farm assessment tools to estimate greenhouse gas emissions

Farm assessment tools are designed to calculate a complete GHGE net balance of a farm. There are several tools available. All seek to achieve good quality of results with the minimum of measurements required to achieve an acceptable level of accuracy. Tools are designed of take account of existing data that are available for each farm. In general, these tools are designed for non-modellers, mainly farmers and policymakers. This means, the use of the tools should be simple and that all the data for the calculation should be easily available, while still providing good-quality results.

Tools typically include a database in the background to provide missing information and to translate available data into model inputs. However, these data bases differ and their use can be separated into two different tool structures. One option is to provide all relevant environmental information in a background database to run a detailed (Tier 3) process-based model. However, the quality of the results will depend on the quality of the available environmental data in the provided data base, which may restrict this method to those areas or countries that are able to provided these data.

One example is the Comet Farm Tool that was developed for the USA (Paustian et al., 2017). This tool allows users to locate and mark the target area by using an interactive map, with all required environmental data extracted to provide model input data in the background. Agronomic details are added by the user. As these are standard farm-specific information (e.g. field management activities; number of animals), no additional measurements are required.

The other option is to use simpler approaches to quantify GHGEs (Tier 1 and Tier 2). The Cool Farm Tool (Hillier et al., 2011) is based on this concept. In this case, some data may be derived from an underlying data base but, for some values, default data are used. Corresponding carbon footprint data or emission factors are used to quantify the GHGE for these values (e.g. the diesel consumption for machinery). However, these data include some uncertainty as the factors are derived for the county scale, while they are applied on the farm scale. The Cool Farm Tool is applied globally and includes all farm activities and is applicable for all kind of farms. There are some limitations for specific livestock.

In addition to farm tools, there are life cycle assessment (LCA) tools that include all relevant sources of emissions for a final product into account considering the entire production chain. For a food product, this includes the production on farm (direct and indirect emissions), imported goods or services (e.g. fertiliser or rented machinery) and emissions after transport from farm gate, including impacts of processing, packaging and transport to retailer. Some LCA models restrict the scope up to the farm gate, as this can be accounted for as the carbon footprint associated with the production of the goods on the farm. LCA studies can be separated into consequential studies, that quantify environmental changes due to changes in demand or production and attributional studies that quantify the impact on the environment (de Vries et al., 2020).

LCA tools simulate the emissions based on indicators, generic factors and simple equations. As with the farm tools, there are global or local factors or empirical equations that are developed based on local data sets in the background that translate management data into emissions. Environmental aspects are considered but in a simplistic way. LCA models depend on the quality of the data set used in their development. System conditions (and assumptions made) that a LCA tool is applied to should be compatible with the data set that was used for its development. This means that, for LCA tools developed in a particular country, there is a requirement to prove their applicability in other geographic areas. Examples of country-specific LCA models are the ANCA tool (The Netherlands) (de Vries et al., 2020), CAP2'ER (France), Carbon Audit (Ireland) and Bovid-CO_2 (Spain) (O'Brien et al., 2020).

LCA approaches can also give misleading results if applied to organic agriculture if originally designed using conventional production data sets. The LCA will change as the processes to produce that project change, potentially rendering the original calculation invalid. Additionally, key aspects of sustainable agriculture (e.g. soil health or biodiversity) might be ignored in some LCA models (van der Werf et al., 2020), which is a problem faced by models in general.

Energy and fuel use on the farm also contributes to the overall GHG balance. Specific energy and fossil fuel use can be calculated for individual farms. However, there are no generally accepted measurement methods to quantify emissions from these sources and gathering information is potentially time consuming. The simple use of emission factors (Tier 1 approach) provides results, as derived by the IPCC (2006). This is the standard method also for farm tools or LCA tools. Fossil fuel use can be differentiated between diesel and petrol. Some models can translate different field management practices into a carbon emission equivalent (Hillier et al., 2011; Lal, 2004). Renewable energy production on the farm can contribute to the overall farm carbon footprint as negative emissions within the overall balance. The energy produced can be directly translated to saved emissions. However, wind turbines or in-field solar arrays may affect the types of land use. In the case of these technologies, simple use of accepted factors is often sufficient to provide a suitable estimate for the contribution to the farm GHGE budget.

9 How do different methods contribute to net zero?

There is an increasing demand by different governments and non-governmental organisations (NGOs) to reduce GHGEs and achieve net-zero targets. To achieve these targets requires inventories to quantify and track actual emissions. Emissions inventories are typically on a national scale, while GHGE programmes may target particular regions or individual farms. While all the methods discussed are used, some methods are more favoured by farmers than others.

Tools like the Cool Farm Tool or the Comet Farm Tool were developed to be used by farmers and designed based on existing data that farmers have available. These tools allow farmers to estimate GHGEs and impacts of any management changes so that farmers can then develop farm-specific strategies to reduce GHGEs. This may then allow them to access government or other schemes incentivising practices to achieve net zero or the opportunity for carbon certificate trading (Oldfield et al., 2022).

Central to incentive schemes or carbon certificate trading is a measurement, reporting and verification (MRV) system. These systems allow a robust quantification of changes in GHGEs (Paustian et al., 2019; Smith et al., 2020). While several MRV protocols already exist, a standardised protocol with clear guidelines is still missing (Kuhnert et al., 2022). The crucial problem with MRV systems is quantifying changes in GHGEs over time. Accurate measurements are costly, time and labour intensive and require expert knowledge. Carbon certificate trading platforms e.g. do not have the capacity to undertake measurement themselves, leaving it as a potential cost to the farmer. However, reducing a requirement for ongoing measurement makes it harder to track progress accurately (Kuhnert et al., 2022).

Besides quantification of changes in GHGEs, reported data require verification. As additional on-site measurements are often impractical, modelling or remote sensing provides the most promising options for verification. Both methods show great potential, but still rely on measurements for calibration. These two methods are also useful options for the quantification of large-scale measurements exceeding the field scale, which make them suitable approaches for inventories or monitoring of large farms and regional emissions. However, for farm- or field-specific emissions, these methods are potentially less accurate, especially in accounting for small-scale changes at the field level. As there is still a gap between the most accurate and the most affordable and practical approaches for farmers, future research needs to provide appropriate solutions to reduce and close this gap.

10 Future developments in measuring greenhouse gas emissions at the farm scale

There is an increased interest in reducing GHGEs or achieving negative emissions on farms. Current solutions still lack quality to provide effective support for MRV systems (either too costly, not accurate enough, or both). Future developments will focus partly on the technological developments to find cheaper solutions for accurate measurements and developing systems that combine the advantages of low-cost modelling and remote sensing with the accuracy of on-site measurements. The short-term modelling will most likely used for estimations. Further development of the models depends on available data sets which are still limited. More data of a sufficient temporal and spatial resolution and representing a wide range of different environmental conditions are required to provide a robust database. This will be of special relevance for increased use of machine learning algorithms to develop improved models to quantify emissions going forward.

In the long run, remote sensing will be probably a likely solution for farm-specific measurements of emissions from croplands and grasslands as well as monitoring of grazing animals. One potential approach is inversion measurements (or top-down measurements) using remote sensing combined with modelling (bottom-up approaches). These methods show the potential to support a monitoring and verification support (MVS) for GHGEs (Ciais et al., 2017). While this method is already showing good results at the landscape scale (Ricaud et al., 2021), the method requires further development to provide robust results at the field or farm scale.

11 References

Albanito, F., McBey, D., Harrison, M., Smith, P., Ehrhardt, F., Bhatia, A., Bellocchi, G., Brilli, L., Carozzi, M., Christie, K., Doltra, J., Dorich, C., Doro, L., Grace, P., Grant, B., Léonard, J., Liebig, M., Ludemann, C., Martin, R., Meier, E., Meyer, R., De Antoni Migliorati, M., Myrgiotis, V., Recous, S., Sándor, R., Snow, V., Soussana, J. F., Smith, W. N. and Fitton, N. (2022). How modelers model: The overlooked social and human dimensions in model intercomparison studies. *Environmental Science and Technology* 56(18), 13485–13498.

Arndt, C., Leytem, A. B., Hristov, A. N., Zavala-Araiza, D., Cativiela, J. P., Conley, S., Daube, C., Faloona, I. and Herndon, S. C. (2018). Short-term methane emissions from 2 dairy farms in california estimated by different measurement techniques and US

Environmental Protection Agency inventory methodology: A case study. *Journal of Dairy Science* 101(12), 11461–11479. https://doi.org/10.3168/jds.2017-13881.

Aubinet, M., Grelle, A., Ibrom, A., Rannik, Ü., Moncrieff, J., Foken, T., Kowalski, A. S., Martin, P. H., Berbigier, P., Bernhofer, C., Clement, R., Elbers, J., Granier, A., Grünwald, T., Morgenstern, K., Pilegaard, K., Rebmann, C., Snijders, W., Valentini, R. and Vesala, T. (2000). Estimates of the annual net carbon and water exchange of forests: The euroflux methodology. *Advances in Ecological Research* 30, 113–175.

Aubinet, M., Vesala, T. and Papale, D. (2012). *Eddy Covariance: A Practical Guide to Measurements and Data Analysis*. Dordrecht, Heidelberg, London, and New York: Springer.

Baldocchi, D., Falge, E., Gu, L., Olson, R., Hollinger, D., Running, S., Anthoni, P., Bernhofer, C., Davis, K., Evans, R., Fuentes, J., Goldstein, A., Katul, G., Law, B., Lee, X., Malhi, Y., Meyers, T., Munger, W., Oechel, W., Paw, K. T., Pilegaard, K., Schmid, H. P., Valentini, R., Verma, S., Vesala, T., Wilson, K. and Wofsy, S. (2001). FLUXNET: A new tool to study the temporal and spatial variability of ecosystem-scale carbon dioxide, water vapor, and energy flux densities. *Bulletin of the American Meteorological Society* 82(11), 2415–2434.

Barchyn, T. E., Hugenholtz, C. H. and Fox, T. A. (2019). Plume detection modeling of a drone-based natural gas leak detection system. *Elementa: Science of the Anthropocene* 7, 41.

Behjati, M., Noh, A. B., Alobaidy, H. A. H., Zulkifley, M. A., Nordin, R. and Abdullah, N. F. (2021). LoRa communications as an enabler for internet of drones towards large-scale livestock monitoring in rural farms. *Sensors* 21(15), 5044.

Bekele, W., Guinguina, A., Zegeye, A., Simachew, A. and Ramin, M. (2022). Contemporary methods of measuring and estimating methane emission from ruminants. *Methane* 1(2), 82–95.

Berardi, D., Brzostek, E., Blanc-Betes, E., Davison, B., DeLucia, E. H., Hartman, M. D., Kent, J., Parton, W. J., Saha, D. and Hudiburg, T. W. (2020). 21st-century biogeochemical modeling: Challenges for century-based models and where do we go from here? *Global Change Biology Bioenergy* 12(10), 774–788.

Ciais, P., Palmer, P., Scholze, M., Kentarchos, A., Brunhes, T., Dolman, H., et al. (2017). *An Operational Anthropogenic CO2 Emissions Monitoring & Verification System: Baseline Requirements, Model Components and Functional Architecture*. Commission European; Joint Research Centre.

de Vries, M., van Dijk, W., de Boer, J. A., de Haan, M. H. A., Oenema, J., Verloop, J., et al. (2020). *Calculation Rules of the Annual Nutrient Cycling Assessment (ANCA) 2019: Background Information about Farm-Specific Excretion Parameters (Update of ANCA Report 2018)*. Wageningen: Wageningen Livestock Research.

Eggleston, H. S., Buendia, L., Miwa, K., Ngara, T. and Tanabe, K. (2006). *2006 IPCC Guidelines for National Greenhouse Gas Inventories*. Japan: IPCC.

Falge, E., Baldocchi, D., Olson, R., Anthoni, P., Aubinet, M., Bernhofer, C., Burba, G., Ceulemans, R., Clement, R., Dolman, H., Granier, A., Gross, P., Grünwald, T., Hollinger, D., Jensen, N., Katul, G., Keronen, P., Kowalski, A., Lai, C. T., Law, B. E., Meyers, T., Moncrieff, J., Moors, E., Munger, J. W., Pilegaard, K., Rannik, Ü., Rebmann, C., Suyker, A., Tenhunen, J., Tu, K., Verma, S., Vesala, T., Wilson, K. and Wofsy, S. (2001). Gap filling strategies for defensible annual sums of net ecosystem exchange. *Agricultural and Forest Meteorology* 107(1), 43–69.

Foken, T., Aubinet, M. and Leuning, R. (2012). The eddy covariance method. In Aubinet, M., Vesala, T. and Papale, D. (Eds.), *Eddy Covariance: A Practical Guide to Measurement and Data Analysis*. Dordrecht: Springer Netherlands, 1–19.

Frank, S., Havlik, P., Soussana, J., Levesque, A., Valin, H., Wollenberg, E., Kleinwechter, U., Fricko, O., Gusti, M., Herrero, M., Smith, P., Hasegawa, T., Kraxner, F. and Obersteiner, M. (2017). Reducing greenhouse gas emissions in agriculture without compromising food security? *Environmental Research Letters* 12(10), 105004.

Fronzek, S., Pirttioja, N., Carter, T. R., Bindi, M., Hoffmann, H., Palosuo, T., Ruiz-Ramos, M., Tao, F., Trnka, M., Acutis, M., Asseng, S., Baranowski, P., Basso, B., Bodin, P., Buis, S., Cammarano, D., Deligios, P., Destain, M., Dumont, B., Ewert, F., Ferrise, R., François, L., Gaiser, T., Hlavinka, P., Jacquemin, I., Kersebaum, K. C., Kollas, C., Krzyszczak, J., Lorite, I. J., Minet, J., Minguez, M. I., Montesino, M., Moriondo, M., Müller, C., Nendel, C., Öztürk, I., Perego, A., Rodríguez, A., Ruane, A. C., Ruget, F., Sanna, M., Semenov, M. A., Slawinski, C., Stratonovitch, P., Supit, I., Waha, K., Wang, E., Wu, L., Zhao, Z. and Rötter, R. P. (2018). Classifying multi-model wheat yield impact response surfaces showing sensitivity to temperature and precipitation change. *Agricultural Systems* 159, 209–224. https://doi.org/10.1016/j.agsy.2017.08.004.

Gålfalk, M., Nilsson Påledal, S. and Bastviken, D. (2021). Sensitive drone mapping of methane emissions without the need for supplementary ground-based measurements. *ACS Earth and Space Chemistry* 5(10), 2668–2676.

Goopy, J. P., Woodgate, R., Donaldson, A., Robinson, D. L. and Hegarty, R. S. (2011). Validation of a short-term methane measurement using portable static chambers to estimate daily methane production in sheep. *Animal Feed Science and Technology* 166–167, 219–226. https://doi.org/10.1016/j.anifeedsci.2011.04.012.

Goopy, J. P., Chang, C. and Tomkins, N. (2016). A comparison of methodologies for measuring methane emissions from ruminants. In *Methods for Measuring Greenhouse Gas Balances and Evaluating Mitigation Options in Smallholder Agriculture*. Cham: Springer, 97–117. https://doi.org/10.1007/978-3-319-29794-1_5.

Hillier, J., Walter, C., Malin, D., Garcia-Suarez, T., Mila-i-Canals, L. and Smith, P. (2011). A farm-focused calculator for emissions from crop and livestock production. *Environmental Modelling and Software* 26(9), 1070–1078.

Hutchinson, G. L. and Livingston, G. P. (2001). Vents and seals in non-steady-state chambers used for measuring gas exchange between soil and the atmosphere. *European Journal of Soil Science* 52(4), 675–682.

IPCC (2014). *Climate Change 2014: Synthesis Report: Contribution of Working Groups I, II and III to the Fifth Assessment Report of the Intergovernmental Panel on Climate Change, Core Writing Team*, R. K. Pachauri and L. A. Meyer (Eds.). Geneva: IPCC.

IPCC (2021). Summary for policymakers. In Masson-Delmotte, V., et al. (Eds.), *Climate Change 2021: The Physical Science Basis. Contribution of Working Group I to the Sixth Assessment Report of the Intergovernmental Panel on Climate Change*. Cambridge: Cambridge University Press.

Johnson, K. A. and Johnson, D. E. (1995). Methane emissions from cattle. *Journal of Animal Science* 73(8), 2483–2492.

Jonker, A. and Waghorn, G. C. (2020). *Guidelines for Use of Sulphur Hexafluoride (SF6) Tracer Technique to Measure Enteric Methane Emissions from Ruminants*. No. MPI

Technical Paper. No. 2020/06. New Zealand: New Zealand Agricultural Greenhouse Gas Research Centre.

Johnson, K., Huyler, M., Westberg, H., Lamb, B. and Zimmerman, P. (1994). Measurement of methane emissions from ruminant livestock using a sulfur hexafluoride tracer technique. *Environmental Science and Technology* 28(2), 359–362.

Jose, V. S., Sejian, V., Bagath, M., Ratnakaran, A. P., Lees, A. M., Al-Hosni, Y. A. S., Sullivan, M., Bhatta, R. and Gaughan, J. B. (2016). Modeling of greenhouse gas emission from livestock. *Frontiers in Environmental Science* 4, 27.

Kuhnert, M., Vetter, S. H. and Smith, P. (2022). Measuring and monitoring soil carbon sequestration. In Rumpel, C. (Ed.), *Understanding and Fostering Soil Carbon Sequestration*. Philadelphia, USA: Burleigh Dodds Science Publishing, 305–322.

Kumari, S., Fagodiya, R. K., Hiloidhari, M., Dahiya, R. P. and Kumar, A. (2020). Methane production and estimation from livestock husbandry: A mechanistic understanding and emerging mitigation options. *Science of the Total Environment* 709, 136135.

Kupper, T., Hani, C., Neftel, A., Kincaid, C., Buhler, M., Amon, B. and VanderZaag, A. (2020). Ammonia and greenhouse gas emissions from slurry storage - a review. *Agriculture, Ecosystems and Environment* 300.

Lal, R. (2004). Carbon emission from farm operations. *Environment International* 30(7), 981–990. https://doi.org/10.1016/j.envint.2004.03.005.

Livingston, G. P., Hutchinson, G. L. and Spartalian, K. (2006). Trace gas emission in chambers. *Soil Science Society of America Journal* 70(5), 1459–1469.

Madsen, J., Bjerg, B. S., Hvelplund, T., Weisbjerg, M. R. and Lund, P. (2010). Methane and carbon dioxide ratio in excreted air for quantification of the methane production from ruminants. *Livestock Science* 129(1–3), 223–227. https://doi.org/10.1016/j.livsci.2010.01.001.

Maier, M., Weber, T. K. D., Fiedler, J., Fuß, R., Glatzel, S., Huth, V., Jordan, S., Jurasinski, G., Kutzbach, L., Schäfer, K., Weymann, D. and Hagemann, U. (2022). Introduction of a guideline for measurements of greenhouse gas fluxes from soils using non-steady-state chambers. *Journal of Plant Nutrition and Soil Science* 185(4), 447–461.

Mauder, M., Foken, T., Aubinet, M. and Ibrom, A. (2021). Eddy-covariance measurements. In Foken, T. (Ed.), *Springer Handbook of Atmospheric Measurements*. Cham: Springer International Publishing, 1473–1504.

Montanholi, Y. R., Odongo, N. E., Swanson, K. C., Schenkel, F. S., McBride, B. W. and Miller, S. P. (2008). Application of infrared thermography as an indicator of heat and methane production and its use in the study of skin temperature in response to physiological events in dairy cattle (Bos taurus). *Journal of Thermal Biology* 33(8), 468–475.

Moureaux, C., Ceschia, E., Arriga, N., Béziat, P., Eugster, W., Kutsch, W. L. and Pattey, E. (2012). Eddy covariance measurements over crops. In Aubinet, M., Vesala, T. and Papale, D. (Eds.), *Eddy Covariance: A Practical Guide to Measurement and Data Analysis*. Dordrecht: Springer Netherlands, 319–331.

Munger, J. W., Loescher, H. W. and Luo, H. (2012). Measurement, tower, and site design considerations. In Aubinet, M., Vesala, T. and Papale, D. (Eds.), *Eddy Covariance: A Practical Guide to Measurement and Data Analysis*. Dordrecht: Springer Netherlands, 21–58.

O'Brien, D., Herron, J., Andurand, J., Caré, S., Martinez, P., Migliorati, L., Moro, M., Pirlo, G. and Dollé, J. B. (2020). Life beef carbon: A common framework for quantifying grass and corn based beef farms' carbon footprints. *Animal* 14(4), 834–845. https://doi.org/10.1017/S1751731119002519.

Ogink, N. W. M., Mosquera, J., Calvet, S. and Zhang, G. (2013). Methods for measuring gas emissions from naturally ventilated livestock buildings: Developments over the last decade and perspectives for improvement. *Biosystems Engineering* 116(3), 297–308. https://doi.org/10.1016/j.biosystemseng.2012.10.005.

Oldfield, E. E., Eagle, A. J., Rubin, R. L., Rudek, J., Sanderman, J. and Gordon, D. R. (2022). Crediting agricultural soil carbon sequestration. *Science* 375(6586), 1222–1225.

Paustian, K., Lehmann, J., Ogle, S., Reay, D., Robertson, G. P. and Smith, P. (2016). Climate-smart soils. *Nature* 532(7597), 49–57.

Paustian, K., Easter, M., Brown, K., Chambers, A., Eve, M., Huber, A., et al. (2017). *Field- and Farm-Scale Assessment of Soil Greenhouse Gas Mitigation Using COMET-Farm*.

Paustian, K., Collier, S., Baldock, J., Burgess, R., Creque, J., DeLonge, M., Dungait, J., Ellert, B., Frank, S., Goddard, T., Govaerts, B., Grundy, M., Henning, M., Izaurralde, R. C., Madaras, M., McConkey, B., Porzig, E., Rice, C., Searle, R., Seavy, N., Skalsky, R., Mulhern, W. and Jahn, M. (2019). Quantifying carbon for agricultural soil management: From the current status toward a global soil information system. *Carbon Management* 10(6), 567–587.

Petropoulos, G., Carlson, T. N. and Wooster, M. J. (2009). An overview of the use of the SimSphere Soil Vegetation Atmosphere Transfer (SVAT) model for the study of land-atmosphere interactions. *Sensors* 9(6), 4286–4308.

Ramírez-Restrepo, C. A., Barry, T. N., Marriner, A., López-Villalobos, N., McWilliam, E. L., Lassey, K. R. and Clark, H. (2010). Effects of grazing willow fodder blocks upon methane production and blood composition in young sheep. *Animal Feed Science and Technology* 155(1), 33–43. https://doi.org/10.1016/j.anifeedsci.2009.10.003.

Reichstein, M., Falge, E., Baldocchi, D., Papale, D., Aubinet, M., Berbigier, P., Bernhofer, C., Buchmann, N., Gilmanov, T., Granier, A., Grünwald, T., Havránková, K., Ilvesniemi, H., Janous, D., Knohl, A., Laurila, T., Lohila, A., Loustau, D., Matteucci, G., Meyers, T., Miglietta, F., Ourcival, J., Pumpanen, J., Rambal, S., Rotenberg, E., Sanz, M., Tenhunen, J., Seufert, G., Vaccari, F., Vesala, T., Yakir, D. and Valentini, R. (2005). On the separation of net ecosystem exchange into assimilation and ecosystem respiration: Review and improved algorithm. *Global Change Biology* 11(9), 1424–1439.

Ricaud, P., Attié, J., Chalinel, R., Pasternak, F., Léonard, J., Pison, I., Pattey, E., Thompson, R. L., Zelinger, Z., Lelieveld, J., Sciare, J., Saitoh, N., Warner, J., Fortems-Cheiney, A., Reynal, H., Vidot, J., Brooker, L., Berdeu, L., Saint-Pé, O., Patra, P. K., Dostál, M., Suchánek, J., Nevrlý, V. and Zwaaftink, C. G. (2021). The monitoring nitrous oxide sources (MIN2OS) satellite project. *Remote Sensing of Environment* 266, 112688. https://doi.org/10.1016/j.rse.2021.112688.

Rochette, P. (2011). Towards a standard non-steady-state chamber methodology for measuring soil N2O emissions. *Animal Feed Science and Technology* 166–167, 141–146. https://doi.org/10.1016/j.anifeedsci.2011.04.063.

Rodríguez, A., Ruiz-Ramos, M., Palosuo, T., Carter, T. R., Fronzek, S., Lorite, I. J., Ferrise, R., Pirttioja, N., Bindi, M., Baranowski, P., Buis, S., Cammarano, D., Chen, Y., Dumont, B., Ewert, F., Gaiser, T., Hlavinka, P., Hoffmann, H., Höhn, J. G., Jurecka, F., Kersebaum, K. C., Krzyszczak, J., Lana, M., Mechiche-Alami, A., Minet, J., Montesino, M., Nendel, C., Porter, J. R., Ruget, F., Semenov, M. A., Steinmetz, Z., Stratonovitch, P., Supit, I., Tao, F., Trnka, M., de Wit, A. and Rötter, R. P. (2019). Implications of crop model ensemble size and composition for estimates of adaptation effects and agreement of recommendations. *Agricultural and Forest Meteorology* 264, 351–362. https://doi.org/10.1016/j.agrformet.2018.09.018.

Roth, L., Camenzind, M., Aasen, H., Kronenberg, L., Barendregt, C., Camp, K. H., Walter, A., Kirchgessner, N. and Hund, A. (2020). Repeated multiview imaging for estimating seedling tiller counts of wheat genotypes using drones. *Plant Phenomics* 2020, 3729715.

Shen, Y., Xiao, Z., Wang, Y., Yao, L. and Xiao, W. (2022). Multisource remote sensing based estimation of soil NOx emissions from fertilized cropland at high-resolution: Spatio-temporal patterns and impacts. *Journal of Geophysical Research: Atmospheres* 127(20), e2022JD036741.

Smith, P., Soussana, J. F., Angers, D., Schipper, L., Chenu, C., Rasse, D. P., Batjes, N. H., van Egmond, F., McNeill, S., Kuhnert, M., Arias-Navarro, C., Olesen, J. E., Chirinda, N., Fornara, D., Wollenberg, E., Álvaro-Fuentes, J., Sanz-Cobena, A. and Klumpp, K. (2020). How to measure, report and verify soil carbon change to realize the potential of soil carbon sequestration for atmospheric greenhouse gas removal. *Global Change Biology* 26(1), 219–241.

Tedeschi, L. O., Abdalla, A. L., Alvarez, C., Anuga, S. W., Arango, J., Beauchemin, K. A., Becquet, P., Berndt, A., Burns, R., De Camillis, C., Chará, J., Echazarreta, J. M., Hassouna, M., Kenny, D., Mathot, M., Mauricio, R. M., McClelland, S. C., Niu, M., Onyango, A. A., Parajuli, R., Pereira, L. G. R., Del Prado, A., Paz Tieri, M., Uwizeye, A. and Kebreab, E. (2022). Quantification of methane emitted by ruminants: A review of methods. *Journal of Animal Science* 100(7).

van der Werf, H. M. G., Knudsen, M. T. and Cederberg, C. (2020). Towards better representation of organic agriculture in life cycle assessment. *Nature Sustainability* 3(6), 419–425.

Vetter, S. H., Abdalla, M., Kuhnert, M. and Smith, P. (2022). Soil carbon sequestration and biochar. In Bui, M. and Mac Dowell, N. (Eds.), *Greenhouse Gas Removal Technologies* (1st ed.). London: Royal Society of Chemistry, 194.

Wallach, D. and Thorburn, P. J. (2017). Estimating uncertainty in crop model predictions: Current situation and future prospects. *European Journal of Agronomy* 88, A1–A7. https://doi.org/10.1016/j.eja.2017.06.001.

Wohlfahrt, G., Klumpp, K. and Soussana, J. (2012). Eddy covariance measurements over grasslands. In Aubinet, M., Vesala, T. and Papale, D. (Eds.), *Eddy Covariance: A Practical Guide to Measurement and Data Analysis*. Dordrecht: Springer Netherlands, 333–344.

Chapter 16

The role of standards and certification in delivering safe, sustainable and socially responsible agri-food supply chains

Louise Manning, University of Lincoln, UK

1 Introduction

Land use has been linked to the production of food, feed, fibre and fodder for humans since human life began, but particularly with the beginnings of agriculture around 12 000 years ago. Since then, agriculture has gone through various phases of development. Early civilisations such as Ancient Egypt (3000–1000 BCE) developed domestication and selective breeding of crops and livestock, preparation of land for crops through ploughing (using oxen), use of organic fertilisation, irrigation techniques and improved storage and transport (e.g. using wheeled carts), generating a sufficient food surplus to support large populations, urban centres, greater division/specialisation of labour and complex social and government structures.

A second major phase of development was the agricultural revolution in Europe and the United States in the eighteenth and nineteenth centuries. This was brought about by innovations such as: improved drainage, land reclamation and conversion of permanent pasture to arable land; developments in technology, including the development of the horse-drawn seed drill, corn reaper and threshing machine, and improvements to the traditional plough; introduction of new crops (such as turnips, clover and potatoes) and animal breeds (mainly cattle and sheep); development of the Norfolk Four Course Rotation with reduced fallows; enclosure of medieval open field systems, which allowed more efficient utilisation of land; increased use of organic pesticides,

http://dx.doi.org/10.19103/AS.2024.0133.33

the development of the first synthetic fertilisers; and the development of more sophisticated market infrastructure (including roads, inland waterways and rail networks to transport food more easily to urban centres). The Agricultural Revolution led to a doubling of cereal yields in the UK during the 1700s, and almost tripled yields and farm-labour productivity by 1870 (Mazoyer and Roudart, 2006).

A third major phase of development was the 'Green Revolution' after the Second World War. This was characterised by the development of more productive crop varieties and animal breeds (underpinned by new breeding techniques based on a better understanding of genetics), the widespread use of synthetic fertilisers and pesticides, increasing mechanisation (e.g the replacement of horses by tractors), as well as growing specialisation and concentration of production at farm, regional and national levels. These innovations meant that between 1950 and 1984, world grain production increased by 160%. This period also saw the growth of increasingly sophisticated global agri-food supply chains often dominated by a relatively small group of multinational food manufacturers and retailers in developed countries with strong leverage over their farmer suppliers (often smallholder farmers in developing countries) (Kilby, 2019).

More recently, new technologies such as global positioning systems (GPS), a new generation of proximal and remote sensor technologies, as well as developments in data processing have led to a new era of precision farming. Current developments include the advent of autonomous vehicles, robotics and the application of artificial intelligence and machine learning. With livestock systems, game-changing technologies have been the access to antibiotics, vaccines and medication and the introduction of mechanised milking parlours through to robotic milkers, artificial insemination, and the development of large-scale, intensive poultry meat and egg production within highly controlled indoor environments. Large-scale abattoirs have also led to greater efficiency through the mechanisation of slaughter and processing of animals. The food production industry has continued to use science and technology innovations to mechanise, automate and create smart food production processes assuring the consistency and safety of products.

It is easy to see the development of agriculture in primarily technological terms, as a series of innovations that have allowed food production to continue to feed a rapidly growing population. The daily number of available calories in the United Kingdom (UK) for instance, increased from around 1500 in 1270 to 3340 in 2018. The continued improvements in the efficiency of modern agricultural production can be seen in the fact that, since 1961, the land used globally for agriculture has only increased by 7% whilst the global population has increased by around 250%, with the land used per person halving from 1.43 ha to 0.63 ha (Ritchie, 2021). The increasing capacity and capability of the food system has been a factor in increasing average life expectancy from 32 years in 1900 to 71 years by 2021 (Dattani et al. 2023a). Average child mortality has fallen from 50% pre-1700 to 4.3% and in some countries is 0.4% today (Dattani et al. 2023b).

However, although this technological achievement is a key part of the story, the science and technology narrative neglects other key dimensions: namely political, economic, social and environmental drivers. Indeed, the development of farming and food systems has been characterised as a series of food 'regimes', each with its own technological, political, social and environmental characteristics and impacts (both positive and negative) (Watson, 2024). Agri-food supply chains affect many actors, from farmers and rural communities who may struggle to make a living in a competitive, global marketplace to consumers who may not always have access to a supply of affordable, nutritious food to maintain a healthy diet. The latter has been highlighted by the COVID-19 pandemic, where estimates in 2021 suggest that 161 million additional people experienced chronic undernourishment in 2020 with those facing chronic hunger rising to between 720 and 811 million people in 2020 (FAO et al., 2021). The growth of global agriculture and the development of more intensive methods of production has also resulted in growing concerns about animal welfare and wellbeing, pollution from fertilisers and crop protection products which has damaged the environment and threatened human health, as well as greenhouse gas emissions and changes in land use which make agriculture a significant contributor to climate change and biodiversity loss. As an example, 10 000 years ago, 57% of the world's habitable area (10.6 billion hectares) was covered by forest. One-third of the forested land has been lost, half of this area in the last 100 years (Ritchie, 2021). Two-thirds of wild grassland, savannah and shrubs have been converted to crop production or managed grazing land, with around 1% of land being urbanised. As Ritchie states, 'humanity's biggest footprint is due to what we eat, not where we live' (Table 1).

Considering the global biomass of mammals in tonnes of carbon (174 million) whilst humans account for one-third of the total (34%), livestock accounts

Table 1 The expansion of agricultural land. Adapted from Ritchie (2021)

Timeframe	Global population (millions)	Land use (%)			
		Forest	Wild grassland and shrubs	Grazing land	Crops
10 000 years ago	–	57	42	0	0
5000 years ago	50	55	44	0	0
1700	603	52	38	6	3
1900	1650	48	27	26	8
1950	2500	44	15	31	12
2018	7634	38	14	31	15

for 62% (cattle 34%, pigs 12%, buffalo 5%, horses, camels and asses 4%, sheep 3%, goats 3%), with wild mammals only 4% (Bar-On et al. 2018; Ritchie, 2022). Domesticated poultry comprise 71% of global bird biomass (7 million tonnes of carbon) with wild birds being the rest (Bar-On et al. 2018). These impacts have raised serious questions about the direction that agriculture has taken over the last century.

Humans are inventive, but the application of technology and the systems and processes that are employed in food, fodder, fuel and fibre production from field to fork is framed not just by what we can do and how we do it, but also by what we should do, or what we ought to do. It requires anticipating and reacting to the (sometimes unintended) consequences of a particular innovation. PAS 440:2020 (p. 2) defines responsible innovation as, 'careful consideration of, and action to address, the potential impacts of introducing a new product, service, process or business model'. This will include whether the outcomes of modifying or replacing existing products, services, processes or business models provides better outcomes, and for whom. Inclusion in the context of innovation is described as the opening up of 'visions, purposes, questions, and dilemmas to broad, collective deliberation through processes of dialogue, engagement, and debate, inviting and listening to wider perspectives from publics and diverse stakeholders' (Owen et al. 2013, p. 38). Inclusive innovation considers the aspects and impacts of a particular innovation for all stakeholders, in all anticipated situations, with a reflective and a reflexive feedback loop when situations or ethical framings change, what has been described as reflexive governance architectures (Manning et al. 2023). Thus, inclusive innovation requires dynamic, participatory and anticipatory governance that shapes the direction and context of the innovation (Owen et al. 2013), leading to more robust, relevant and desirable outcomes (Fraaije & Flipse, 2020).

This chapter reflects on the role of standards in helping to ensure we deliver safe, sustainable and socially responsible agri-food supply chains, now and into the future. It looks at how standards have evolved and ways they might be improved to deliver the changes we need.

2 The evolution of standards

Almost as soon as food systems developed, there was a recognition of the need for standards and regulation, beginning with rules relating to issues such as weights and measures, food safety and adulteration in order to prevent unsafe and unfair practices affecting consumers. These were mainly restricted to food safety and quality regulations enacted by governments. Supranational organisations such as the United Nations Food and Agriculture Organisation (FAO), Codex Alimentarius Commission (a joint initiative between FAO and the World Health Organisation (WHO) and the World Trade Organisation

(WTO) set standards for how global agri-food supply chains should interact sustainably. The Codex Alimentarius international food standards for products, guidelines and codes of practice, e.g. Hygienic Practice Guidance all contribute to social elements of sustainability such as food safety, food quality and fairness in supply chains (Codex, 2024).

More recently, the standards landscape has become much more complex. The complexity of modern agri-food supply chains, with a host of intermediaries separating farmers from the final consumer, often located in differing jurisdictions, has significantly complicated standard setting as well as monitoring and ensuring compliance. As food supply chains have become more complicated, multi-layered, globalised and in some ways become more opaque, consumer trust has moved from an early relational trust in food suppliers (where customers interacted directly with a known local network of suppliers), 'I know you so I trust you', to institutional trust, 'I see the certification or the claims and so I trust you' (Manning & Kowalska, 2021). They argue, 'Whilst institutional trust can develop consumer confidence in regulatory, inspections and market systems of enforcement, and surveillance; it provides confidence by default, primarily via the absence of incidents' (Ibid. p. 2). Institutional trust has led to the use of validation and verification processes to underpin standards and certification.

The thematic scope of standards has also now expanded to cover issues such as social fairness and ethical trading (e.g. to protect the interests of vulnerable actors in agri-food supply chains such as smallholders), animal welfare, environmental sustainability, biodiversity loss and nature-related risks, as well as setting out standards for particular kinds of production system (e.g. organic production or integrated farm management which are both perceived to have environmental and other benefits). The expanding scope of standards affecting food supply can be seen in the concept of social responsibility which has been increasingly applied to business operations. BS EN ISO 26000:2020 (p. 3) defines social responsibility as: 'responsibility of an organisation for the impacts of its decisions and activities on society and the environment, through transparent and ethical behaviour that [firstly] contributes to sustainable development, including health and the welfare of society, takes into account the expectations of stakeholders, is in compliance with applicable law and consistent with international norms of behaviour; and is integrated throughout the organisation and practised in its relationships.' With increasing societal concern expressed around aspects such as biodiversity loss and global warming, and with regard to their social responsibility, organisations operating in food supply chains are addressing nature-related risks as part of their environmental and social governance (ESG) strategies. This means that these aspects are being embedded more deeply into certification processes and the development of international standards such as the ISO 14067:2018

Greenhouse gases – carbon footprint of products – Requirements and guidelines for quantification and Greenhouse Gas (GHG) Protocol Product Life Cycle Accounting and Reporting Standard (2011).

Finally, the range of actors involved in setting and monitoring standards affecting agri-food supply chains has proliferated, from official national, regional (e.g. the EU) and international bodies to non-government organisations (NGOs), the food industry itself and, more recently the banking and investment sector concerned to meet ESG standards and manage climate-related investment risks. There is now a plethora of both compulsory and voluntary standards that impact on, or will impact on, food supply chains and land-based businesses involved in fibre, fuel and fodder production. The WTO and Codex have been instrumental in the harmonisation of public food standards to allow for cross-border food trade. However, harmonisation of private, voluntary standards has been more limited as vested interest, market positioning and competitive advantage affect how far those setting and using such standards may wish to cooperate (Schrobback et al. 2023).

3 Measuring compliance and improving performance

Compliance is either the action, or the physical status, of complying with a standard that can be a regulatory requirement (law, or legal standard), a normative or a contractual requirement (Manning, 2020). Professional standards, protocols and codes of conduct are developed in agri-food supply chains against which organisational compliance can be measured. These standards and protocols reflect regulatory, legal, market and/or normative institutional-based standards relating to food safety, food quality or sustainability. They can promote business-to-business (B2B) or business-to-consumer (B2C) trust (Bachmann and Inkpen, 2011; Nooteboom, 2019). Supply chain governance systems, whether driven by public or private actors, extend beyond market collaboration, driving structures and processes to limit non-conformance that could affect trust, brand reputation or regulatory compliance (Abebe et al, 2017). However, embedding cross-supply chain governance structures, such as certification schemes demonstrating compliance with a standard, comes with implementation costs which can be a barrier to adoption, particularly in low-margin supply chains (Abebe et al, 2017; Manning, 2024).

Measuring compliance against a set of prescribed normative standards and, as a result, being able to certify that a product meets these standards also does not necessarily embed performance improvement. For example, certifying that a carbon footprint calculation has been undertaken in line with prescribed methods to provide a carbon dioxide equivalent (CO_2e) calculation is only the first step in providing a baseline that informs a decarbonisation strategy. Unless performance improvement is embedded in the criteria within the standard, product certification does not necessarily drive better practices. Continuous improvement requires an organisation not only to comply with prescribed requirements (defined by themselves, or others), but also to implement management systems that support improvement of processes and operational efficiency (Aven & Krohn, 2014). Delivering real improvement requires that characteristics such as integrity, respect, trust, honesty and a genuine desire for change are embedded within organisational practices (Manning, 2020). Paine (1994, p. 111) states that: 'While compliance is rooted in avoiding legal sanctions, organisational integrity is based on the concept of self-governance in accordance with a set of guiding principles.... [the task] .. is to define and give life to an organisation's guiding values, to create an environment that supports ethically sound behaviour, and to instill a sense of shared accountability among employees.' The development of normative sustainability standards in supply chains is seeing a greater focus on integrity, ethical behaviour and accountability. It is crucial that these values are embedded in an organisation rather than seen as optional extras to a primary focus on financial performance. Lingham et al. (2022) have introduced the concept of non-financial values associated with foods:

- *Intrinsic value* related to the compositional nature of the food;
- *Product-related value* reflecting how the food was produced in terms for example of sustainability standards and practices;
- *Supply-related value* which relates to integrity and fairness in the supply chain through to the final purchase by the consumer; and
- *Emotion-cultural value,* i.e. the emotional and cultural aspects of food beyond its nutritional value.

These 'alter-values' support a holistic concept of value and organisational performance based on respect for food, the land, the people and the livestock that have produced it. A quote attributed to the celebrated management thinker W. Edwards Deming (2018) is, 'It is wrong to suppose that if you can't measure it, you can't manage it – a costly myth.' Deming was highlighting that focussing too much on measurement can obscure what good management should be. For example, rapid and deep decarbonisation in food systems requires the integration of policy, regulations, market normative standards, organisational strategies and associated innovation, technologies and infrastructure and ultimately behavioural change (Geels et al. 2017). It is behavioural change, overcoming 'lock-in', that is most important and difficult to deliver (Geels et al. 2017; Erickson et al. 2015). The role of certification as a means to overcome lock-in and barriers to change towards more sustainable, green practices has been explored in a number of sectors including forestry (Susskind et al. 2020). It is important however to consider how certification standards are designed, otherwise, the standard itself, if it is too restrictive and prescriptive, could drive singular approaches to compliance that in themselves reduce resilience, especially adaptive capacity and the ability to deliver the same outcomes in the diverse range of settings and organisations in which they need to be applied. This requires thinking in certification standards development to transition from being activity to outcomes-based.

4 Conclusion

Resilience has been described in many ways in general, and implicitly within this chapter. In the context of improving standards and certification of agri-food supply chains, resilience relates to how standards themselves promote the ability of an organisation or the wider food supply chain to not just 'bounce back' to a steady state, but to 'bounce forward' to a new status that is shaped by the evolving geopolitical and socio-economic environment in which they operate (Shadbolt et al. 2013; Manning, 2023). Agility, and in particular adaptive capacity, underpins resilience in an uncertain or changing world. Regulatory and normative standards need to promote agility within agri-food supply chains to extend beyond coping strategies and survival mechanisms to consider how organisations and supply chains can thrive and evolve during a period of rapid change, volatility and uncertainty (Kowalska et al. 2023). Operational agility needs to be linked to sustainable practices, ensuring that economic, environmental and social performance can be embedded within work practices and more widely in normative standards and certification schemes. Bringing new toolkits and decision-support tools to market that will support sustainable practices and increase visibility and transparency within food supply chains is essential.

5 Where to look for further information

The following sources and links can provide useful information:

- BRGS (see https://www.brcgs.com).
- Codex Alimentarius (see https://www.fao.org/fao-who-codexalimentarius/en/).
- GFSI (see https://www.isealalliance.org).
- GlobalG.A.P (see https://www.globalgap.org/).
- ISEAL Alliance (see https://www.isealalliance.org).
- Sedex (see https://www.sedex.com).

6 References

Abebe, G. K., Chalak, A. and Abiad, M. G. (2017). The effect of governance mechanisms on food safety in the supply chain: evidence from the Lebanese dairy sector. *Journal of the Science of Food and Agriculture* 97(9), 2908–2918.

Aven, T. and Krohn, B. S. (2014). A new perspective on how to understand, assess and manage risk and the unforeseen. *Reliability Engineering and System Safety* 121, 1–10.

Bachmann, R. and Inkpen, A. C. (2011). Understanding institutional-based trust building processes in inter-organizational relationships. *Organization Studies* 32(2), 281–301.

Bar-On, Y. M., Phillips, R. and Milo, R. (2018). The biomass distribution on Earth. *Proceedings of the National Academy of Sciences of the United States of America* 115(25), 6506–6511.

BS EN ISO 26000:2020. *Guidance on Social Responsibility*. British Standards Institute, London.

Codex (2024). About Codex alimentarius. Available at: https://www.fao.org/fao-who-codexalimentarius/about-codex/en/ Retrieved March 10, 2024.

Dattani, S., Rodés-Guirao, L., Ritchie, R., Ortiz-Ospina, E. and Roser, M. (2023a). Life expectancy. Available at: https://ourworldindata.org/life-expectancy Retrieved March 10, 2024.

Dattani, S., Spooner, F., Ritchie, H. and Roser, M. (2023b). Child and infant mortality. Available at: https://ourworldindata.org/child-mortality Retrieved March 10, 2024.

Deming, W. E. (2018). *The New Economics for Industry, Government, Education*. MIT Press, Cambridge, MA.

Erickson, P., Kartha, S., Lazarus, M. and Tempest, K. (2015). Assessing carbon lock-in. *Environmental Research Letters* 10(8), 084023.

FAO, IFAD, UNICEF, WFP and WHO (2021). State of Food Security and Nutrition in the World: transforming food systems for resilient food security, nutrition and access to health diets for all. Available at: http://www.fao.org/publications/sofi/2021/en/

Fraaije, A. and Flipse, S. M. (2020). Synthesizing an implementation framework for responsible research and innovation. *Journal of Responsible Innovation* 7(1), 113–137.

Geels, F. W., Sovacool, B. K., Schwanen, T. and Sorrell, S. (2017). Sociotechnical transitions for deep decarbonization. *Science* 357(6357), 1242–1244.

ISO 14065:2020. General principles and requirements for bodies validating and verifying environmental information. Available at: https://www.iso.org/standard/74257.html Retrieved 14 March 2024.

Kilby, P. (2019). *The Green Revolution*. Routledge, London, UK.

Kowalska, A., Lingham, S., Maye, D. and Manning, L. (2023). Food insecurity: is Leagility a potential remedy? *Foods* 12(16), 3138.

Lingham, S., Manning, L. and Maye, D. (2022). Reimagining food: readdressing and respecting values. *Sustainability* 14(12), 7328.

Manning, L. (2020). Moving from a compliance-based to an integrity-based organizational climate in the food supply chain. *Comprehensive Reviews in Food Science and Food Safety* 19(3), 995–1017.

Manning, L. (2023). Being resilient in challenging times in food supply chains. *Engineering Proceedings* 40(1), 1.

Manning, L. (2024). Operationalising integrity within Supply Chains. In: Kaptein, M. (Ed.). *A Research Handbook on Organisational Integrity*. Edgar Elgar, Cheltenham.

Manning, L., Brewer, S., Craigon, P. J., Frey, J., Gutierrez, A., Jacobs, N., Kanza, S., Munday, S., Sacks, J. and Pearson, S. (2023). Reflexive governance architectures: considering the ethical implications of autonomous technology adoption in food supply chains. *Trends in Food Science and Technology* 133, 114–126.

Manning, I. and Kowalska, A. (2021). Considering fraud vulnerability associated with credence-based products such as organic food. *Foods* 10(8), 1879.

Mazoyer, M. and Roudart, L. (2006). A history of world agriculture. Earthscan. Routledge, London, UK.

Nooteboom, B. (2019). Uncertainty and the economic need for trust. In: Sasaki, Masamichi (Ed.). *Trust in Contemporary Society*. Brill, Netherlands, pp. 60–73 .

Owen, R., Stilgoe, J., Macnaghten, P., Gorman, M., Fisher, E. and Guston, D. (2013). Responsible innovation. A framework for responsible innovation. In: Owen, R. J., Bessant, J. R. and Heintz, M. (Eds). *Responsible Innovation: Managing the Responsible Emergence of Science and Innovation in Society* (vol. 104). Wiley, Chichester, pp. 27–50. doi:10.1002/9781118551424.ch2.

Paine, L. S. (1994). Managing for organisational integrity. *Harvard Business Review* 72(2), 106–117.

Pas 440:2020 (2020). *Responsible Innovation – Guide* (vol. 440). British Standards Institute. Available at: https://pages.bsigroup.com/l/35972/2020-03-17/2cgcnc1?utm_source=pardot&%23x00026;utm_medium&%23x0003D;email&%23x00026;utm_campaign&%23x0003D;SM-STAN-LAU-PAS-PAS440-2003 Retrieved March 10, 2024.

Ritchie, H. (2022). Wild mammals make up only a few percent of the world's mammals. Available at: https://ourworldindata.org/wild-mammals-birds-biomass Retrieved March 10, 2024.

Ritchie, H. (2021). The world has lost one-third of its forest, but an end of deforestation is possible. Available at: https://ourworldindata.org/world-lost-one-third-forests Retrieved March 10, 2024.

Schrobback, P., Zhang, A., Loechel, B., Ricketts, K. and Ingham, A. (2023). Food credence attributes: a conceptual framework of supply chain stakeholders, their motives, and mechanisms to address information asymmetry. *Foods* 12(3), 538.

Shadbolt, N., Olubode-Awosola, F. and Rutsito, B. (2013). Resilience to 'Bounce without breaking'in New Zealand Dairy Farm Businesses. *International Facility Management Association 19 Transforming Agriculture*, 2, 1–14.

Susskind, L., Chun, J., Goldberg, S., Gordon, J. A., Smith, G. and Zaerpoor, Y. (2020). Breaking out of carbon lock-in: Malaysia's path to decarbonization. *Frontiers in Built Environment* 6, 21.

The Greenhouse Gas Protocol (2011). *Product Live Cycle Accounting and Reporting Standard* ISBN 9781569737736. Available at: https://ghgprotocol.org/product-standard. Accessed 14 March 2024.

Watson, D. (2024). *Transforming Food Systems: the Quest for Sustainability*. Burleigh Dodds Science Publishing, Cambridge, UK.

Index

www.ingramcontent.com/pod-product-compliance
Lightning Source LLC
LaVergne TN
LVHW010620300926
849162LV00001B/12

* 9 7 8 1 8 0 1 4 6 4 5 1 2 *